Nuclear Decom 2001

Conference Organizing Committee

Jim Jones
Chairman

Mike Brewin
RWE NUKEM Limited

Keith Collett
RWE NUKEM Limited

Bob Forgan
BNES

Michael Grave
Rolls-Royce Nuclear Engineering Services Limited

John Lee
Magnox Electric plc

Roy Manning
UKAEA

Graham McEwan
Athene Engineering Limited

Andrew McIntyre
NNC Limited

Robin Sellers
British Nuclear Fuels plc

Fred Sheil
British Nuclear Fuels plc

IMechE
Conference Transactions

International Conference on

Nuclear Decom 2001
Ensuring Safe, Secure, and Successful Decommissioning

16–18 October 2001
Commonwealth Conference and Events Centre, London, UK

Organized jointly by
the Nuclear Power Committee of the Power Industries Division of the Institution of Mechanical Engineers (IMechE) and British Nuclear Energy Society (BNES)

IMechE Conference Transactions 2001–8

Professional Engineering Publishing

Published by Professional Engineering Publishing Limited for The Institution of Mechanical Engineers, Bury St Edmunds and London, UK.

First Published 2001

This publication is copyright under the Berne Convention and the International Copyright Convention. All rights reserved. Apart from any fair dealing for the purpose of private study, research, criticism or review, as permitted under the Copyright, Designs and Patents Act, 1988, no part may be reproduced, stored in a retrieval system, or transmitted in any form or by any means, electronic, electrical, chemical, mechanical, photocopying, recording or otherwise, without the prior permission of the copyright owners. *Unlicensed multiple copying of the contents of this publication is illegal.* Inquiries should be addressed to: The Publishing Editor, Professional Engineering Publishing Limited, Northgate Avenue, Bury St Edmunds, Suffolk, IP32 6BW, UK. Fax: +44 (0) 1284 705271.

© 2001 The Institution of Mechanical Engineers, unless otherwise stated.

ISSN 1356–1448
ISBN 1 86058 329 6

A CIP catalogue record for this book is available from the British Library.

Printed by The Cromwell Press, Trowbridge, Wiltshire, UK

The Publishers are not responsible for any statement made in this publication. Data, discussion, and conclusions developed by authors are for information only and are not intended for use without independent substantiating investigation on the part of potential users. Opinions expressed are those of the Author and are not necessarily those of the Institution of Mechanical Engineers or its Publishers.

Sponsors

ALSTEC systems technology

RWE Solutions
RWE NUKEM Limited

Co-sponsors

JSME

IEE

THE INSTITUTION OF CIVIL ENGINEERS

Related Titles of Interest

Title	Editor/Author	ISBN
Improving Maintainability and Reliability Through Design	G Thompson	1 86058 135 8
Engineers' Guide to Pressure Equipment	C Matthews	1 86058 298 2
Handbook of Mechanical Works Inspection – a Guide to Good Practice	C Matthews	1 86058 047 5
IMechE Engineers' Data Book – Second Edition	C Matthews	1 86058 248 6
Integrity of High-temperature Welds	IMechE/IOM	1 86058 149 8
Assuring its Safe: Integrating Structural Integrity, Inspection, and Monitoring into Safety and Risk Assessment	IMechE Conference	1 86058 147 1
Nuclear Decommissioning '98	IMechE Conference	1 86058 151 X
Radioactive Waste Management 2000 – Challenges, Solutions, and Opportunities	IMechE Conference	1 86058 276 1
Storage in Nuclear Fuel Cycle	IMechE Conference	0 85298 998 9
Boiler Shell Weld Repair – Sizewell 'A' Nuclear Power Station	IMechE Seminar	1 86058 244 3
Flaw Assessment in Pressure Equipment and Welded Structures	IMechE Seminar	1 86058 284 2
Materials and Structures for Energy Absorption	IMechE Seminar	1 86058 321 0
Recent Advances in Welding Simulation	IMechE Seminar	1 86058 310 5
Remanent Life Prediction	IMechE Seminar	1 86058 154 4

For the full range of titles published by Professional Engineering Publishing contact:

Sales Department, Professional Engineering Publishing Limited, Northgate Avenue, Bury St Edmunds, Suffolk, IP32 6BW UK

Tel: +44 (0)1284 724384 Fax: +44 (0)1284 718692
Website: www.pepublishing.com

Contents

Regulation, Compliance, and International Co-operation

C596/034/2001	**BNFL's response to the Health and Safety Executive (HSE) Team Inspection Report on the control and supervision of operations at Sellafield** P Bigg	3
C596/035/2001	**Decommissioning – a regulatory view** S J Blakeway and F E Taylor	13
C596/020/2001	**Dounreay – life after audit** P Welsh	23

Finance and Policy

C596/036/2001	**Decommissioning of nuclear installations in the research framework programmes of the EC** H Bischoff	31

Liabilities Management

C596/005/2001	**The management of nuclear liabilities** R M Sellers and D R T Warner	41
C596/010/2001	**Management of liabilities for later decommissioning of nuclear facilities in Germany** P Petrasch, R Paul, and J-P Luyten	49
C596/044/2001	**Software tools for environmental restoration – the nuclear site end game** C R Bayliss, G J Coppins, and M Pearl	61

Public Perception

C596/037/2001	**Public opinion on nuclear power** B Ingham	73
C596/038/2001	**Public acceptance and the development of long-term radioactive waste management policy** C Curtis	81
C596/039/2001	**Public requirements in nuclear waste management** R E J Western	95

Innovation

C596/002/2001	**The potential role of high-power lasers in nuclear decommissioning** L Li	103

C596/021/2001	**Innovations in nuclear decontamination techniques** K Riley, G Fairhall, I D Hudson, and F Bull	123
C596/001/2001	**The Dounreay PFR liquid metals disposal project** B Burnett	139
C596/012/2001	**The RASP – safe size reduction of herogeneous objects** S Bossart, S Rosenberger, H-U Arnold, and M J Sanders	153
C596/013/2001	**Copper cable recycling system – value gained from the recycling of copper from decommissioning projects** R Meservey, C Conner, S Rosenberger, and M J Sanders	161

Learning from Other Industries

C596/024/2001	**Market forces facing the management of environmental liabilities** R Fouquet	171
C596/007/2001	**Health and safety aspects of decommissioning offshore facilities** M Lunt	183
C596/026/2001	**Leveraging military technology for nuclear decommissioning** T Young	195

Contract Management

C596/043/2001	**Contract strategy selection** R D Nicol	207
C596/008/2001	**Partnering experience at BNFL** T J Carr	217
C596/017/2001	**WOMAD – the use of NEC–ECC (New Engineering Contract – Engineering Construction Contract) on a major decommissioning contract** S J Parkinson, K P Gregson, and A Staples	223

Implementation

C596/033/2001	**The Greifswald decommissioning project – strategy, status, and lessons learned** H Sterner and D Rittscher	235
C596/016/2001	**Retrieval of intermediate level waste at Trawsfynydd nuclear power station** S Wall and I Shaw	255
C596/015/2001	**Windscale advanced gas-cooled reactor decommissioning – hot gas manifold dismantling strategy and tooling** G J Walters, S J Batchelor, and M J Steele	265

C596/018/2001	Decommissioning the Berkeley vaults – remote operations P K J Smith, R A Peckitt, and N P Salmon	277
C596/011/2001	Big rock point successfully employs innovative process for fuel pool cleanout K Forrester	287
C596/040/2001	Aspects of decommissioning the UTR-300 reactor at the Scottish Universities Research and Reactor Centre H M Banford, R D Scott, I Robertson, J D Allyson, and T M McCool	303
C596/022/2001	Assessment of the costs of decommissioning the nuclear facilities at Risø national laboratory K Lauridsen	315
C596/045/2001	Robotic decommissioning of a Caesium-137 facility A Murray	325
C596/004/2001	Implementation of NECSA's nuclear liability management programme with special reference to plant decommissioning P J Bredell	337
C596/030/2001	IAEA views on issues, trends, and development in decommissioning of nuclear facilities and member states' experience M Laraia	345
C596/041/2001	Progress in decommissioning of a major alpha processing facility at AWE Aldermaston J Starkey	355
C596/014/2001	Soil sorting gate equipment for site remediation B G Christ, K Froschauer, and G G Simon	369
C596/042/2001	Decommissioning of the nuclear fuel storage ponds at Berkeley Power Station D M Williams	375

Technical Visit

| C596/099/2001 | JET decommissioning project
K A Wilson | 387 |

Additional Paper

| C596/028/2001 | Regulation of decommissioning projects in a Scottish context
H S Fearn, J Gemmill, and M Keep | 405 |

Authors' Index   415

Regulation, Compliance, and International Co-operation

C596/034/2001

BNFL's response to the Health and Safety Executive (HSE) Team Inspection Report on the control and supervision of operations at Sellafield

P BIGG
BNFL plc, Sellafield, UK

1. INTRODUCTION

BNFL's Sellafield operation is the largest nuclear site in the UK. Operations embrace spent nuclear fuel reprocessing, treatment and storage of radioactive wastes and the generation of steam and electricity from four Magnox reactors. In addition a number of older facilities are in the process of being decommissioned. The workforce, comprising of;

    6300 BNFL employees
    1200 agency staff
    3500 contractors and support staff

operate a continuous shift system, working around the clock, 365 days a year, making a significant contribution to the clean up and de-commissioning of nuclear operations world wide. BNFL Sellafield operation is a major employer in the North West and supports several thousand other jobs in West Cumbria.

It is important to understand that the overall safety and environmental performance of Sellafield over the last two decades is generally good. In recent years the trend has continued to improve and the HSE report concluded Sellafield was safe but action was required to make improvements in stated areas.

Following an apparent increase in the number of incidents in the early part of 1999, allied to concerns about progressive reductions of the workforce resulted in the Nuclear Installations Inspectorate (NII) to carry out a Team Inspection of BNFL's arrangements for the control and supervision of operations at Sellafield. The Inspection was carried out in September 1999.

On the 18th of February 2000 the Health and Safety Executive (HSE) published their findings in the report "HSE team inspection of the control and supervision of operations at BNFL's Sellafield site" (Reference 1).

The publication of the report inevitably generated a very high degree of interest from all stakeholders and particularly the media. This provided the Sellafield site management team one of the major challenges it had ever faced. In order to reform BNFL, Norman Askew the newly appointed Chief Executive outlined the key objectives to drive the changes necessary to safeguard the future:

- Safety of the Public and workforce as being the number one priority and

- Restoring confidence in BNFL's ability to deliver high quality goods and services.

BNFL formally published the company's response "Going forward safely" (reference 2) in April 2000 and made major changes to the management of the company.

"Going forward safely" outlined the key activities and milestones comprising the work program designed to address the shortcomings identified by the NII at the time of the Team Inspection.

In response to the HSE report BNFL undertook to re-enforce and demonstrate Sellafield was safe and to:

- Improve the safety management system and management structure.

- Improve resource availability.

- Improve independent inspection, audit and review.

Subsequently BNFL has made a significant investment to deliver a major transformation program to re-visit and raise standards across the whole Site, demonstrating to all stake holders that the company was committed to continual improvement of the environment, health, safety and operating practices.

2. TEAM INSPECTION PROGRESS TO DATE (GENERAL)

In addition to the appointment of a new Chief executive and Sellafield Site Director, BNFL established a core project team to develop detailed action plans to address the recommendations in the Team Inspection report to assure consistent implementation and co-ordinate the linkages and interdependencies between the recommendations. The basis of the approach was designed to secure the highest standards of safe operation at Sellafield.

The implementation of the work program over the last 18 months to achieve the Site re-organisation with a strong focus on quality, consistency, control and clarity of accountability and supervision has proved to be an extremely demanding and stretching task. Not only have the 28 recommendations proved to be highly integrative and far more resource intensive than initially envisaged, but the added burdens of recruiting (800 people at all levels) and ensuring plant in non steady states of production remain compliant have tested the resolve of everyone within the company.

Throughout the project BNFL management have remained committed to their Vision:

- To be a responsible licensee, meeting all legal obligations.
- To reduce hazard potential both on and off the Site.
- To be acknowledged by customers and stakeholders as a Site that values quality in what we do and what we produce.
- To operate in an environmentally responsible manner meeting all our discharge authorisation and consent requirements and working towards HMG" OSPAR commitments.
- To make our plants and processes more robust in order to establish safe, stable and predictable plant operation.
- To introduce better systems and processes to the Site on sensible timescales; in a phased manner; in away that people can absorb the change so that it becomes embedded; and without creating risk.
- To have a skilled motivated and responsible workforce, well led by developing and sustaining management capability at all levels.

Despite the pressures and huge scope of the work, the company has achieved considerable successes in responding to the NII report and the vision has been a key focus in driving change.

The main themes of work outlined in the HSE report are summarised in table 1.

Common Theme	*Area Examined*	*Recommendation*
Safety Management Systems	Corporate Policy	1
	Safety documentation	2
	Management Structure	3
	Independent inspection audit and review	4
	Implementation at Sellafield	5
Safety Management Practices	Management tasks	6,7,8
	Management availability and visibility	9,10
	Role of Duly Authorised Persons	11,12
	Safety Monitoring	13
	Operational Experience Feedback	14,15,16
Management of Change	Baseline for safe operation	17,18,19
	Management of Change Procedure	20
Control and Supervision	Plant Operations	21,22,23
	Maintenance	24,25
Safety Culture		26,27
Recent BNFL Safety Initiatives		28

Table 1 – Main Themes of HSE Report

BNFL's response document "Going forward safely" (Reference 2), outlines the key deliverables and initially planned timescales for responding to the recommendations.

3. PROGRESS ON DELIVERABLES (OVERALL)

Good progress has been made against all of the 28 recommendations. The majority of key deliverables (milestones) have been completed to programme as indicated in Figure 1. A list of progress against each of the 28 recommendations is given in Appendix 1. So far the following actions have been completed:

- 32 out of 41 key deliverables completed to date.
- 7 out of 28 recommendations closed, as indicated in Appendix 1.
- Resources have been assessed across the Site and recruitment undertaken to strengthen the organisation at all levels.
- 545 vacancies filled to date (168 craft, 283 process and 94 management/technical).
- Baseline resource assessments have been completed for the whole of Sellafield
- The Management of change procedure has been implemented to control personnel moves.
- The new organisation has been designed and is currently being implemented.
- The implementation of the single management system has commenced and is being progressively rolled out across the Site.
- The Independent Compliance Advisors and Independent Inspectors are fully operational.
- A single Company EH & S policy statement is now in place.
- A UK Environmental, Health and Safety manual has been produced and issued.
- Improved safe systems of work arrangements have been implemented across the Site.
- Improvements have been made to the labelling of isolation points with over 100,000 valves on the Site being relabelled.
- Over 10,000 personnel have been trained and have in new 'standards and expectations'.
- Improvements in the training and assessment of key personnel have been introduced.

Overall, substantial progress is being made against the key findings of the NII report and demonstrates the commitment and hard work of everyone involved. Achieving the key deliverables in accordance with planned milestones involved Senior Project Managers working closely with representatives from each operating unit and specialists to define the deliverables and develop practical implementation plans and support materials.

BNFL has a full time project team of 30 working on team inspection issues and over 300 full time equivalent people involved across the Site.

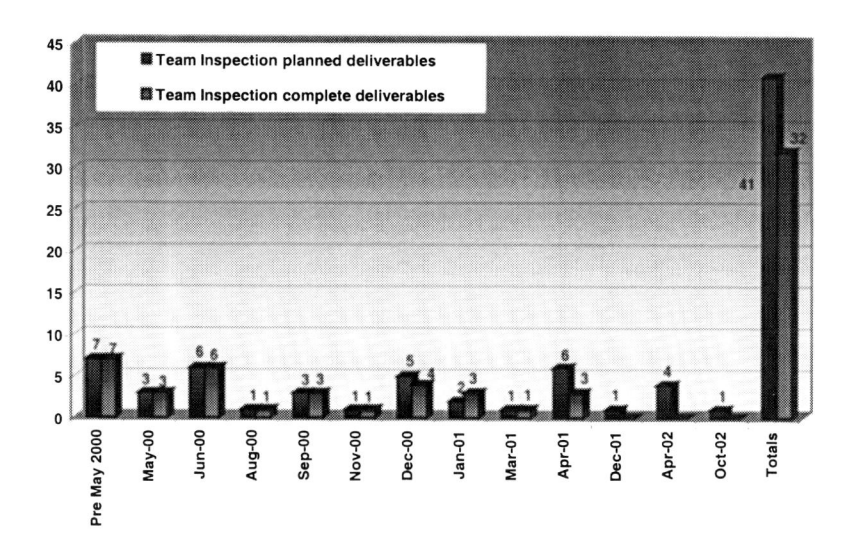

Figure 1 – Progress on Key Deliverables

Within this paper it is not possible to provide detail on all aspects of Team Inspection, and therefore concentrate on providing an insight into the progress that has been achieved overall and in detail by:

The implementation of the new organisation structure.	- Recs 3, 7 & 17.
Communication of Standards and Expectations	- Rec 6.
Independent Inspection, Audit and Review of Health and Safety	- Rec 4.

Work in the above areas continues to reveal the large scope of work to be completed and the intense pressure people within Sellafield are experiencing to achieve planned objectives.

3.1 Sellafield New Organisation

In order to address the NII recommendations substantial changes have been made at board level and clear reporting lines and accountabilities have been established and cascaded to operating units (including specific safety accountabilities) for key post holders. Individual specific responsibilities are currently been incorporated and finalised within each post holders Role Description.

The Site, at the time of the NII report was operating as a group of individual business units. This has been addressed, recognising interdependencies between business units and the strive to provide common site standards to ensure Sellafield operates as an integrated Site.

Currently the new organisation is in the process of implementation and work is underway to build on the new structure by implementing clear standards and responsibilities impacting all staff on Site, introducing new Safe Systems of Work and verifying compliance to new arrangements.

3.2 Standards and Expectations

BNFL have produced a list of 10 clearly stated standards and expectations for the way people should work. They were developed in partnership with management and employee representatives and are illustrated in Figure 2.

Key to effective implementation is the engagement of the whole work force in exhibiting behaviours that underpin the standards.

Over 10,800 people at Sellafield have been trained in these standards through a series of group awareness briefings supported by a video, encouraging ownership, involvement and compliance and developing a commitment to "professional general conduct".

Secondly, small group discussions were successful in focusing on local issues and established the differences between good and poor service provision. Individuals were encouraged to take away and implement one personal action in their workplace.

Regular audits take place to ensure these standards are maintained and become embedded within the organisation.

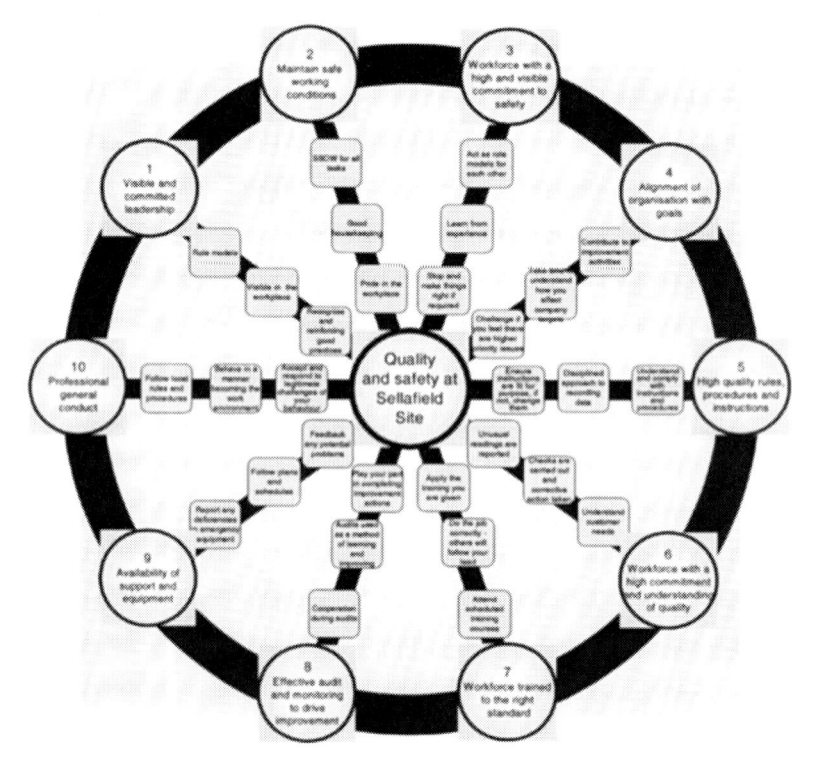

Figure 2 – Standards and Expectations

3 Audit and Inspection

Significant progress has also been achieved as a result of the success of the appointed Independent audit and Inspection Teams. Independent Compliance Advisors (reporting to the Director of Operations via the Head of Environment, Health and Safety) and Corporate Independent Inspectors (reporting to the Chief Executive via the Head of UK EH&S and the Head of Corporate safety management and assurance) were appointed in response to NII recommendation 4.

A team of six Independent Compliance Advisors (ICA's) were appointed at Sellafield. The team is made up of experienced operations staff. Their role is:-

- To provide assurance to the Director of Operations that "Compliance" is being achieved on Site.
- To provide advice to Sellafield employees on how to achieve compliance with Site procedures and legislation.

This essentially involves 20% inspection and 80% advice, mentoring to help employees comply with Sellafield requirements.

The team of 10 Corporate Independent Inspectors (CII's) cover the whole of BNFL's UK operations and were chosen primarily from external candidates. Their role is:-

- To provide assurance to the corporate centre that Health and Safety policy is being implemented throughout the company.

CII's effectively spend 80% of their time Inspecting (verifying compliance) and 20% providing advice. They act as the Company's internal inspectors.

The ICA's and CII's have been successful in working with operating units and are now a respected catalyst, key to encouraging improvement, raising standards of performance and the helping teams to adopt best practice across the Site.

Despite the above early success, BNFL are committed to developing the role of the ICA's in order to improve the ways of working with the CII's and other auditors to avoid "Audit Overload", be more demanding of operating units and foster working relationships with other Regulatory bodies.

4 CONCLUSIONS

Team Inspection work, consistent with the successes and challenges outlined in the above progress summaries has proved to be a demanding and resource intensive challenge. Success to date demonstrates BNFL's commitment and resolve to work towards implementing safe systems of work to world class standards. BNFL is committed to working to their Vision, systematically closing out NII recommendations and continuously improving operations demonstrating the highest standards of safety to all stakeholders.

REFERENCES

1 "HSE team inspection of the control and supervision of operations at BNFL's Sellafield site".
2 "Going forward safely" BNFL's response to the HSE team inspection report on the control and supervision of operations at Sellafield.

Appendix 1 – Team Inspection Recommendation Status

Recommendation 1 Health and safety policy • A revised corporate health and safety policy has been issued, signed by the Chief executive.	**Closed out by NII**
Recommendation 2 Safety management systems • A UK EH&S manual has been produced and issued.	**Closed out by NII**
Recommendation 3 Safety management structure • The new Chief Executive has made significant changes to the organisation at the most senior level to ensure responsibilities and accountabilities are clear. A Director of Operations for Sellafield has been appointed and his top management team established.	**Closed out by NII**
Recommendation 4 Inspections and audits • A team of corporate inspectors has been recruited and have been in post for 6 months. A Sellafield team of independent compliance advisors have been in post for 9 months. The WANO peer review has been adopted and an assessment is planned for 2002.	3 of 3 Key deliverables complete
Recommendation 5 Standardise detailed working arrangements • A programme for the review and development of a single set of Sellafield site arrangements has been prepared. A number of revised documents have been issued and are being implemented across the Site.	Key deliverables in progress
Recommendation 6 Performance standards • Standards and Expectations have been defined and introduced to all employees across Sellafield.	2 of 3 key deliverables complete
Recommendation 7 Organisational structure • A new organisation structure and accountabilities have been defined for all management positions on the Site. Implementation of these changes is in hand.	3 of 3 key deliverables complete
Recommendation 8 Day to day operations control • Action has been taken to ensure that BNFL personnel are directly responsible for all work undertaken by a contractor on the Site. Additional controls have been implemented on the deployment of contractors.	Complete
Recommendation 9 Manage safe operations of plant • Targets for the time plant managers spend on safety related issues have been set. Work is in hand to reallocate priorities to ensure that they can be achieved.	1 of 2 key deliverables complete

Recommendation 10 Systematic assessment • A review of safety work load of staff has been undertaken in line with the response to recommendation 7.	2 of 2 key deliverables complete
Recommendation 11 Safety related training • A consistent process for identifying, assessing and recording training needs and delivery has been defined. Work is currently being taken to implement this across the Site.	1 of 2 key deliverables complete
Recommendation 12 Appointment of DAPs and SQEPs • A consistent process for assessing and appointing Duly Authorised Persons and others with safety duties has been defined. Work is currently being undertaken to implement this across the Site.	1 of 1 key deliverables complete
Recommendation 13 Proactive monitoring • Proactive monitoring has been introduced across the Site. A review of its implementation is currently being carried out.	1 of 1 key deliverables complete
Recommendation 14 Record and investigation of incidents and recommendations • Improved arrangements for reporting, recording and investigation of incidents have been implemented across the Site.	Complete
Recommendation 15 Learning from experience • Revised arrangements for learning from experience have been issued. Following concerns expressed by NII, an independent review of LFE at Site has been undertaken. The results of this review are currently being evaluated and an action plan established.	1 of 1 key deliverables complete
Recommendation 16 Exchange of information re LFE • Formal arrangements for the exchange of information with other chemical and nuclear operators have been established.	1 of 1 key deliverables complete
Recommendation 17 Baseline resources • The baseline resource assessment for the Sellafield site has been completed.	**Closed out by NII**
Recommendation 18 Minimum staffing levels • Arrangements for establishing minimum staffing levels have been established. These levels have been included as part of the baseline assessment work.	2 of 2 key deliverables complete
Recommendation 19 Key roles/vacancies • Key roles have been identified as part of the baseline assessment, but further work is required.	**Closed out by NII**
Recommendation 20 Management of change procedure • Revised management of change arrangements have been implemented across the Site.	**Closed out by NII**
Recommendation 21 Safety envelope • Based on best industry practice, BNFL has developed systems for confirming that the plant is operating within its safety case at all times. These are being introduced in the near future on a trial basis.	1 of 2 key deliverables complete

Recommendation 22 Deviation from normal operations • BNFL has specified processes for better control of operations with particular reference to deviations from normal conditions.	As Recommendation 21 above
Recommendation 23 Labelling of plant and equipment • A programme for implementing consistent arrangements for plant labelling and plant configuration control has been developed together with the provision of up to date drawings. Work against this programme is on target	1 of 2 key deliverables complete
Recommendation 24 Safe systems of work site wide • A revised safe system of work procedure has been issued and implemented across the Site. Key personnel have been trained in the operation of this system.	1 of 2 key deliverables complete
Recommendation 25 Plant handover/handback • A revised system of isolations, handover and handback has been issued. Training of key personnel in the operation of this system has commenced.	Complete by December 2001
Recommendation 26 Safety culture/behaviour • A bench mark assessment of the safety culture/behaviour at Sellafield has been completed. An action plan to address areas of concern has been developed.	1 of 1 key deliverables complete
Recommendation 27 Review resource availability/management action matching words • This will be one of the last recommendations that can be closed out as it relies on the majority of other recommendations.	See other recommendations
Recommendation 28 Review of safety initiatives • A review of initiatives has been completed and this has resulted in a number being terminated whilst others have been rationalised.	**Closed out by NII**

© With Author 2001

C596/035/2001

Decommissioning – a regulatory view

S J BLAKEWAY and **F E TAYLOR**
HM Nuclear Installations Inspectorate, Health and Safety Executive, Bootle, UK

ABSTRACT

Decommissioning is the final stage in the life of a facility and is regulated in a similar manner to the operational phase. The nuclear regulatory regime is described including the new regulations on Environmental Impact Assessment for reactors. The Health and Safety Executive has recently issued guidance to its inspectors on the regulation of facilities being decommissioned. A number of aspects of this guidance are discussed including strategies, timetables, safety cases, management and organisation and finally delicensing.

1. INTRODUCTION

Decommissioning is the set of actions taken at the end of a nuclear facility's operational life to take it permanently out of service with the ultimate aim of making the site available for other purposes. As far as the regulator is concerned the expectation is that the process should have adequate regard for the health and safety of workers and the public and the protection of the environment. The actions taken should systematically and progressively reduce the level of hazard on a site, and will in most cases include the physical dismantling of the facilities. It is not necessarily a single step process and may involve work being done in stages over a significant period of time.

From a nuclear licensing point of view decommissioning is a continuation of the period of operation and therefore the existing regulatory framework, which is generally non-prescriptive, is applied to this type of work. The Health and Safety Executive (HSE) also takes account of Government policy, which was last stated in 1995 (1). As the nuclear industry matures many facilities have reached the end of their lives and have either been decommissioned or their decommissioning is being planned. There are many challenges to both the industry and the regulator as these complex projects get under way especially since many of the older plants were not designed with decommissioning in mind. This paper discusses some aspects of the regulatory framework and our expectations of nuclear licensees which have recently been made available as guidance to HSE inspectors.

2. NUCLEAR REGULATORY REGIME

The main legislation governing the safety of nuclear installations is the Health and Safety at Work etc Act 1974 and those sections of the Nuclear Installations Act 1965 (as amended) which are relevant statutory provisions. Under the Nuclear Installations Act no site may be used for the purpose of installing or operating any nuclear installation unless the operator holds a valid licence from HSE. Once a licence has been issued, there is a continuing period of responsibility of licensees under the Nuclear Installations Act throughout operation and decommissioning, until there is no longer any danger from ionising radiations from the site.

HSE may attach conditions to nuclear site licences which appear to it to be necessary or desirable, in the interests of safety, or with respect to the handling, treatment and disposal of nuclear matter. These standard licence conditions are essentially non-prescriptive and many of them require the licensee to make and implement adequate arrangements to address safety issues (2). The regime is flexible in that it can be used for all stages of the life of the plant and each licensee can develop arrangements which best suits its business. Consequently these conditions apply equally to decommissioning and operating sites and form a continuous process of regulation "from cradle to grave". The arrangements that a licensee develops to meet the requirements of the licence conditions constitute elements of a safety management system. HSE reviews the licensees' arrangements to check that they are clear and unambiguous and address the main safety issues adequately. It also carries out regular inspections to check compliance.

A particular requirement, under these conditions is that the licensee should produce and implement decommissioning programmes. The current expectation is that outline decommissioning programmes should be available for plants at the design stage and more detailed programmes would normally be produced some time before cessation of use of the plant. HSE has the option of approving these programmes which then cannot be changed without the approval of HSE. In addition, the arrangements for decommissioning should, where appropriate, divide decommissioning into stages and HSE has the power to specify that the licensee shall not proceed to the next stage without its consent. These powers under the licence can be used by HSE to give a high degree of regulatory control over decommissioning. Also of relevance to decommissioning are several licence conditions which relate to the management of radioactive waste and other material on the site, for which HSE has regulatory responsibility.

During the decommissioning period, of course all other health and safety legislation will apply and some will be prominent as the plant state changes and projects commence. The Ionising Radiations Regulations 1999 will remain important as it will be necessary to ensure that doses remain as low as reasonably practicable during different operations. Projects will need to be managed under the Construction (Design and Management) Regulations 1994. Other regulations such as the Provision and the Lifting Operations and Use of Work Equipment Regulations 1998 and Lifting Equipment Regulations 1998 will also be relevant.

This paper covers HSE's regulatory responsibilities, however it should be noted that discharges to the environment and the disposal of radioactive waste are regulated under the Radioactive Substances Act 1993 by the Environment Agency and the Scottish Environment Protection Agency. Close liaison is maintained between the HSE and the two environment agencies in addition to specific statutory consultation arrangements. This is to ensure that all aspects are properly regulated and licensees are not subject to conflicting requirements.

Licensees are also required to make submissions under Article 37 of the Euratom Treaty describing the potential radiological impact of decommissioning on other European Union States. These are Government documents and the environment agencies generally manage their preparation.

3. ENVIRONMENTAL IMPACT ASSESSMENT

The recently amended European Directive on Environmental Impact Assessment covers decommissioning projects associated with nuclear reactors above 1 kW continuous thermal load. The aim is to ensure that the possible environmental effects of decommissioning reactors are properly considered before the project commences. The licensee should also consider the impacts of different ways of implementing the options. It is recognised that it is difficult to assess environmental impact in detail over the very long time span of some of these projects. Nevertheless the licensee is expected to identify the more significant effects.

This Directive has been implemented in the UK by the Nuclear Reactors (Environmental Impact Assessment for Decommissioning) Regulations 1999 which are administered by HSE. They apply both to new decommissioning projects for nuclear reactors and nuclear power stations and also to changes in existing projects which may have a significant adverse environmental effect. Licensees will need to produce an Environmental Statement describing the Environmental Impact Assessment to support their applications for consent from HSE to commence the project. The public is given access to the Environmental Statement and is invited to make representations. Before granting consent, HSE is required to consult with a wide range of interested parties and must be satisfied that an adequate Environmental Impact Assessment has been performed. A decommissioning project is deemed to have started when action has been taken to permanently disable the reactor so it cannot return to service. Defuelling, as long as it is carried out using the normal operating procedures, is not included in the project.

An optional stage in the process is that a licensee can request a pre-application opinion from HSE of the proposed content of the Environmental Statement. This gives consultees the opportunity to comment and to suggest topics which should be considered. In particular local consultees are able to provide information on local issues. This process has been undertaken for the BNFL (Magnox Generation) stations at Bradwell and Hinkley Point A. HSE recognised that the scoping document could not give a complete picture of the intent but came to the view that BNFL would be covering most issues in its Statement. Therefore, taking account of the comments from consultees, HSE has provided its opinion on issues that need to be covered that were not explicitly described in the document. In particular it pointed out the need for BNFL to cover the whole project and discuss the impact of options. The full reports are expected later in the year will be sent out for wide consultation. HSE will assess the Statements taking into account the comments received and, if satisfied grant consent for the projects to commence.

4. GOVERNMENT POLICY

Government policy on radioactive waste management was reviewed in 1994/95 and the conclusions of that review were set out in "Review of Radioactive Waste Management Policy, (Cm 2919)" (1). This includes policy concerning decommissioning which has been taken into account by HSE in developing guidance for its inspectors. Following the publication of the report of the House of Lords Select Committee Enquiry into Nuclear Waste Management (3), issued in March 1999, the government has announced a review of this policy.

The OSPAR/Sintra agreement (4), which the government signed in July 1998, commits the UK to a progressive and substantial reduction of the radioactivity in liquid discharges by adopting best available techniques, such that additional concentrations in the marine environment above historic levels, are close to zero by 2020. This agreement may have an impact on existing disposal routes and requirements for discharges during decommissioning.

5. DECOMMISSIONING GUIDANCE

As explained earlier the licensing regime is goal setting rather than prescriptive and therefore HSE does not generally issue guidance to the nuclear industry on its expectations. However it has published the Safety Assessment Principles (5) used by its inspectors in assessing licensees' proposals. It also issues more detailed guidance to its inspectors. Over the last few years HSE has been developing and consolidating its guidance on radioactive waste management and decommissioning taking account of Government policy. Because it realised that this was a topic of wide interest, HSE discussed the draft guidance with interested organisations and presented papers at conferences on its emerging thinking. It found this to be a very useful exercise. This work is now complete and the guidance has been placed on the Nuclear Safety Directorate section of the HSE website. The guidance will be reviewed regularly in the light of experience with its use.

The guidance starts with four fundamental expectations for the decommissioning of nuclear facilities, which, should be met so far as is reasonably practicable, they are:

i) In general, decommissioning should be carried out as soon as it is reasonably practicable, taking account of all relevant factors.

ii) Hazards associated with the plant or site should be reduced in a progressive and systematic manner.

iii) Full use should be made of existing routes for the disposal of radioactive waste.

iv) Remaining radioactive material and radioactive waste should be put into a passively safe state for interim storage pending future disposal or other long-term solution.

It then goes on to discuss a number of aspects in more detail and some of these are described below.

5.1 Strategic Planning

Many decommissioning projects are large and complex, hence good strategic planning is essential to ensure that the work proceeds efficiently and the resultant radioactive wastes are managed effectively. The strategy should identify the extent of a licensee's liabilities, describe the means of dismantling each part of the facilities and the management of all the radioactive material and waste. The timescales over which the different stages of decommissioning will take place should be defined. The strategy should be linked to, or integrated with, the strategy for the management of existing radioactive waste and waste which is produced during decommissioning. If a licensee is responsible for a number of nuclear sites then it may be appropriate to provide a corporate strategy supported by series of site-specific strategies.

In selecting a preferred strategy, licensees should demonstrate that they have examined a full range of options covering different timescales, technical factors, social factors and financial factors. Licensees should demonstrate that their strategy is consistent with Government policy and consider the extent to which it is consistent with the concepts of sustainable development and the precautionary principle. It should be clear how the costs of implementing the strategy have been estimated and how the appropriate funds will be provisioned.

Licensees' strategies should take into account uncertainties in the future especially since decommissioning of nuclear facilities will continue for many decades in to the future. The next 50 to 100 years will undoubtedly see many social, political and environmental changes. Over recent years the perception of risk by society has changed and people are becoming more averse to those risks which are imposed upon them, are unevenly distributed, affect future generations or the environment. Other trends such as stricter regulation of radiation exposure, radioactive waste disposal, increased regulation from Europe may place further responsibilities on licensees in the future. The effects of climate change may challenge the design bases on which the plant and structures were originally designed.

In particular estimates of costs will be increasingly uncertain as they are projected into the future. As well as unforeseen influences and events that can lead to additional costs, there are a number of costs which are virtually certain to rise in the future such as those associated with waste management and disposal. Similarly, there is an uncertainty associated with the projection of investment returns into the future to demonstrate adequate financial provision.

5.2 Quinquennial Reviews of Decommissioning Strategy

HSE, in consultation with the environment agencies, has been carrying out quinquennial reviews (QQRs) of licensee's decommissioning strategies as requested by the Government (1). In its review HSE assesses the strategy as to determine whether it is technically feasible and in addition consistent with existing legislation, safety and environmental requirements. HSE will seek to satisfy itself that the strategies remain soundly based as circumstances change. The review also examines the financial provisions to determine whether they are adequate to meet the costs of the decommissioning strategy, and also flexible enough to take account of changes and do not foreclose the option of carrying out decommissioning on an earlier timescale should that be required.

HSE expects licensees to produce a publicly available document setting out their decommissioning strategy, and HSE will in turn publish the conclusions of its review. The first round of QQRs is well underway and it is covering all the UK licensees.

HSE recognises that licensees have undertaken substantial studies to support their decommissioning strategies. However, the area is complex and so it is not surprising that as a result of these reviews HSE has identified the need for further work in areas such as in the identification of liabilities, the clarification of responsibilities and the justification of the timing of dismantling.

5.3 Decommissioning Timetables

One of the key elements that will emerge as a result of strategic planning is a decommissioning timetable and a number of factors will influence the timescale over which decommissioning takes place. Some of these are discussed below. It also needs to be recognised that if necessary, in the interests of safety, HSE may require decommissioning to be completed on an earlier timescale than originally planned.

Government policy (1) states that decommissioning of nuclear facilities should be done as soon as it is reasonably practicable, taking account of all relevant factors and this is recognised in HSE's first fundamental expectation. The next expectation is that the hazards associated with shutdown facilities will be reduced in a progressive and systematic manner, over an appropriate period of time. The rate at which work proceeds will be determined by several factors, each of which will exert a particular influence, and licensees will need to demonstrate that they have considered and balanced these influences in reaching and justifying their proposals on a case-by-case basis. The Government's expectation is that timetables will be developed that were acceptable to the regulators. It follows therefore that decommissioning may proceed as a continuous activity, or if appropriate, as a series of sequential stages, the end result of each stage being a significant reduction in hazard. The order, timing and extent of each stage will be influenced by the hazard posed by a particular plant on a site with the priority being on those with the highest hazard or risk. It is recognised that in some circumstances, actions may be required which temporarily increase risk to enable hazard reduction to take place and this will clearly need to be justified.

Another of HSE's fundamental expectations is that full use should be made of existing routes for the disposal of radioactive waste. At present routes are only available for low level wastes but it is expected that at some time in the future routes will become available for other wastes and HSE will expect licensees to make use of them. In the meantime other wastes, which are potentially mobile should be removed and processed, for passive safe storage (6) pending disposal. In general priority should be given to the most significant hazards which are likely to be represented by the most active materials, radioactivity that is in a mobile form, and waste that is inadequately characterised, packaged, contained in nuclear facilities that do not meet modern standards, or that are deteriorating.

HSE considers that decommissioning techniques are sufficiently developed and proven for licensees to undertake decommissioning of facilities in the short term but recognises that there are potential safety benefits from deferring dismantling of some installations containing short-lived radionuclides. In the particular case of reactors, where the principal form of radioactivity is Co-60 with a half life of 5.3 years, HSE recognises that there are safety benefits from deferring final reactor dismantling for several decades in order to reduce operator and public radiation exposure, and also in reducing the quantity and possibly the classification of radioactive waste that is produced. However, the benefits in radiological safety and waste generation from deferral diminish with time and there will be a point at which there is little further gain. On the other hand, there are nuclear chemical facilities contaminated with plutonium, where the in-growth of americium leads to a radiological hazard that increases with time, and there is a detriment from delaying decommissioning.

The benefits of delay must be balanced against those of placing any hazardous material into a state of passive safety as soon as practicable. In addition, if periods of deferral are proposed, licensees will need to demonstrate that they can ensure long-term safety by maintaining an appropriate organisation, supporting infrastructure and corporate memory. There is also the risk that the physical structures will degrade leading to an increased possibility of leakage of radioactive materials.

In terms of costs, early dismantling of facilities will be more expensive initially, but this must be compared to the long-term costs of deferral, because of the need to maintain the facility in a safe condition, the licensees' infrastructure, the increasing risk of unforeseen routes for environmental releases and other uncertainties associated with the future. In addition, where licensees use discounting techniques to compare costs incurred at different times this inevitably favours the options for delay, and HSE believes that such arguments should be used with caution in justifying the timescales for decommissioning.

The *Precautionary Principle* describes a philosophy for addressing potentially serious risks subject to high scientific uncertainty, particularly where they are in the environmental field and where there are risks that could affect future generations. It basically prescribes that as uncertainty increases then emphasis should increase on reducing the hazard by cost-effective means.

In HSE's view, early decommissioning, particularly where there are few safety benefits from deferral, is more in line with the Government policy of sustainable development and the use of the precautionary principle. In addition, it is noted that, internationally, the approach is for early decommissioning; this is in marked contrast to the approach of some UK licensees.

5.4 Safety Cases

The nuclear site licence requires licensees to produce safety cases for the operation of facilities at all stages of their life, including decommissioning. Indeed the current expectation is that the safety aspects of decommissioning should be considered when the plant is designed. The licence conditions also require that the safety case should be regularly reviewed and HSE's expectation is that a major review should be carried out every 10 years. In addition during decommissioning, the safety case should be updated when necessary to reflect the impact of modifications to the facilities and to address the changing nature of the hazard. Some decommissioning activities may temporarily result in an increase in the risk in order to achieve an overall reduction in the hazard. In such instances, the safety case must identify the impact of the changes and demonstrate that the risks remain at an acceptable level whilst the work is undertaken. Examples of such activities could include partial dismantling of structures, removal of systems, post-operational clean out and radioactive waste retrieval and processing.

Many buildings, structures or equipment may have to remain in place for long periods of time during the decommissioning of a nuclear facility. In such cases the assessment of the continuing safety of the nuclear facility involves determining the current physical condition and establishing how it will change in the future. The safety case should describe the arrangements for the continued surveillance, maintenance and monitoring of the facilities that will ensure that any unexpected degradation will be detected. Similarly, adequate arrangements should be made for detecting leakage of radioactivity.

5.5   Management and Organisation

A fundamental basis of UK nuclear regulation is that the licensee bears the sole and absolute responsibility for safety and that this cannot be delegated to another party. Significant changes in organisational structure and personnel are likely to occur during decommissioning. One of the conditions attached to the site licence requires licensees to have arrangements for management of organisational change. These should be used in advance of shutdown and throughout the decommissioning process which, in some cases, may last for a very long period time. The licensee is expected to have in place an appropriate management structure and numbers of staff, both to fulfill the key roles associated with licensees' responsibilities under the site licence and to provide the general infrastructure to support the project.

Contractors are of course used widely in decommissioning and contribute valuable skills and expertise. HSE has been considering the implications of the use and control of contractors, and other arrangements such as partnering, and has presented its views on a number of occasions (7, 8, 9). In this context, the term intelligent customer has been in use for several years, and essentially it means that a licensee must take steps to retain an adequate capability within its own organisation to understand the nuclear safety requirements of all of its activities, and also those of any contractors; to take responsibility for managing safe operation; and to set, interpret and ensure the achievement of safety standards.

In carrying out decommissioning work there are practical benefits, in terms of doses and costs, to be gained from using workers from the former operating team who have good knowledge of the facilities and the site. If decommissioning is deferred, then it is unlikely that this resource will still be available when dismantling starts. In that case, licensees must make arrangements to capture and maintain the knowledge base, and to assemble and train teams of workers to undertake activities when required, which might be a considerable time into the future.

5.6   Delicensing

The ultimate end point of the decommissioning process is the return of the site to unrestricted use and its release from regulatory control or delicensing. According to the Nuclear Installations Act a nuclear licensee's period of responsibility for the site only ceases when HSE has given notice in writing that, in its opinion, there has ceased to be any danger from ionising radiations on the site or the licence has been transferred to another corporate body. It is clear that the intention of the Nuclear Installations Act is that sites should be able to be delicensed but there is no definitive guidance on the appropriate criteria. It should be noted that there is no facility in UK law for releasing the site for restricted use only.

HSE has delicensed a number of sites in the past. These were mostly research facilities and each case was assessed on its merits. In the most recent cases it was reasonably straightforward to demonstrate that the levels of radioactivity were indistinguishable from background. HSE is currently developing generic criteria for delicensing and a number of papers have been published describing its emerging thinking (10, 11). This will assist in its consideration of applications from more complex sites. In developing criteria HSE will need to take into account forthcoming regulations on radioactively contaminated ground being prepared by the Department of Environment, Transport and the Regions and the Scottish Executive.

As a first step the licensee will be expected to undertake a thorough survey of the site and identify radioactive material including any in structures or in the ground. HSE will generally wish to have an independent survey undertaken on its own behalf before being satisfied the site can be delicensed.

It is HSE's view that it would not be able to delicence a site if it fell above the exemption limits of other legislation such as the Radioactive Substances Act 1993 or the Ionising Radiations Regulations 1999. This implies, for example, that the concentration of most radioactive materials in soil would need to be less than 0.4 Bq/g. In addition HSE would expect the licensee to consider reducing levels further where this is reasonably practicable. HSE believes this approach is consistent with its general approach to risk reduction described in the document "Reducing Risk, Protecting People" (12).

6. SUMMARY AND CONCLUSIONS

The flexible nature of the nuclear licensing regime enables HSE to regulate decommissioning as a stage in the life cycle of the plant. New regulations on the need for an Environmental Impact Assessment before the start of a reactor decommissioning project have recently come into force. Two major projects, at Bradwell and Hinkley Point A are starting to go through the process. HSE has recently issued detailed guidance to its inspectors which has been placed on its website. This will assist a consistent approach to the regulation of facilities being decommissioned.

© Crown Copyright 2001

REFERENCES

1. Department of the Environment *Review of radioactive waste management policy: Final conclusions* Cm2919 HMSO 1995 ISBN 0 10 1291922.

2. Health and Safety Executive *Nuclear site licences under the Nuclear Installations Act 1965 (as amended): Notes for applicants* HSE Books 1994 ISBN 0 7176 0795X.

3. The Government Response to the House of Lords Select Committee Report on the Management of Nuclear Waste, DETR, October 1999.

4. "OSPAR Convention: Convention for the Protection of the Marine Environment of the North-East Atlantic - the United Kingdom became a signatory at Sintra, Portugal during the Ministerial Meeting of the Oslo/Paris ("OSPAR") Commission in July 1988.

5. Health and Safety Executive *Safety Assessment Principles for Nuclear Plants,* HSE Books 1998, ISBN 0-11-882043-5.

6. L Williams, D Mason, S Blakeway and C Snaith *A regulatory view of the long term passively safe storage of radioactive waste in the UK* Proceedings of an international conference on Safety of Radioactive Waste Management, IAEA Cordoba, 13-17 March 2000, IAEA 2000 ISBN 92 0 101700 6.

7. F E Taylor and D Turton *Regulatory Requirements for the use of Contractors on Nuclear Licensed Sites* Nuclear Energy, 1998, 37, No 1, 55-58.

8. F E Taylor and A Coatsworth *Partnering in the Nuclear Industry. A Regulatory Perspective* Proceedings of an International Conference in Alliancing, INE, Erskine, 9-11 June 1999.

9. F E Taylor *Regulatory View of the Nuclear Licensee as an Intelligent Customer* Nuclear Energy, 2000, 39, No 3, June, 175-178.

10. I F Robinson *De-licensing Nuclear sites in the UK* P-9-133 in Proceedings of the 10th International Congress Site International Radiation Protection Association May 14-19 2000 Hiroshima, Japan.

11. I F Robinson *A Nuclear Inspector's perspective on decommissioning at UK Nuclear Sites*, J Radiol. Prot. 1999 Vol. 19 No. 3 p203-212.

12. Health and Safety Executive *Reducing Risk, Protecting People* Discussion Document, HSE Books 1999.

© Crown Copyright 2001

C596/020/2001

Dounreay – life after audit

P WELSH
UKAEA, Dounreay, UK

INTRODUCTION

A joint Health and Safety Executive/Scottish Environmental Protection Agency (HSE/SEPA) team carried out an Audit of the management of safety at UKAEA Dounreay in June 1998. The Audit report was published on 1st September 1998 and contained 143 recommendations for action by UKAEA, including the preparation of an Action Plan. On 30th November '98 UKAEA published 'The Way Forward' which contained UKAEA's initial response to the Audit and the Action Plan for addressing the audit recommendations.

UKAEA have taken the opportunity provided by the Audit to review all their systems and processes, not only addressing the NII recommendations but looking forward to the business needs of the future and making the changes necessary to allow site restoration to progress in a timely manner.

It would be tedious in the extreme to describe changes made to the business against each recommendation. Instead this paper summaries a number of areas where key improvements have been made:

	No. Recommendations
• Management and Organisation	24
• Human Resource	7
• Safety Culture	16
• Safety Cases	26
• Safety Management Systems	23
• Integrated Waste Management and Decommissioning Strategy	46

90 of the 143 recommendation have now been completed and 74 of these have been formally closed-out by the NII, the rest being under review at the present time. The remaining 53 recommendations are of a long term or strategic nature. For each of these recommendations a position statement has been written and the action transferred to a monitorable programme such as those for plant upgrades or safety case delivery. NII and SEPA will continue to monitor these programmes as part of normal regulatory business.

MANAGEMENT AND ORGANISATION

The main finding of the HSE/SEPA audit was that organisational changes made within the UKAEA had so weakened the management and technical base at Dounreay that it was not in a good position to tackle its principal mission, i.e. the decommissioning of the site. The issues were those of control, e.g. the degree of contractorisation, the management of contractors and the use of contractors in safety related posts, and, of having sufficient in house skills and experience to be able to act as an intelligent customer.

UKAEA has addressed the issue of control by:

- Dismantling the Management Support Contract. The contract ceased to exist earlier in the year and all senior managers at Dounreay are now UKAEA staff.

- Introducing a new management structure, designed to ensure that UKAEA is in direct control of all key safety related processes and that all responsibilities and accountabilities are clearly understood and encapsulated in the site management procedures. The structure has a strong functional basis with the site's key roles of waste management and decommissioning at its centre. These groups are supported by three service groups, Business Support, Engineering and Safety and Environment.

- Introducing clear visibility of how UKAEA complies with each licence condition and who is responsible for ensuring compliance with each criterion.

- Strengthening the teams responsible for management of contractors, clearly defining the circumstances under which a contractor can be used and the skills and experience required of the contractor, and strengthening the policy on the control of contractors.

To be able to concentrate on achieving its aim of 'restoring our environment' without continual distraction Dounreay needs the support of the public, the regulators, government and its staff. Whilst improved communications with stakeholders, better management of the media and consultation all have a part to play, obtaining widespread confidence ultimately depends upon continuously delivering a high quality performance. To effectively deliver the site forward programme, including in the short term responding satisfactorily to the Audit and lifting the NII Direction Dounreay, requires efficient processes and an effective organisation.

To this end, Dounreay has:

- Provided a clear vision of its future business direction. The first step in clarifying the site's vision was the decision to stop future commercial work, thus enabling the site to concentrate upon the core activities of waste management and decommissioning. The second stage was the establishment of high level site goals, objectives and targets relating to Safety and Environment, People and Business delivery, to which all staff could relate, and which lead to a unity of focus across the site.

- Introduced arrangements to measure achievement, whether personal or site, against stretching, challenging and realistic targets. Where possible the site now measures itself against nationally or internationally recognised standards, e.g. ISRS, IIP accreditation, ISO 14001 and ISO 9001 and uses these standards to monitor continuous improvement.

- Revised the management structure so that it aligns more closely with the site objectives and complementary business plans for operating and service groups.

HUMAN RESOURCE

Human Resource issues are especially important at Dounreay, not least because of its remote location and the executive team take a strong lead on ensuring adequate numbers of competent staff are available to undertake the work of the site. A key part of this work involves succession planning. Training needs are identified on an individual basis and harmonised with UKAEA and site management training requirements into a Site Training Plan. All training on the site is co-ordinated by a new central training organisation.

To ensure a sufficient supply of suitably qualified staff to support the future programme. Dounreay has put in place:

- A forward resource plan that identifies the staff required to support the future programme, in terms of numbers and disciplines.

- Post profiles that identify the experience skills, training and qualifications needed for each post.

- Identification of the competence and qualifications possessed by individual staff.

- Derivation of training, succession, staff development and recruitment plans.

The information is fed into a database, incorporating a set of human resource management tools, to allow mangers to identify training needs for their staff and manage the release of staff for training.

The process provides managers with the information they need to manage future manpower requirements, including ensuring that their staff are suitably qualified and experienced for the tasks expected of them. A list of safety related posts has been compiled as a subset of the total list of posts, and include those posts which are required to meet statutory obligations, both nuclear and non nuclear.

SAFETY CULTURE

The issues of safety culture are the most difficult to address within any organisation. The way people behave and react to a given situation is influenced by a huge variety of factors, not all of them organisational. Dounreay has taken the view that culture can most readily be influenced by setting standards for the organisation, monitoring compliance to the standards and by rewarding examples of good practice.

Dounreay has established an integrated programme of activities directed at improving safety culture. The programme is identified in the Site Business Plan; targets are set and levels of achievement are regularly monitored. The activities include:

- Increasing management presence in the workplace.

- Regular safety road shows addressing safety both in the workplace and at home. The road shows have been well attended and have highlighted many issues that people may not otherwise been aware of, for example a recent seminar focused on the dangers of carbon monoxide.

- Encouraging and providing support for visits by staff to other sites and companies.

- Providing operational feedback, particularly with respect to incidents both internal and external, with emphasis placed on lessons that can be learnt.

- Operating a Team Brief system whereby a prepared brief, together with any local issues is delivered on a weekly basis by first line supervisors to their work teams. The brief is aimed at improving communication within the organisation, so that information is delivered by supervisors rather than through the grapevine.

- Encouraging staff that identifying and reporting items of concern, e.g. unsafe situation and practices in their working environment.

- Strengthening and expanding the safety representatives network, with increase emphasis on contractor's safety representatives.

SAFETY CASES

The Audit expressed concerns that Dounreay's Safety Cases were largely written when the site was licensed in 1990 and were dependent primarily upon Probabilistic Risk Assessment (PRA); and the programme of reviewing the cases after five years had fallen behind schedule.

In response Dounreay has trebled the in house resources dedicated to both writing and project managing the production of safety cases and to the support of projects requiring safety documentation.

Safety Case programmes for operational, decommissioning and engineering projects have been established. The work within the programmes has been prioritised primarily to ensure safety cases are available for waste facilities as they are brought back on line, to demonstrate shutdown safety conditions and to support site decommissioning priorities.

The safety cases being produced are essentially deterministic, based on the identification of layers of protection, but with PRA playing a supporting role. The key outputs from the safety cases are the limits and conditions of operations, which are incorporated in plant operating and maintenance instructions. The engineering substantiation of safety related plant against its safety related requirement plays a prominent part in the safety case process and can lead to significant engineering programmes to support plant operations.

The end product of these programmes will be robust safety cases that have been implemented with documentation clearly defining the safety envelope, Maintenance Instruction and Testing requirements and necessary improvements to safety related plant. The first of the revised cases to be produced are for D9867 a solid ILW store and D1208 a highly active liquor treatment plant. These modern standards safety cases will be living documents and will be maintained to incorporate any modifications that arise in the future.

SAFETY MANGEMENT SYSTEMS

The Audit drew attention to the need for UKAEA and Dounreay to revise their existing safety and environmental management systems to be in line with modern practice, i.e. to develop a comprehensive and structured management system which includes all aspects of the business, including safety and environment. The system has been redesigned around the HSG65 guidance for Successful Health and Safety Management. As a result:

- Dounreay is integrating its safety, quality and environmental systems into a single loss control system.

- Links between the top-level corporate procedures/policies to the Dounreay Management Systems Manual and to lower level instructions have been clarified and strengthened.

- Documentation has been reviewed to remove duplication, inconsistencies and to ensure that it fits within the integrated system.

- The roles and responsibilities for compliance with individual site licence conditions have been clearly identified and documented.

- Emergency arrangements have been reviewed and training in command and control principles carried out.

- Codes of practice have been introduced to better control contamination at source.

- New approaches to facilitate incident reporting and improve how modifications are dealt with have been developed and implemented.

INTEGRATED WASTE MANAGEMENT AND DECOMMISSIONING STRATEGY

The Audit questioned the adequacy of strategies and plans that were in place for future decommissioning and waste management work at Dounreay and recommended that UKAEA should produce a detailed, substantiated integrated decommissioning and waste management plan which clearly identified each step to final decommissioning of the Dounreay site.

Dounreay has produced the Site Restoration Plan for the site, which was delivered to the regulators in September last year. It is a document which sets out the way all the waste streams on site will be treated and how and when each facility and building on the site will be decommissioned. The Plan is by definition highly integrated, the way in which you treat one waste stream can effect the decommissioning of other plants and waste stores. As part of the treatment of the waste streams existing facilities need to be operated for a number of years and a number of new plants need to be built. The site infrastructure needs to be able to support these operations and an important part of the Plan is the estates and utilities section which ensures that the site infrastructure develops to allow the Plan to be delivered.

The regulators are broadly content with the plan and we expect that they will have completed their detailed assessment by the autumn of this year. The Site Restoration Plan it sets out the work to be done and a method of achieving it, along with key dates which will allow the regulators to monitor the process.

CONCLUSION

Significant effort has been expended in restructuring the business to meet the needs of the future. The basic principles of control (both internally and of contractors), co-operation, communication and competence are at the heart of a strong safety management system, and have been fundamental to the changes that have been implemented.

Control has been improved through the new business structure, rationalisation of written procedures, the strengthening of key teams on the site and the introduction of clear targets and goals and targets as part of the business planning process. The launch of the Dounreay Site Restoration Plan has given clear focus to the work on site and integrates all the site activities.

Communication has been improved through the strengthened management presence in the workplace, more emphasis has been placed on the team brief system and regular feedback on safety topics. The arrangements for consultation with safety representatives has also been revised to bring them in line with industry best practice and to include contractor's safety representatives as well as UKAEA employees.

More emphasis has been placed on the competence of employees rather than looking just at the training they have received. Competence based training is being developed in many areas, in particularly through the use of Scottish Vocational Qualifications for Operations and Decommissioning.

A measure of the success in improving the safety systems can be found in the increasing International Safety Rating System score (ISRS), which has increased from Level 5 to Level 7 since we were first audited in November 1999. The next formal audit will take place in the autumn and Dounreay are striving to achieve Level 8.

The Dounreay Site Restoration Plan provides a strategic framework for the delivery of long term audit recommendations and to takes the site forward to allow it to meet its new mission of site restoration.

© With Author 2001

Finance and Policy

C596/036/2001

Decommissioning of nuclear installations in the research framework programmes of the EC

H BISCHOFF
European Commission, Nuclear Fission and Radioprotection, Brussel, Belgium

Decommissioning is the final phase in the lifecycle of a nuclear installation and is to be considered part of a general strategy of environmental restoration after the final suspension of the industrial activities. At present, over 110 nuclear facilities[*] within the Union are at various stages in the decommissioning process and it is forecast that at least a further 160 facilities will need to be decommissioned over the next 20 years (within the present 15 Member States). Enlargement of the Union would contribute to a rapid increase in the number of nuclear facilities to be decommissioned (at least 50 facilities).

Since 1979, the European Commission's DG Research has conducted four successive five-year research and development programmes on the decommissioning of nuclear installations performed under cost-sharing contracts with organisations within the European Union. The main objective of these programmes was, and still is, to establish a scientific and technological basis for the safe, socially acceptable and economically affordable decommissioning of obsolete nuclear installations.

These programmes were carried out by public organisations, research institutes and private companies in the Member States under shared-cost contracts and concerted actions. The main objectives of these activities were to strengthen the scientific and technical knowledge in this field, with a particular view to enhance safety and environmental protection aspects, minimising the occupational exposures and dismantling costs as well as the radioactive waste arisings.

Since 1979, more than 60 Mio € have been spent on:
- the development of decontamination and dismantling techniques for different kinds of nuclear installations
- technologies for waste minimisation, such as melting of steel components
- the development of decommissioning strategies and management tools

---

[*] nuclear power plants, fuel cycle facilities, particle accelerators and nuclear research installations

- the development of remote handling systems for high activated components (TELEMAN programme)
- development of planning and management tools for decommissioning projects

In the beginning of the 90s, four pilot decommissioning projects were chosen to compare the differences in the approach of:

- a fuel processing plant (AT1 in La Hague)
- a gas-cooled reactor (WAGR in Windscale)
- a boiling water reactor (KRB-A Gundremmingen in Germany) and
- a pressurised water reactor (BR-3 in Belgium)

Five years ago, a VVER type reactor (Greifswald in Germany) has been added to this list of pilot decommissioning sites.

The **WAGR** dismantling, for instance, served as a bridgehead for the future dismantling of graphite gas-cooled reactors. It was an extremely important textbook case, which rightly used the most modern techniques, thus enabling the choice of the scenario which is best suited to lower, the doses received by the operators, the costs, and the volumes of the wastes.
Operations to remove the reactor internals were undertaken with the use of innovative dismantling techniques involving amongst others:

- Computer-controlled Remote Dismantling Machine using stereoscopic television cameras to assist in the dismantling process;
- Acoustic-cleaning of electrostatic pre-filters.
- Ultra-violet laser to decontaminate vital parts of the machine before maintenance;
- Video-gamma-camera to identify and sort radioactive materials and hot spots;

The dismantling of the **BR3** in Belgium concentrated successfully on developing dry and underwater cutting techniques for the highly activated core internals. The **Greifswald** decommissioning project, one of the largest in the world, started the stage 3 dismantling of five commercial VVER-440 reactors in Greifswald and one VVER-70 reactor in Rheinsberg. The remote-controlled dismantling of the first reactor pressure vessel and reactor internals, using a new developed robotic system, will start in 2001.
In the **KRB-A** (Germany), a 250 MWe Boiling Water Reactor, the dismantling of the core internals, the heat exchanger, the activated concrete bio-shield and the reactor pressure vessel was finished.
The **AT1** reprocessing plant in France has successfully completed its decommissioning period and the site is currently being cleaned up for further use.

Within the EC programme, two databases on decommissioning have been created:

- EC DB TOOL for collecting technical performance data and
- EC DB COST for collecting data on waste arising, doses etc.

Both are now being merged into one database, EC DB NET, which is available on the internet (so far only for members of the project group).

The interest from the IAEA and the OECD/NEA and EC in the development of a common understanding of the decommissioning process led to the creation of a list of *Standardised Decommissioning Cost Item Definitions* (INCOSIT), another project under FP-4, to ease a world-wide comparability and transferability of data on decommissioning.

With this set of standardised decommissioning cost items it should be possible to create a common tool for the calculation of whole decommissioning projects, regardless of the type of reactor or the chosen method of dismantling. Under FP-5, a first benchmark exercise on decommissioning costs of VVER reactors will be executed. Similar activities, using the same list, are currently under work in the IAEA and OECD/NEA.

With the support of the EC, conferences, workshops and seminars were held on:
- Melting of dismantling steel
- Decommissioning strategies
- The use of databases and
- Dismantling techniques

Under the 4th Research Framework Programme of the EC (1994–98), a 20-year period of EC funded research activities in the decommissioning field was concluded, which has been qualified as essential in that sector. It can be stated that most of the dismantling techniques and technologies involved in the decommissioning process have reached industrial stage.
A large number of Final Reports and publications on various aspects of decommissioning are available at our EC service or from the relevant authors.

The activities in the decommissioning which are supported in FP-5 are clearly shifting from research on technology to

– dissemination of results from former research activities
– exchange of experience and provision of training
– collection of relevant data from decommissioning projects as well as
– development of decision-supporting and management tools
– integration of the needs of the candidate countries.

The current Work programme in Nuclear Fission Research supports the creation of *networks* to "exchange of information between national and Community sponsored research; promote exchange and feedback between the research and user communities; achieve consensus or a common understanding on key technical/scientific issues; identify research needs and develop strategies for how they can be addressed, promote training activities within a specific area, etc."[1]

And in the Communication of the Commission on the European Research Area (Oct.2000), which will be created within the period of FP-6 (2002-2006), the Commission proposes European Networks of Excellence around special areas of interest.

For this purpose the Commission decided to support the creation of a **Thematic Network on Decommissioning** (www.ec-tnd.net) as an effective instrument for facilitating these objectives.

This network is in line with the Commission's current and future intentions of interconnecting individual, national and European initiatives in a certain field and has the ability to serve as a forum for extended exchange of experience and the integration of future members from the Eastern Europe. It shall involve research facilities, the decommissioning industry, ongoing and future decommissioning projects as well as authorities and regulators.

---

[1] Nuclear Energy Work Programme 1998 to 2002, Official Journal L26,1999

It is foreseen to provide free access to the database EC DB NET for the members of the network, with the objective, on one side, to disseminate collected experience from different decommissioning projects, but also to receive more data to improve the usability of the database. An extended set of data and a large number of clients are an indispensable condition for a longstanding existence of a database.

Besides the Thematic Network and the database, there are 3 more projects, which receive substantial financial support from the Commission under FP-5.
The **IRDIT** project (Innovative Remote Dismantling Techniques) is dealing with a special aspect of the dismantling process – the remote dismantling of the high-activated reactor pressure vessel (RPV) and its internals. For this aim, two projects were chosen:
- The remote cutting of the Russian-type VVER reactor pressure vessels of unit 1 and 2 of the NPP Greifswald (Germany) and
- The cutting of the RPV and internals of the BR3 pressurised water reactor in Mol (Belgium).

The **Standardised Decommissioning Cost Estimation of VVER NPP** is a benchmark exercise which is based on the Standardised Decommissioning Cost Item List and shall comprise the real and the planning data of several VVER plants in Germany, Eastern Europe and NIS.

The **Compendium on the state-of-the-art in Decommissioning** shall provide the practical to date experience and knowledge in all aspects of decommissioning - starting with the planning and licensing and finishing with waste management and site release for further use - which has been collected during the last 20 years of the EC research in this field. It is intended to serve as a handbook on decommissioning and shall give practical guidance on all the questions future decommissioning projects might face.

As a result of the latest call under the EC research programme, two consecutive **Training Courses in Nuclear Decommissioning** will be organised by a consortium of experienced decommissioners. These courses shall comprise all main aspects of decommissioning and will also give the possibility to extend the theoretical part with a practical course in one of the partners facilities. The first course will start most probably in the second half of 2002.

With these projects, supported under the current research programme of the Commission, it is intended to create a network of excellence in nuclear decommissioning and to keep and enhance the high level of European expertise and competitiveness in this field.

In recent times a number of decommissioning activities have been started by other international organisations like IAEA or OECD/NEA, as well as in other services of the EC, like DG Environment or DG Enlargement (cost studies, Environmental Impact Assessments etc.). The Commission supports the idea of a close co-operation with the different international bodies in this field and a complementary approach to solve open questions.

As an example could serve the co-operation in the field of cost estimation methodologies, where the NEA (Liaison Committee on Decommissioning), the IAEA (cost studies on VVER) and the EC/DG Environment (decommissioning cost handbook) have recently initiated complementary investigations focusing on different cost aspects of decommissioning.

Where does the Commission still see issues to discuss and to work on during the oncoming years in the decommissioning field?

Cost aspects of decommissioning

The cost of decommissioning should reflect all activities of the decommissioning process, starting with the planning and licensing and post-operation and finishing with radioactive waste management and site clearance.

If the decommissioning is deferred for an extended period of time, also surveillance and security of the facility should be taken into account. According to the legal framework, a mechanism has to be established before operation in order to secure the funds needed for the decommissioning of each facility.

However, for plants that were constructed in earlier nuclear programmes, i.e. in the 50s and 60s or under different legal frameworks as in the Eastern European Countries, funds are often limited, which may have impact on the chosen decommissioning strategy.

The decommissioning projects have demonstrated that costs can be managed. Comparisons of individual cost estimates for specific facilities have shown relatively high variations, however, which result mainly from the use of different cost estimation methodologies, using different data requirements.

This could be one field for comparison, analysis and discussion.

A second issue in the decommissioning field is the:

Radiation safety during decommissioning

In 1996 the European Commission has issued the Directive on Basic Safety Standards (Directive 96/29/EURATOM).

It has introduced a series of new measures to improve the protection of the health of workers and the general public. For this purpose, the Directive reduced the dose limits and contains explicit provisions for intervention situations. It also structures the concept of clearance and exemption for materials containing radioactivity.

To advise the Member States on the implementation of the Directive, DG Environment has issued several publications, i.e.

- definition of clearance levels
- the ALARA concept in Decommissioning
- radiological protection criteria for the clearance of buildings
- calculation of individual and collective doses from the recycling of metals from
- the dismantling of nuclear installations
- practical use of the concepts of clearance and exemption.

There are a number of Final Reports from EC-funded research projects, which deal with radiological aspects in decommissioning as well.

There is however a need for clarification and coherence of the current system with aspects such as optimisation, dose limits, triviality, public and environmental protection.

With respect to operator safety during interventions in hazardous environments, such as areas with alpha contamination, development of safe and comfortable and, in the same time, cost-effective protective clothing equipment is needed. That concerns also the efficiency of protective clothing as well as the biological and physical monitoring of the operator.

Another important aspect is the investigation of

Environmental Impact of Decommissioning

In January, a workshop was organised by the Commission on "Current Regulatory Status on the EU Member States and Applicant Countries" concerning "Environmental Impact Assessment for Decommissioning of Nuclear Installations" (EIA).
The study, which was initiated and financed by DG Environment, aimed at reviewing the requirements of the relevant EIA Directive of the European Commission (97/11/EC) and to provide guidelines for their application to the specific issue of decommissioning.
Special consideration has been given to public involvement in this process.

As there is presently limited experience in applying EIA to the decommissioning of nuclear power plants in either the EU or in the applicant countries, it is believed that there is a need for discussion and exchange of experience on this item.

In this context should be stressed the extended work in environmental questions of decommissioning which has been done by DG Environment during recent years.

One of the main issues in decommissioning is

Waste treatment and unconditional release of dismantling waste

The management of large volumes of materials arising from the decommissioning of nuclear facilities represents one of the major tasks to be undertaken and one of the most substantial cost fractions. According to existing experience, less than 1% of the materials produced will be managed as low and intermediate level radioactive wastes.

Various international organisations, as the IAEA and the EU, have issued a number of release recommendations, relating to exemption and clearance criteria, i.e. the present EC recommendation on unconditional clearance of scrap metal (EC Radiation Protection 89, Recommended radiological protection criteria for the recycling of metals from the dismantling of nuclear installations, 1998)

In addition, each Member State has its own strategies and policies about waste management, including the material release criteria.

Harmonisation of the waste management practices and of the clearance levels among the member states or rather world-wide would be beneficial not only in terms of equivalent levels of safety in waste management and disposal, but also in the minimisation of wastes through release and recycling.

The newly created Thematic Network could promote the creation of a broadly accepted and coherent system of release criteria and associated regulations for recycling and reuse of materials from decommissioning, considering all aspects of global optimisation.

Socio-economical, political and public perception issues are the major non-technical problems influencing decommissioning projects. They should be addressed as early as possible in the conceptual phase of a decommissioning project.
In case of an early shut down of a plant, the questions will gain even higher importance.

As some of the Eastern European operators will have to face the challenges of an early shut down, it is most valuable to have the possibility to gain from outside experience to build a strategy on employment of redundant staff, educational and training programmes and site development and reuse.

In this respect should be mentioned the successful strategy on the reuse of the Greifswald decommissioning site where currently 1.5 Billion DM foreign investments are under negotiation (which, by the way, supports the strategy of immediate dismantling and site clearance).

Public perception is one of the main issues related to nearly all activities in the nuclear field, and there is room for strong improvement. Therefore the current co-operation projects under FP-5 should provide a forum for the exchange of experience and the start of new initiatives.

In this respect, the EC supported Decommissioning web-site (http://www.ec-decom.be) should be mentioned, which provides an opportunity for all interested parties working in the nuclear decommissioning to provide data and information and make it an interactive forum to communicate with the public.

Conclusion

The European Commission recognised very early the need for research, development and demonstration of an effective and safe decommissioning of nuclear installations after completing operation. With more than 140 nuclear power plants and almost the same amount of research reactors within the Member States there was a clear need for a programme on decommissioning and dismantling of those installations.

Relating to the results of a 20-year research and development programme, comprising all aspects of a decommissioning project including the management and treatment of dismantling waste, this programme contributed significantly to the fact that the European nuclear industry currently is probably one of the few industries that has demonstrated to be able to manage successfully the end-of-life of their installations. It can be stated that Decommissioning and Dismantling of nuclear installations have reached industrial stage and is a mature technology.

In order to disseminate the accumulated know-how and improve the exchange of information within the participating organisations, databases have been set up, a thematic network, which is open for all interested parties, has been created and a compendium on the state-of-the-art knowledge in Decommissioning and Dismantling is under way.

This wide dissemination of knowledge and best practice in the decommissioning through international co-operation, networking, training activities, conferences and workshops integrating future Member States of the European Union should provide the basis to keep and enhance the existing high level of expertise in this field.

However, there is still room for discussion and improvement, especially in strategy and management, reduction of waste arising, harmonisation in recycling and reuse of materials, free release levels as well as on assessment of environmental impacts and public perception issues.

Despite the common opinion that the EC should further decrease the funding of *research* activities in the nuclear decommissioning, there is also a common attitude to continue the support for dissemination of best practice and accumulation of knowledge within networks of excellence and through co-operation with international organisation which are active in this field.

References

1. *Research and Training Programme in the Field of Nuclear Energy*, 1998 to 2002, Official Journal, L 26, Febr. 1999

2. *Remote Dismantling Operations – WAGR Decommissioning*, T Benest et al., UKAEA Windscale, Final Report to EC contract No FI4D-CT95-0006

3. *A Proposed Standardised List of Items for Costing Purposes in the Decommissioning of Nuclear Installations*, Interim Technical Document, OECD/NEA-IAEA-EC 1999

4. *The BR3 Pressurised Water Reactor Pilot Dismantling Project*, Final Report, V Massaut et al., EUR 18229, 1998

5. *Study into the Development of a Decommissioning Planning Tool*, Final Report to contract FI4D-CT96-0008, Garner/Colquhoun/Elder, NLM Co Ltd., 1999

6. *Decommissioning of Nuclear Plants and Related Waste*, Euroelectric, June 2000

7. *Recycling and Reuse of Radioactive Material in the Controlled Nuclear Sector*, Hopkinson/Bishop/Cross/Harrison, AEA Technology, EUR 18041, 1998

8. IAEA TECDOC-1032, *Factors for formulating strategies for environmental restauration*, July 1998

9. IAEA-TECDOC-1133, *The decommissioning of VVER type nuclear power plants*, January 2000

10. IAEA-TECDOC-855, *Clearance levels for radionuclides in solid materials – application of exemption principles*, 1996

11. *Decommissioning experience in the European Union to date*, V Massaut et al., paper presented at the EURADWASTE'99 conference, 15-18 Nov. 1999 in Luxembourg

12. *Proceedings of the second workshop "Decommissioning of Nuclear Installations – Technical Aspects"*, Mol 8 –10 June 1999, EC Doc. XII-217-99

13. *Proceedings of the first workshop "Decommissioning of Nuclear Installations – Strategy and Planning"*, Heringsdorf, 21-22 September 1998, EC Doc.XII-25-99

14. *Proceedings of the joint NEA/IAEA/EC workshop on "Regulatory Aspects of Decommissioning"*, Rome, 19 –21 May 1999, ISBN 8-448-0033-0

15. *Managing Radioactive Waste in the European Union*, Publication of the EC, DG Research, 1999

© European Commission 2001

Liabilities Management

C596/005/2001

The management of nuclear liabilities

R M SELLERS
Liabilities Unit, BNFL, Sellafield, UK
D R T WARNER
Liabilities Unit, BNFL, Berkeley, UK

ABSTRACT

Many industrial processes require activities to be undertaken after operations have terminated. Determining the funding required for these is a complex task forming an essential part of "liabilities management". This paper sets out to explain certain aspects of liabilities management with particular reference to BNFL's management of its nuclear liabilities.

The focus will be on the nature of the liabilities, the activities to be undertaken in their discharge and how provisions are derived. Liabilities management is not, however, simply about providing funds, but also ensuring value for money in discharging the liability, and reducing costs wherever possible. BNFL's approaches on these will be outlined.

1 INTRODUCTION

There can be few industrial processes that do not require certain activities to be undertaken after the main income-generating phase has come to an end. These activities may include managing any unwanted by-products formed in the process, the demolition and disposal of the process plant and the remediation of the ground on which the process plant was built. Determining how much money should be set aside to pay for all these obligations may itself be a complex task and forms a key part of what has become known as "liabilities management".

Historically most nuclear liabilities on BNFL's sites were the financial responsibility of its customers but following integration in 1998 with the former nuclear generator, Magnox Electric plc, BNFL inherited both financial and physical responsibility for all of the liabilities from the magnox fuel cycle. This included taking over ten additional nuclear licensed sites including operating power stations, decommissioning sites and active facilities at Berkeley Centre.

This paper describes the physical liabilities for which BNFL is responsible and explains how it sets about providing an estimate of the cost of managing its liabilities and how it takes into account uncertainties in the estimate.

2 WHAT ARE LIABILITIES?

Financial Reporting Standard 12 (FRS12) defines liabilities as *'Obligations of an entity to transfer economic benefits as a result of past transactions or events'*. Put more simply, a liability is an obligation to do some work, or spend some money, in the future as a result of past or present activities. Timing of the work or expenditure is determined by safety, environmental, strategic and commercial considerations.

BNFL's nuclear liabilities have been generated and continue to arise as a consequence of all its activities in the manufacture of products in the nuclear fuel cycle. The most important activities include:

- decommissioning (decontamination, dismantling and demolition of redundant reactors, chemical plants and other nuclear facilities);
- waste management (retrieving, treating and packaging of wastes plus subsequent storage and disposal);
- remediation of contaminated ground (on which reactors and other plant have stood) and site termination; and
- reprocessing of fuel from BNFL's own reactors (most notably of "final cores").

The total liabilities, as stated in BNFL's accounts for the 1999/2000 FY were £34.2Bn. This figure includes some £11Bn of liabilities which BNFL manages (and provides cost estimates for) on behalf of MoD and UKAEA.

Table 1 Breakdown of BNFL Liabilities

	Undiscounted (£Bn)	Discounted @ 2.5% (£Bn)
Waste Management (incl. Disposal)	11,941	6124
Decommissioning	19,640	7930
Reprocessing	2661	1944
Total	**34,242**	**15998**

Source: BNFL Annual Accounts 2000

Nuclear liabilities are by their very nature uncertain. FRS12 requires that a provision is recognised for a liability of uncertain amount or timing if:

- the entity has a present obligation as a result of a past event,
- it is probable that a transfer of economic benefits will be required to settle the obligation, and
- a reliable estimate can be made of the amount of the obligation. For the latter a best estimate is required.

In the rare cases that a reliable estimate cannot be made, it recommends that the liability is disclosed as a *contingent liability* (that is to say, it will only be defined adequately by some future event, referred to in FRS12 as an *obligating event*), and no provision has to be made. It is open to debate whether any nuclear liabilities fall into this category, though the termination of complex sites such as Sellafield or the disposal of wastes certainly come close. In practice BNFL has sought to make such situations manageable by defining a set of assumptions which allow a reliable estimate of the cost to be made.

FRS12 also addresses the issue of timescales in considering what provision should be made for a particular liability, in that it recognises the time value of money and recommends that the provision should be the "present value of the expenditures". It goes on to say that the discount rates should reflect current market assessments of the true value of money and risks specific to the liability where these have not already been taken into account. In other words, prudent discounting is permissible.

3 HISTORICAL DEVELOPMENT

The increases in liabilities over the period of operation of BNFL and its precursor organisations have arisen from three principal causes: business development, changes in regulations and increased recognition.

The UK nuclear industry began with the defence programme in the late 1940s/1950s and the subsequent development of the civil magnox reactor programme. Upon its formation in 1971, BNFL took over responsibility for some of the early defence related plants and, following the integration of Magnox Electric in 1998, BNFL assumed responsibility for the whole UK magnox programme. BNFL's other business areas include the production and reprocessing of oxide fuels, production of fuel from reprocessed material *e.g.* MOX and latterly provision of plant to deal with legacy wastes.

Government policy and regulations have changed as the public perception of environmental issues has developed. This has resulted in changes in the overall philosophy for dealing with wastes. Originally the concept was maximum permitted discharge via authorised disposal routes with minimal pre-treatment. For example liquid reprocessing effluents were discharged to sea with only a short period of interim storage to allow for some radioactive decay. Certain plutonium contaminated wastes were disposed to the deep ocean; a disposal route that has since been closed. ILW and HLW were stored in "raw" form pending future treatment and deep disposal, with little consideration of how wastes were to be retrieved from the early storage facilities

Since the early 1980s BNFL's policy has changed to one of treating wastes as they arise and progressively retrieving and treating the historic "legacy" to a passive safe form. To enable this a major construction programme to provide plant to condition wastes and reduce discharges was started in the mid-1980s. In parallel with this, a start has been made on the decommissioning of early plant and raw waste storage facilities.

Finally, additional liabilities issues have been identified, for example, ground contamination at chemical plant and reactor sites and the whole area of site termination and delicensing.

4 ESTABLISHING A VALUATION OF LIABILITIES

4.1 General Approach
The previous sections have highlighted the scale of the nuclear liabilities on BNFL's sites. This section considers in more detail the basis on which estimates of the cost of discharging these liabilities are made and the other factors and issues which need to be taken into account or which have a bearing on the estimation process.

In general terms the approach taken can be seen as having six distinct components. These are:

1. a knowledge of what the liabilities to be managed are (the inventory);
2. a clear definition of what constitutes the point at which the liability is discharged (the end-point);
3. identification of the activities etc. which need to take place to convert the liability from its current state to the point at which it is discharged (the route-map from (a) to (b));
4. models to convert the activities identified in (c) into costs (cost models);
5. liability cost estimates inevitably have some uncertainties attached to them and these need to be separately assessed (contingency and risk); and
6. Finally, all activities associated with liability discharge must be compliant with existing legislation and conform to other standards as necessary. Mostly, for instance, in the case of safety, these will be factored in to the technical considerations covered by (a), (b) and particularly (c), but where they are not they need to be made explicit to ensure that their consequences are properly taken into account. Similarly, any assumptions made need to be documented (constraints and assumptions).

Where possible, liability cost estimates should be benchmarked against, reflect or build on similar activities already carried out on other, broadly similar, plants.

Funding arrangements and the time value of money are also important considerations that follow from these considerations, but are treated separately.

The following sections amplify these six factors.

4.2 Inventory
Knowledge of the nature of the liabilities to be discharged is a material prerequisite of any liability cost estimate. For wastes, this includes information about their chemical, physical and radiochemical characteristics, for reactors and chemical plants, plans of the facilities, details of their materials of construction, their current condition and the disposition, nature and amounts of any radioactive contamination, for contaminated ground, the nature, quantities

and disposition of radionuclides, the nature of the land and groundwater flows, and so on. Much of this information is available from records, from sampling and analysis, or by calculation (*e.g.* for activation products in reactors), and for the most part is more than adequate for the purposes of developing liability discharge plans and reliable cost estimates. However, there are some uncertainties especially for older plants in which records are incomplete or where sampling or direct inspection is not possible or technically challenging.

4.3 Endpoint

A simple but very effective metaphor for liabilities discharge is that of a journey. Section 4.2 dealt with the liability as it currently exists; this represents the point of departure. The endpoint represents the final destination, the point at which the liability is extinguished, the goal to be achieved. For radioactive wastes the end-point is generally disposal although for other potentially recyclable materials, reuse is a possibility. Similarly, buildings may have their contents dismantled and removed and the structure decontaminated to allow reuse. Alternatively the building may be demolished and the ground on which it stood delicensed. The ability to think through to the end-point is an essential element of liabilities management.

4.4 Route-Map

Continuing the metaphor, the route-map (or strategy) defines the optimum journey for the liquidation of each liability. Just as a number of alternative routes are feasible when travelling so different strategies may exist depending on the constraints applied. Minimum cost is often the main driver but this is often tempered by other requirements and stakeholder expectations.

Once the preferred strategy has been agreed, outline liability discharge plans and detailed project implementation plans and programmes follow. BNFL's major waste management and decommissioning strategic planning reviews have historically been conducted on an annual cycle as part of the business planning round; the eighth iteration has now been completed.

4.5 Cost Models

The approved strategy and outline plan, when applied to the inventory using detailed cost models allows the cost of discharging the liability to be determined. This is the basis of liability estimates and the development of the associated provisions.

4.6 Contingency and Risk

FRS12 requires that the risks and uncertainties that inevitably surround many events and circumstances to be taken into account in valuing a provision. BNFL distinguishes between uncertainty and risk. The inherent uncertainty in estimating costs associated with undertaking the chosen strategy is covered by a contingency allowance. Risk is the net cost (positive or negative) that would result if threats and opportunities were to materialise resulting in different, or additional tasks being required.

4.7 Constraints and Assumptions

Developing a liability cost estimate requires certain assumptions to be made about, say, what technology will be available to discharge the obligation, a particularly important consideration when discharge falls many decades into the future.

Under these circumstances FRS12 says that provision should only reflect such developments if "*there is sufficient objective evidence that they occur*". By the same token, it recommends that possible new legislation should only be taken into account when *"sufficient objective evidence exists that the legislation is virtually certain to be enacted"*.

The timescale over which BNFL will discharge its liabilities extends well over a century into the future. It is inconceivable that some changes will not occur in that time, either in terms of technological advance, business efficiency (which may itself in part be driven by technological advance) or legislation. However, for the purposes of deriving a robust cost estimate, liability discharge strategies and plans are developed on the basis of today's technology, and today's regulatory framework. These are counteracting forces (technological change will lead to a reduction of the costs of liability discharge; regulatory change will generally increase them), and will tend to cancel one another out. By their very natures neither is easy to predict with certainty, but the major areas of potential regulatory change have been identified by analyses carried out within BNFL as having mitigating technical measures.

A variety of other assumptions are made in developing liability cost estimates, and BNFL has been careful to record these. Superficially some, such as plant throughput rates, or drum filling fractions appear relatively minor, though they may have substantial implications for costs; others such as reactor or plant closure dates have more obvious impacts.

The timing of the various activities to be undertaken is also an important. In part it falls out naturally from other considerations such as the availability of waste treatment and packaging capacity or a deep waste repository, but must also reflect the need for some flexibility in certain activities. This is to allowing a degree of work smoothing and, in principle, the elimination of major peaks and troughs in the liability discharge programme. For planning purposes it is assumed, for instance that plutonium-contaminated plant will be dealt with promptly to avoid the additional work needed if levels of the gamma-emitting radioisotope Am-241 are allowed to build up. For plant such as reactors, in which relatively short-lived radioisotopes such as Co-60 are dominant, delay to permit decay makes the tasks associated with demolition substantially easier, and is a key feature of BNFL's current reactor decommissioning strategy.

4.8 Benchmarking and learning from experience
Ideally liability cost estimates should not be developed in isolation but rather should reflect or build on previous experience on similar activities elsewhere. Internal benchmarks are likely to be the most reliable, since these are the ones for which most information is available, although even simple comparisons with estimates from other organisations can be valuable. Too much should not be expected of benchmarks, however; they give some comfort that estimates are similar to (or different from) others, but great precision should not be expected. Their primary merit is in weeding out poorly defined cost estimates, for instance those with poorly defined scope or inadequate technical definition.

5 CATALOGUE

Clearly liabilities management requires the marshalling of substantial volumes of information and BNFL has met this challenge by building an electronic "liabilities catalogue". The four main zones are:

Reviews: allowing access to electronic versions of the main waste management, decommissioning and risk documents and incorporating powerful search facilities.
Graphs: containing a number of standard graphical reports showing, for example, cashflows over time for different decommissioning activity stages. Background descriptions of plants, including photographs and process diagrams, are also included in this zone.
Financial: showing how the published annual accounts data are derived from supporting documentation
Analyse: provides the means for the user to query the underlying database for particular data of interest using an interactive query generator. The results of the query can be presented in tabular format or graphically and can be exported to standard desktop applications for further analysis.

A number of developments are planned for future versions of the catalogue. The underlying database has been linked to a powerful Geographical Information System (GIS) to demonstrate geographical relationships between data. In a further development site photographs have been overlaid and such a system is invaluable for ground contamination studies. For other data the relationship to the overall process flow is more informative than pure geographical information.

6 RECENT DEVELOPMENTS

Following integration with Magnox Electric, BNFL initiated a review of its liabilities with the emphasis particularly on those associated with decommissioning. On the basis of the principles and approaches outlined above, and taking into account the results of recent internal research work, this identified some £7.1Bn (undiscounted) of additional liability. No single factor was responsible for this change, but mostly it reflects increased technical definition of the requirements for decommissioning and dismantling the chemical plants at Sellafield, including such things as penetration of activity into concrete structures, better characterisation of the requirements for the treatment of liquors arising from plant decontamination during "initial decommissioning" and revised plant throughputs based on current experience. These increases fall disproportionately on the older plants arising from the UK defence programme in the 1950s and 1960s. More recent plants built to manage arisings from THORP and the Magnox Reprocessing Plant B205 have been constructed to modern standards and with decommissioning in mind, and these issues do not arise. The bulk of these costs fall well into the future, and in discounted terms have a much smaller impact.

The assumptions underlying the cost estimates were also reviewed and some adjustments made taking into account the latest technical knowledge. Account has also been taken of potential delays to the availability of a deep waste repository and the need for additional waste stores.

6 CONCLUDING REMARKS

In a Conference devoted to decommissioning it is important to remember that liability management covers far more than just this aspect. In the case of nuclear liabilities the full scope encompasses:

- recognition of the potential impacts on the environment that give rise to the liability;
- minimising the actual impact on the environment during plant operation;
- safely storing and disposing of radioactive wastes and remediating facilities used to handle radioactive materials;
- making appropriate provision for future costs.

Thus we can define a *liability lifecycle* :
- up-front recognition that business decisions may result in the *creation* of new or additional liabilities and taking due account of this in the decision-making process;
- control of liability *production* during the design, construction, operation and decommissioning of radioactive processes and plants;
- *transformation* leading to eventual *elimination* of the liability.

Transformation may in turn involve the creation of a new and different liability such that the lifecycle repeats until final remediation, normally disposal or delicensing, is achieved. For example BNFL has constructed an integrated suite of plants to store, retrieve and condition radioactive wastes which will themselves become decommissioning liabilities as soon as active operations commence.

In a very real sense, therefore, liability management permeates the organisation and is not the sole preserve of a few liability specialists.

Overall the purpose of liabilities management may be summarised as to:
- provide *assurance* that liabilities are fully identified, evaluated and provided for;
- ensure that *value for money* is obtained as the liabilities are discharged; and
- secure *continuous improvement* by reviewing performance, capturing learning and feeding back improvements to other projects.

The ultimate goal is to reduce the costs of safely discharging liabilities.

ACKNOWLEDGEMENT

This paper is based on one first given at IBC's 7th International Conference and Exhibition on Decommissioning of Nuclear Facilities, which took place in London on 30th & 31st October 2000.

© BNFL 2001

C596/010/2001

Management of liabilities for later decommissioning of nuclear facilities in Germany

P PETRASCH, R PAUL, and **J-P LUYTEN**
NIS Ingenieurgesellschaft mbH, Germany

In Germany, the legal obligation for decommissioning of nuclear facilities arises from the German Atomic Law. Due to this, German utilities have to fund provisions for the later liabilities, raised over a period of now 25 years. The funds have to be calculated specifically for each individual nuclear facility regarding the individual situation.

The calculation of decommissioning costs for nuclear installations is complex and has take in account many different aspects and inter-related parameters. To deal with this a special software is used, the so called STILLKO. This software enables NIS to calculate the decommissioning costs for different types of nuclear installations on the basis of different boundary conditions.

1 INTRODUCTION

The German nuclear power plants work on a high quality level, and several of them are at the top of the electricity generation ranking list. But of course the authorities and the owners know that the lifetime of a nuclear power plant is limited and that one day decommissioning will come.

Provisions have been made at different levels. In laws and rules such items like the licensing procedures for decommission and the funding of money are regulated, and the utilities have defined technical concepts for the decommissioning and for the waste management.

The present paper describes the decommissioning liabilities at the utilities' side, explaining the concepts and the provisions for the later decommissioning of nuclear power plants which are still in successful operation.

2 KINDS OF DECOMMISSIONING FUNDS

After the planned life time, German nuclear power plants have to be decommissioned and dismantled. The decision on the final shut down is normally in the responsibility of the utility and depends on the technical and economical situation of the plant.

The decommissioning work needs a special decommissioning license as per §7 of the German Atomic Act. Therefore a transition from the operational license state to the decommissioning license state is needed. For some actions, which are needed both for operation and for decommissioning, it is advantageous to handle them under the operational license (e.g. management of the spent fuel assemblies, management of the operational waste, and shut down of circuits). There is no need to apply for a new license if the removal of the spent fuel is already a part of the operational license.

Therefore, as can be seen from Fig. 1, there is a gap between the last shut down and the granting of the decommissioning license. This time period is called the Post Operational Phase.

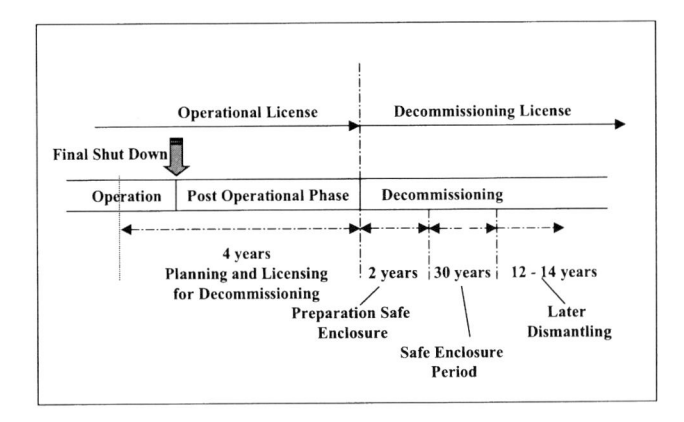

Figure 1: Licensing Situation

Considering these aspects the funds for decommissioning refer to the following subjects:
- Removing and reprocessing of the spent fuel assemblies
- Management and removing of the operational waste and of the mobile reactor pressure vessel internals, i.e. neutron sources, control rods, etc.
- Decommissioning of the plant (this can performed either immediately or later after a safe enclosure period)
- Dismantling of the building structures after clearance of the site.

3 TECHNICAL CONCEPT FOR THE DECOMMISSIONING

It was in the early 70s when the German utilities started to develop decommissioning concepts. From this time up to now the expression "Decommissioning" has been used as a synonym for all the measures within the scope of the special decommissioning license.

The decommissioning and all the concerned actions have to be planned in detail, because nuclear decommissioning means handling of contaminated or activated material, and the expected radiation exposures have to be anticipated. Therefore the safety of the personnel and of the environment are main objects of the decommissioning. All planning work and the later performance are destined to:

- minimise the radiation exposures
- minimise the amount of radioactive waste.

These aims require optimised and qualified techniques and the use of effective radiological protection measures.

The development of the German decommissioning concept started in 1976, and the Basic Report was issued in 1980. All the German nuclear power plants are represented by two so called "Reference Plants", one of the PWR and one of the BWR type. Since that time the Basic Report has been updated periodically to consider new techniques for dismantling and cutting and for waste management as well as for new aspects in the field of licensing procedures. Also practical experiences from real decommissioning projects, like the Niederaichbach total dismantling, Gundremmingen Block A and Würgassen are considered for the update.

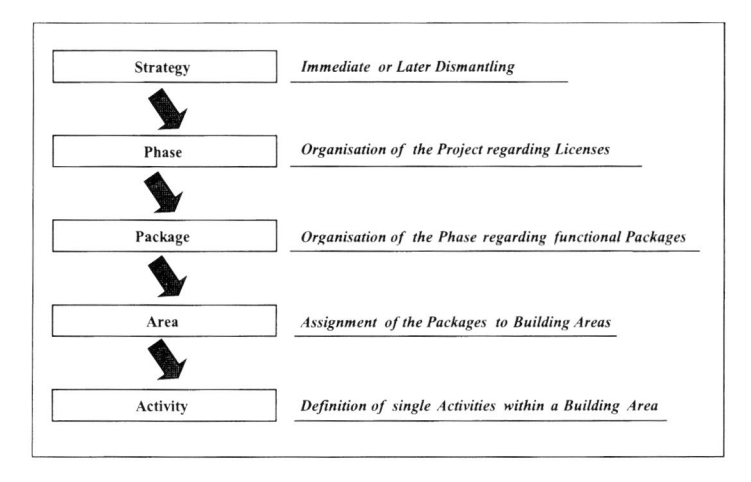

Strategy	Immediate or Later Dismantling
Phase	Organisation of the Project regarding Licenses
Package	Organisation of the Phase regarding functional Packages
Area	Assignment of the Packages to Building Areas
Activity	Definition of single Activities within a Building Area

Figure 2: Organisation of the Decommissioning Project for Calculation

The Basic Report describes the decommissioning of the two Reference Plants in detail and takes into account all the actions and measures during the decommissioning period, including also planning, licensing, operation of the decommissioning site, waste management, etc. The concept is based on the inventory of the Reference Plants and also on a list of necessary measures in a hierarchical model (see Fig. 2).

The hierarchical model guarantees a complete survey on all the measures in a decommissioning project and makes also sure that no important aspect will be forgotten. The possible link between the inventory of the plant and the decommissioning actions is a very important part of the concept enabling a detailed description of the decommissioning expenditures in the field of dismantling, decontamination, release measurements and other waste handling actions. Considering all these aspects the Basic Report for decommissioning in Germany contains four main parts, shown in the Fig. 3.

List of Decommissioning Actions → **Decommissioning Cost Calculation** ← *Waste Management Calculation*

Decommissioning Calculation ↑ ↑ *Time Scheduling*

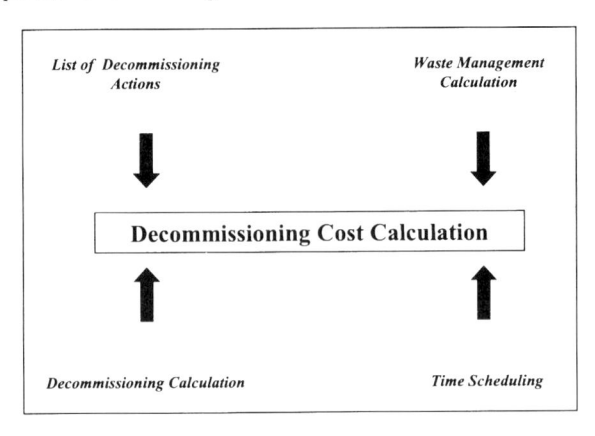

Figure 3: Main Objects of the Decommissioning Calculation

4 DECOMMISSIONING STRATEGIES

As the decision for a decommissioning strategy is in the responsibility of the utilities it is not possible to fix the strategy in the Basic Report. So the Basic Report describes two strategies, the Immediate Dismantling and the Deferred Dismantling after a Safe Enclosure Period of 30 years. Fig. 4 shows the principles of these decommissioning strategies.

In the strategy Deferred Dismantling it is necessary to transfer the plant into safe enclosure condition, this means the systems and components have to be shut down, decontaminated, dried and closed. The plant entrances must be saved against trespassers and the building structures must secure the safe enclosure of the radioactive material.

During the Safe Enclosure Period the plant will be inspected once or twice per year. After a period of (planned) 30 years the plant will be dismantled completely. The conditions for the dismantling work will benefit from the decay of the radioactive nuclides, so the radiation exposure levels will be decreased.

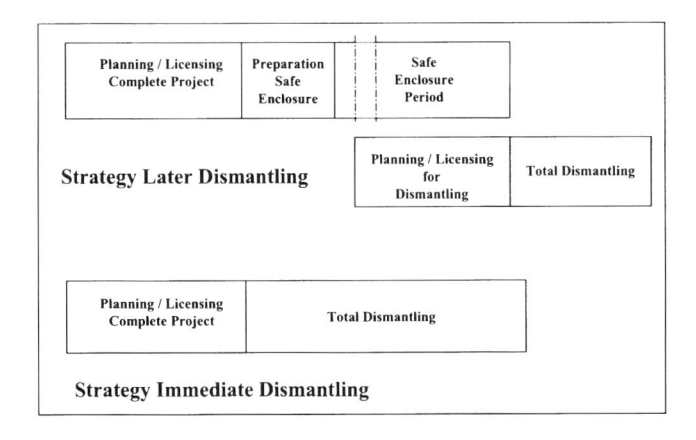

Figure 4: Decommissioning Strategies

The strategy Immediate Dismantling aims for a dismantling at an early stage to use two aspects:

- The staff with the experiences of the plant specifics is still on site and can used for the dismantling work
- The site can used for another power plant or for other industrial purposes.

The immediate dismantling requires more expenditures for radiological protection, shielding, etc. and also a higher quality for the packaging of the radioactive waste material.

5 WASTE MANAGEMENT

Regarding the §9a of the German Atomic Law all the dismantled components and the other radioactive material generated during the decommissioning have to be reused in the industrial cycle or stored in a final repository.

Taking into account the situation in Germany where final repository costs of about 35.000 DM/m³ are expected, plus the costs for the packaging, it is clear that the decommissioning and decontamination work is influenced by the cost. But the waste management and the amount of waste depend also on the release limits which are regulated in the German Radiation Ordinance (update is in preparation at the moment). As a result it is an important goal – also from the financial point of view – to avoid radioactive waste, also the secondary waste that is generated during the decommissioning project.

The components to be dismantled during decommissioning are partially activated by neutron irradiation and partially contaminated through their contact with radioactive media. Under consideration of technical and economical aspects, these components will be safely recovered as residuals or, if this is not possible for technical or economical reasons, they will be properly disposed of as radioactive waste. The masses arising with such decommissioning have been assigned to three groups according to their origin:

primary masses are all equipment, components and buildings existing in the nuclear power plant at the beginning of the decommissioning work and which have to be removed,
secondary masses are articles of consumption which are needed for the dismantling and treatment of the existing primary masses, and
tertiary masses are additional tools and equipment, placed in the controlled area, which are needed for the execution of the decommissioning work.

For each of these mass types a plan has been worked out about how and when they occur, how they can be treated and how the radioactive waste can be packed. The calculation results, given in Fig. 5, show that the predominant part of all the components and systems can be re-used in the industrial material cycle.

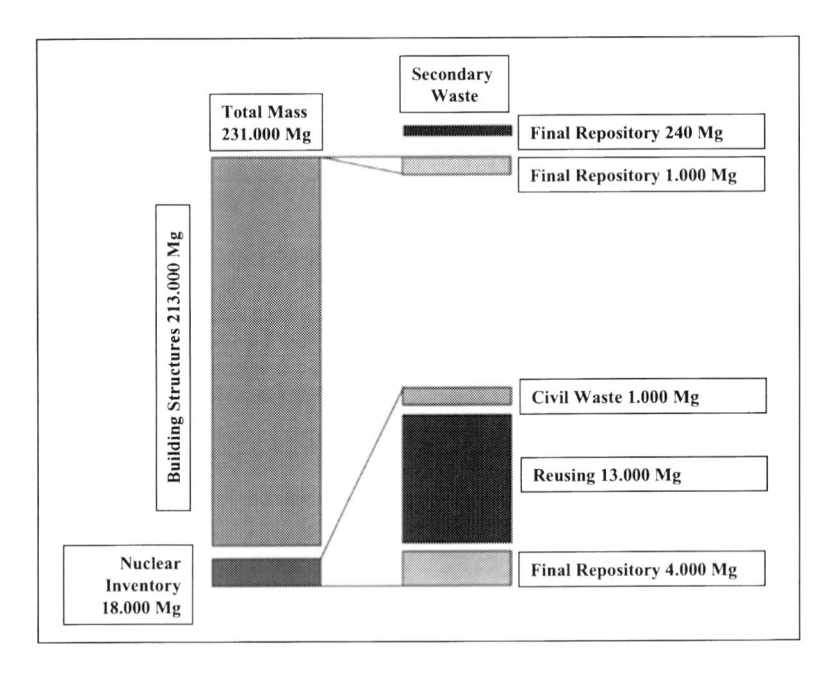

Figure 5: Decommissioning Masses

From the total mass of about 200.000 Mg (Reference Plant for BWR, including the building structures) more than 195.000 Mg can be reused. Less than 3% of the total mass has to be stored in a final repository.

6 BOUNDARY CONDITIONS

The decommissioning cost calculation for funding is subject to site specific boundary conditions. Some of them are very uncertain because the decommissioning work is far in future.

Therefore the decommissioning cost calculation will base on assumptions for possible strategies and scenarios which correspond to the existing situation for:
- Licensing requirements
- Waste management policy, packaging, storage and repository subjected to acceptance criteria of storage and/or repository facilities, release and recycling including melting, decontamination and incineration
- Company policy, personnel organisation on site (use of the staff on site or external companies)
- Use of different technologies and treatments
- Development of different decommissioning plans containing a list of decommissioning activities, description of the main technical items and a time schedule
- Calculation of the decommissioning costs based on calculation of the decommissioning masses, numbers of waste containers, man-power requirements, and others.

Other boundary conditions are:
- The complete nuclear site is considered in the decommissioning plan. Not considered are buildings which are definitely free from contact with radioactivity, e.g. office buildings.
- The work will mainly be carried out by the existing staff on site. External companies can be contracted if it is required for specific decommissioning activities.
- All activities after the final shutdown are included in the decommissioning plan.
- The decommissioning procedures are based on today's technical standard.
- Relevant international rules and regulations, issued by the IAEA and the European Commission, are respected.

7 DECOMMISSIONING CALCULATION

The main topic of the decommissioning calculation is the cost calculation, based on the above mentioned list of decommissioning actions and the plant inventory. But the decommissioning calculation has more than this single aim. Other points of interest are e.g.:
- man-power requirements
- personnel capacity diagrams
- expected radiation exposure
- time schedules
- mass flow diagrams
- cash flows.

7.1 Decommissioning Activities

For each decommissioning activity decisions have to be made on the following items and/or the quantities have to be calculated or estimated:
- employable technology
- necessary tools and equipment
- other articles of consumption
- required manpower, amount of labour and duration of the activity
- costs associated with the activity
- personnel's radiation exposure associated with the activity.

7.2 Costs associated with the Activities

Based upon the calculation results of the decommissioning activities the cost of each activity can be determined. The following types of cost are distinguished:

- personnel costs as a product of the resulting amount of labour and the payroll costs
- costs for equipment such as facilities, machines, special tools and devices (expenses or rent)
- costs for articles of consumption such as tool equipment e.g. saw blades, drilling devices, clothes for the controlled area, acids for the decontamination and others
- fees for the granting of the licence and the fees for the disposal of radioactive wastes
- external costs for all activities which are not taken on the decommissioning site, e. g. incineration or super-compaction of radioactive waste, or melting for unrestricted use
- other costs, e.g. insurance. (proceeds are balanced here as well.)

7.2.1 *Required Manpower and Duration*

The decommissioning activities require qualified personnel that is accustomed or trained to work in the controlled area. The personnel (number and qualification) required for each activity have to be defined considering the following criteria and boundary conditions:

- amount of labour to be expected
- maximum permissible radiation exposure of personnel (ALARA-principle)
- room conditions
- local radiation levels
- employment capability for tools and devices
- already available equipment, e.g. lifting devices.

The man-power needed for the decommissioning project and the needed personnel capacity for different qualifications must comply with the existing staff on site (supposing that the needed staff will be taken from the existing staff).

7.2.2 *Tools and Equipment*

During the decommissioning of a nuclear power plant a large quantity of marketable equipment will be employed. But special equipment and tools will also be needed. In the last few years various R&D projects have been accomplished to develop such special equipment. Many of them have passed their test phase and they are ready for use. The selection of such special equipment is based on previous decommissioning experiences as well as on research results which are taken into consideration and analysed.

7.2.3 *Other Articles of Consumption*

Other articles of consumption are for example:

- cutting gases
- coolants
- clothes for the controlled area
- cover foils, one-way packages, adhesive tapes
- decontamination agents.

The type and the quantity needed depend on the technology employed and on the masses to be dismantled or treated.

7.3 Expected Personnel's Radiation Exposure

The decommissioning activities will be carried out in accordance with the applicable radiological protection regulations and such that the radiation exposure will be as low as possible (ALARA-principle). The radiation exposure to be expected will be calculated prior to carrying out the activities.

The calculation of the decommissioning activities includes the calculation of the collective dose using average dose rate classes assigned to the individual rooms or areas.

7.4 Time schedule

By combining all activities and considering their interdependencies, the duration and course of the whole project is determined in the time schedule.

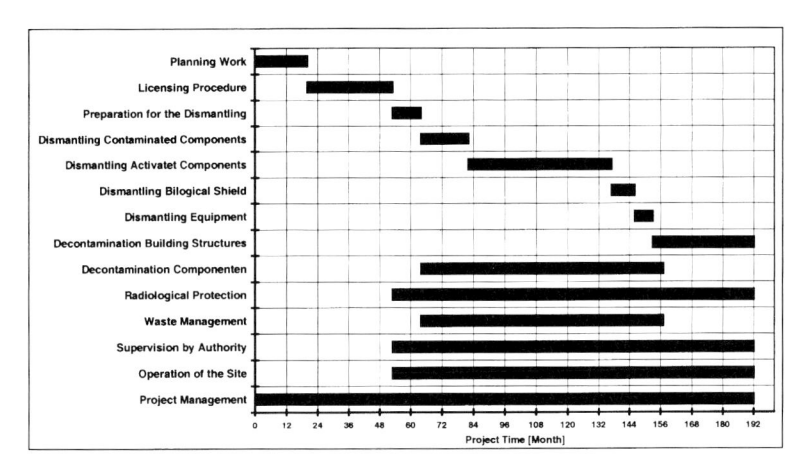

Figure 6: Decommissioning Time Schedule (Example Immediate Dismantling BWR)

7.5 Results of the Decommissioning Calculation

The main results of the decommissioning calculation are the costs, resulting in 305 Mio € (PWR, immediate dismantling) and 330 Mio € (PWR, later dismantling). The results for BWR are given in Table 1 (Price level 2001):

Table 1: Decommissioning Calculation Results

	Immediate Dismantling BWR Reference	Later Dismantling BWR Reference
Costs [Mio €] immediate actions	340	30
Costs [Mio €] 30 years operation SE	-	20
Costs [Mio €] later actions	-	320
Amount of Work [Man-years] immediate actions	1875	250
Amount of Work [Man-years] 30 years operation SE	-	140
Amount of Work [Man-years] later actions	-	1720
Expected Radiation Exposure [Man-Sv] immediate actions	12	1
Expected Radiation Exposure [Man-Sv] later actions	-	9
Duration [Years] immediate actions	12	3
Duration [Years] later actions	-	12

The decommissioning cost contains the following cost categories:

Table 2: Share of the Decommissioning Cost

Cost Category	Immediate Dismantling BWR Reference	Later Dismantling BWR Reference
Staff costs	64%	70%
Investment costs	10%	11%
Costs for waste packages	13%	5%
Consumable costs	4%	3%
Other costs	9%	11%

The decommissioning cost can be assigned to several tasks, e.g. the cost for the treatment of the waste, the cost for the casks, etc. Fig. XX shows the composition of the total decommissioning cost with regard to the main tasks for a BWR Later Dismantling.

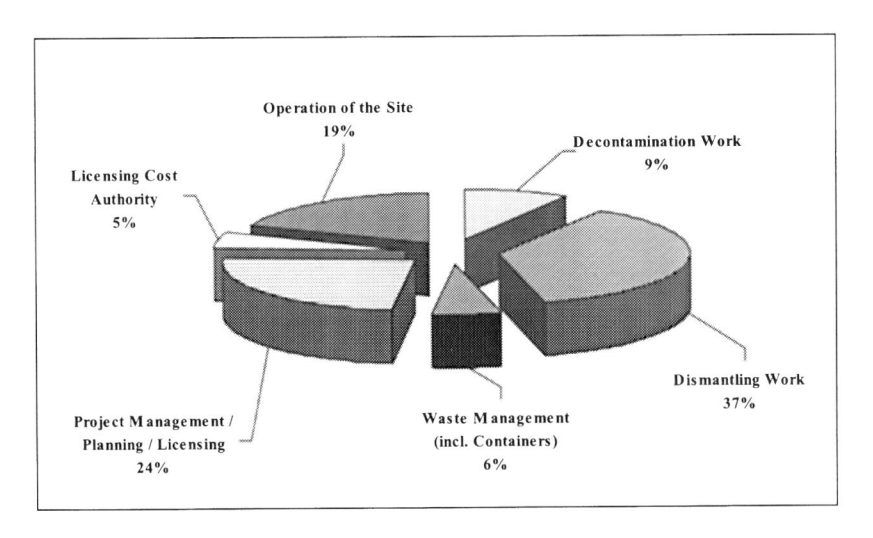

Figure 7: Share of the Decommissioning Cost Concerning Working Packages

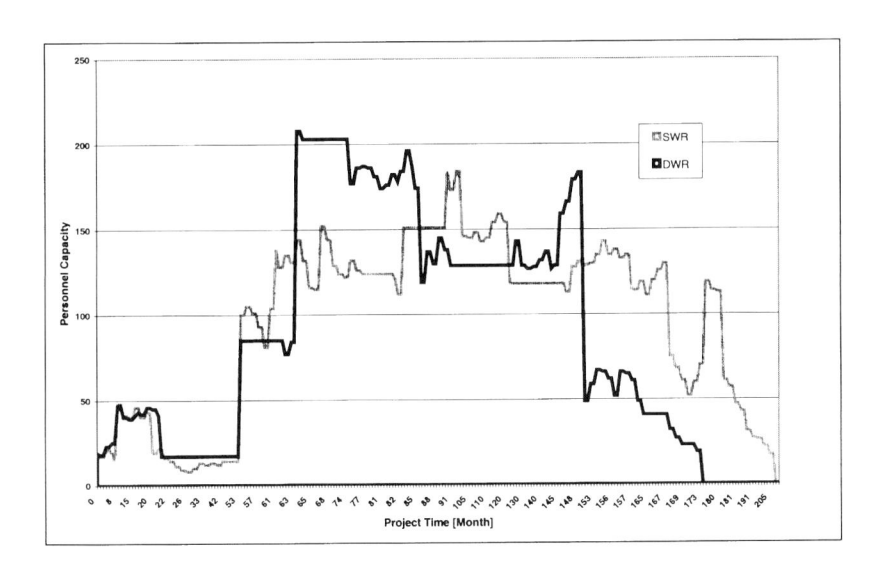

Figure 8: Personnel Capacity

8 COLLECTION OF THE FUND

As mentioned before there is a public interest that nuclear facilities be decommissioned after the termination of their use. This public interest is stated in the German Atomic Law. For the business management of the utility this leads to an uncertain financial commitment.

The obligation for the fund starts with the beginning of the operation, but it is clear – also from the commercial legislation – that not the complete fund is required at this time.

In the early 80s a linear fund collection model was agreed, in accordance with the German Trade Act and with the funding systematic for other industrial facilities. This collection model considers an annual increase of the fund, always based on the actual cost. The following example will explain the model:

- if the decommissioning cost is calculated in the 3rd year of operation, and the result is 400 Mio €, then the fund to be available after these 3 years is: 3/25 of 400 Mio € = 48 Mio €
- suppose that in the 4th year of operation the decommissioning cost is updated to 420 Mio €. Now the fund must be increased to 4/25 of 420 Mio € = 67.2 Mio €
- The increase in the fourth year results to 19.2 Mio €.
- Thus, the increase from the third year to the fourth year contains not only an additional $1/25^{th}$ part of the decommissioning cost, but also the increase of the already collected money to the actual price level.
- After 25 years of collection there is year by year a price increase only (including modifications for technical or other reasons).

This funding model ensures that the fund is always on the actual status.

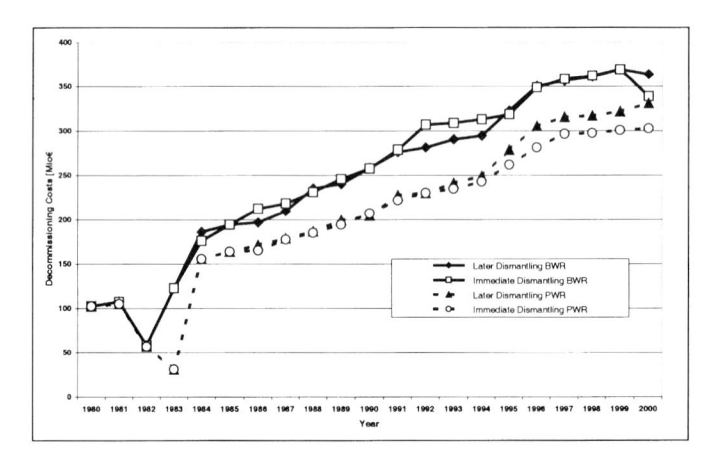

Figure 9: Development of the Decommissioning Cost during the last 20 years

© 2001 NIS Ingenieurgesellschaft mbH

C596/044/2001

Software tools for environmental restoration – the nuclear site end game

C R BAYLISS
Planning Performance and Engineering Division, UKAEA, Dounreay, UK
G J COPPINS and **M PEARL**
Planning Performance and Engineering Division, UKAEA, Didcot, UK

ABSTRACT

UKAEA is progressively delicensing areas of Harwell and Winfrith. To meet the regulatory requirement of demonstrating "no danger", an extensive programme is in place involving historical record review, and radiological and chemical surveys of buildings, drains and land. Consistency of approach and ease of data interpretation is an essential part of this programme and an electronic data capture, interrogation and reporting system coupled to a Geographic Information System has therefore been developed to meet these needs. The bespoke package, called IMAGES – Information Management and Geographical Evaluation System – is discussed in this paper.

1. INTRODUCTION

UKAEA sites generally have complex histories and have been subject to a diverse range of nuclear operations. Most of the nuclear reactors, laboratories, workshops and other support facilities are now redundant and a programme of decommissioning work is in progress. A key objective of the decommissioning programme is the environmental restoration of its sites. This includes assessments of land quality and the management and remediation of contaminated land.

Land quality assessment involves investigation and evaluation of radiological and chemical contamination. The results from this work can then be used to ascertain the most appropriate management and control options to ensure that there are minimum risks to human health and the environment. In addition, where areas of land are being delicensed for redevelopment purposes, it must be demonstrated to the regulators that there is no danger from ionising radiation from the land or buildings.

In view of the amount of information being created that is associated with these activities, a systematic approach is required which not only should ensure consistency in approach, but also should ensure that data is access and revision controlled, and is quality assessed. An electronic system has therefore been developed by UKAEA called IMAGES – Information Management and Geographical Evaluation System to meet these demands. This paper will outline the features of IMAGES and will show examples of its application in the context of delicensing.

2. REQUIREMENTS FOR ENVIRONMENTAL RESTORATION

UKAEA's mission is to restore the environment of its sites in a way that is:

- safe and secure
- environmentally responsible
- value for money
- publicly acceptable

To meet these requirements, decommissioning is being carried out in phases with care and maintenance and care and surveillance periods between stages to allow relatively short-lived radioactivity to decay. This reduces dose levels to personnel and minimises radioactive waste production. Following on from these stages is a post operational phase which corresponds to the point at which the risks to human health and the environment are sufficiently low so that the land can be released for future use. Unconditional release corresponds to meeting the requirement for delicensing.

Because some parts of the site reach the criteria for an "unconditional release" end point earlier than others, land is being released in packages so that an income stream can be generated from the purchase of this land for development. An example of one such area is shown in Figure 1. With so much work being carried out on land quality and contaminated land management and remediation, software tools have been developed to capture and assess the information and to store the outputs. Two of these tools – IMAGES and PRICE Land Characterisation - are outlined in the following sections.

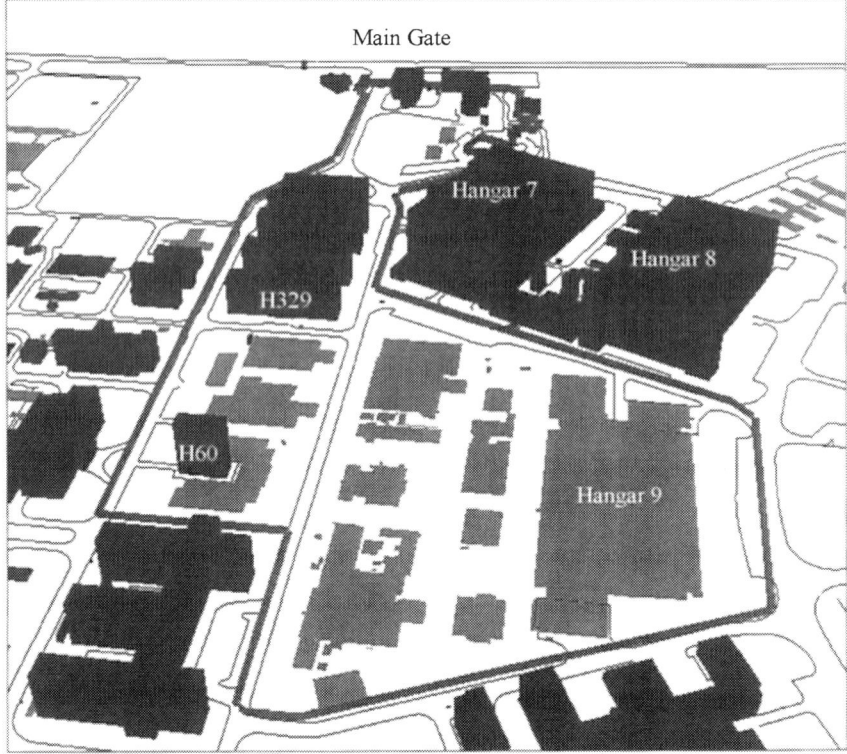

Figure 1 An area of Harwell site showing current and demolished buildings

3. TOOLS FOR ENVIRONMENTAL RESTORATION

The IMAGES system integrates ESRI's ArcView Geographical Information System (GIS) with ESIT's ADMiT data management system.

IMAGES consists of three main elements:

- an ORACLE database system for storage, searching and reporting functionality,
- the ADMiT data management system to capture data from a number of sources in a controlled and auditable way,
- ArcView GIS for geographical analysis and presentation of information.

As a consequence the integrated package enables:

- consistent data capture through a series of input templates
- quality controls on data (e.g. identification of data custodians, data quality marking, revision control, updateability, archivability and traceability)
- data selection through database querying, filtering, and searching
- integration with GIS
- control of data modifications and distribution

At the planning stage of a land characterisation project, the IMAGES system is also being used in conjunction with a cost-estimating tool, also developed by UKAEA. The cost estimating tool is based on UKAEA's PRICE (PaRametrIc Cost Estimating) system. This is normally used for decommissioning cost estimating, but has been modified specifically for land characterisation. Parametric methods of cost estimating are based on the aggregate relationships between cost and physical and performance parameters.

In its simplest form PRICE requires a facility to be divided into simple building blocks or "Components". For each component data is stored on the resources required to undertake unit quantity of that component. This is termed the "Norm" which varies depending on the "Complexity" and "Task" classification attributed to the component. Components can have up to five "Complexity" classifications and three "Task" classifications and thus any one component can have up to 15 "Norm" values.

Cost estimates are created by using the PRICE knowledge base in conjunction with IMAGES and GIS for the area under investigation. Figure 2 presents an example of how spatial database queries can be employed to generate idealised sample strategies, in this instance for roadside survey samples. Such data is typical of that collected and analysed during a land characterisation exercise. The functionality of the GIS can be used to calculate the length of roads within an area and automatically plot survey points at regular intervals along this length, thereby providing unit data for use in PRICE. Other advantages of using this approach include the calculation of physical land areas (e.g. in hectares) and the generation of intrusive survey points using a variety of sample grid patterns.

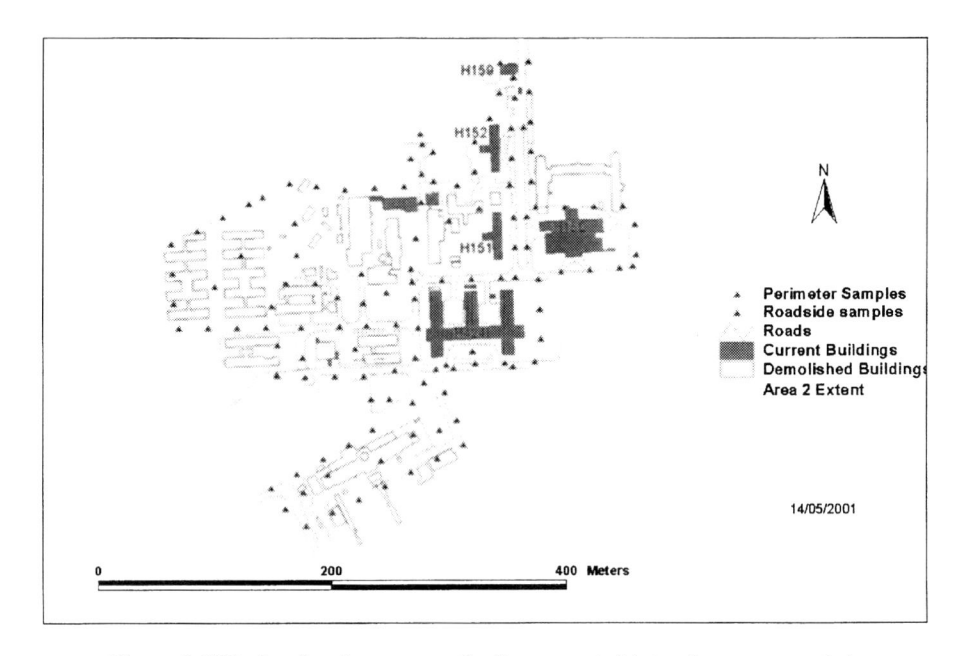

Figure 2 GIS plot showing automatically generated intrusive survey points

3.1 LAND QUALITY MANAGEMENT USING IMAGES

The main interface of the IMAGES system for land quality data is through the hierarchical structure of an assessment database. This includes systematic assessment of land quality, qualitative risk assessment, links to supporting data and any comments and prioritisation of remedial requirements. Much of the supporting data is captured elsewhere in the IMAGES system through a series of standardised input forms. In addition, the assessment can be enhanced using links to GIS where the output from complex database queries can be spatially visualised.

Figure 3 shows the structure of the assessment database. Three levels of information can be stored about the land on a site – Site (regional), Zone (subsection of the site), Area (specific part of the site where land quality is assessed). At the site level is a link to high level strategic documents as well as general documents on site history and geologic setting. Sites are divided into Zones, which contain selected Areas to which detailed land quality information is attached. A qualitative risk assessment section considers sources of contamination, its potential migration and its potential impact to human health and the environment. Following on are recommendations for mitigating measures.

It important that a clear record is kept for various work that is undertaken in Areas as part of the environmental restoration and redevelopment process. The following sections outline some of the detailed information that is collected and accessed through the above structure.

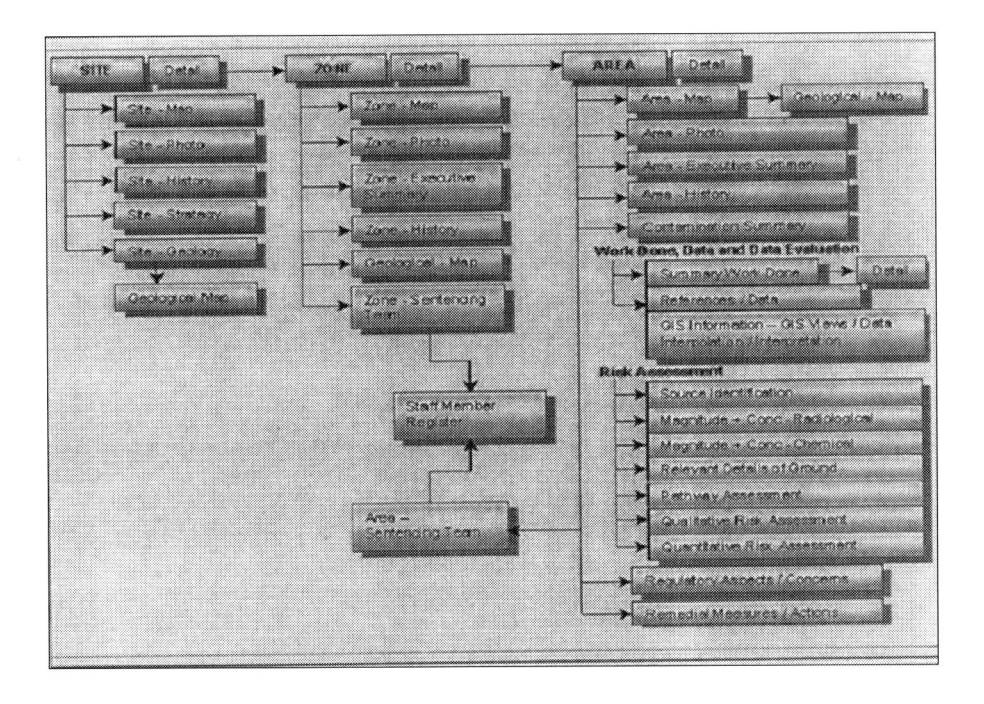

Figure 3 Structure of the IMAGES Assessment Database

3.2 HISTORICAL DATA

Historical evidence is essential to land quality assessment as past knowledge of a site and its activities can be used to guide subsequent surveys. UKAEA has relatively detailed site records which extend back to Royal Air Force usage during World War II. Figure 4 illustrates the format of the forms used for historical data capture. Both current and demolished buildings are recorded along with associated records of any potential contamination that may have resulted from building usage.

Specific documents and photographs can also be attached to provide supporting information and as well as an audit trail back to the original information sources. A link from the "Buildings Module" to ArcView GIS allows useful presentation of historical queries. For example, a database query showing all buildings that contained active laboratories can be viewed in GIS, providing a clear indication of their location on site as well as any spatial relationships that exist with other data held within the system.

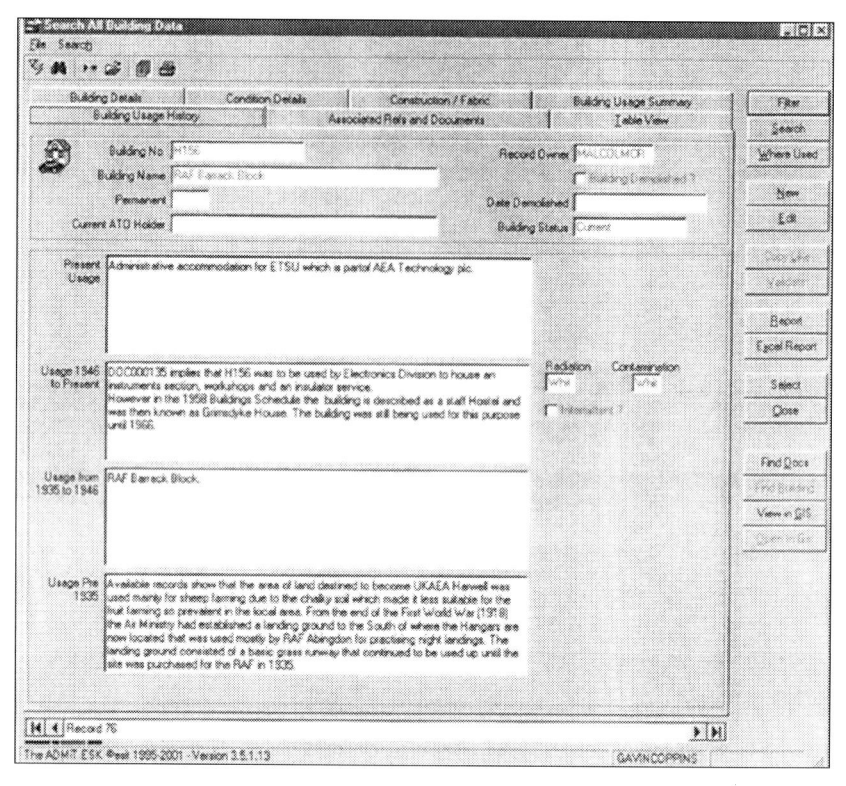

Figure 4 Illustration of the Building History module of IMAGES

3.3 SURVEY INFORMATION

3.3.1 Radiological surveys of buildings
Buildings within certain areas have to be surveyed to ensure that there is no residual contamination. Alpha, beta and gamma surveys are carried out on a room by room basis and the results recorded in a standard electronic form which can be uploaded into IMAGES. Together with the historical evidence and land surveys, this information is being used to support the documentation for delicensing. This data follows the same theme as much of the land quality assessment in that it can be controlled within IMAGES and the results viewed in GIS. In addition, data collected on building histories can also be used to target management and remediation efforts for contaminated land.

3.3.2 Environmental monitoring data (e.g. routine groundwater)
Reassurance environmental monitoring surveys are routinely carried out on all UKAEA sites. Within IMAGES is a database to collect data associated with groundwater quality. GIS can then be subsequently used to assess spatial and temporal variations (Figure 5). This information is also being used to assess the effectiveness of any remedial measures taken, such as hydraulic containment at Harwell.

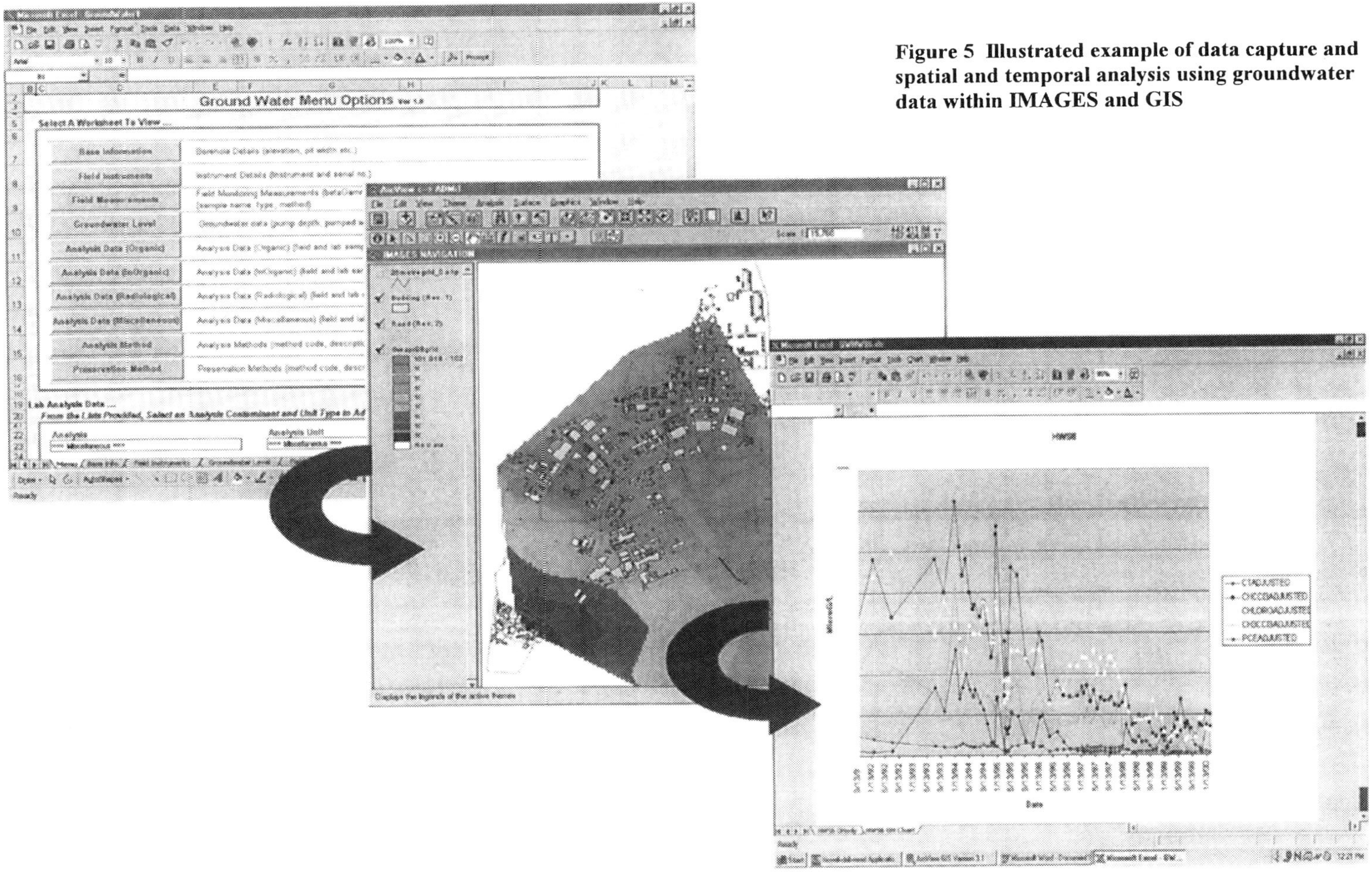

Figure 5 Illustrated example of data capture and spatial and temporal analysis using groundwater data within IMAGES and GIS

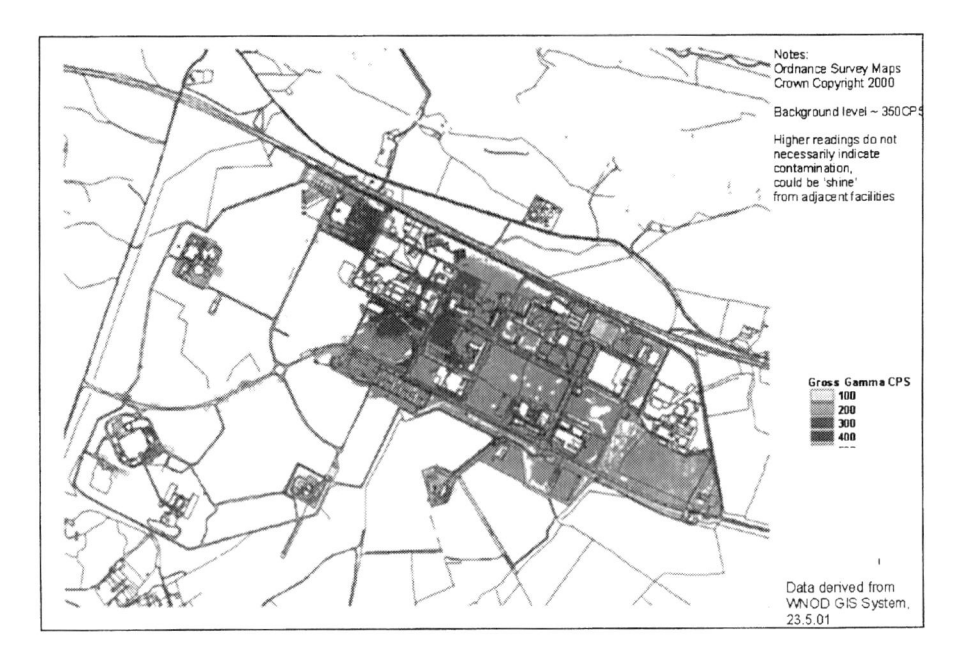

Figure 6 Example data from a GROUNDHOG surface gamma survey

3.3.3 Geophysical surveys
Geophysical techniques can be used to identify subsurface services and potential underground structures. This data can be incorporated into GIS and used to direct the sampling strategy.

3.3.4 Radiological surveys of land surface
A number of techniques can be used to assess contamination at the land surface. The most commonly used technique used on UKAEA sites is GROUNDHOG™ which gives a near total coverage of gamma radiation for the area concerned. An example is shown in Figure 6. Further investigation can then be targeted to areas of elevated activity.

3.3.5 Intrusive land surveys
Where there is the possibility of subsurface contamination and where that subsurface contamination might migrate, intrusive coring or trial pitting is carried out. Standard electronic input forms associated with these types of survey have been developed which can be uploaded into IMAGES. Figure 7 shows information capture, processing and GIS visualisation. By collecting data on a standard input form, consistency of data capture is assured. Completed input forms are logged into IMAGES and quality marked against a set of predefined criteria. The data can then be queried and spatially assessed in GIS – for example, interrogating the analytical data for levels of contamination above a particular level and visualising the result in GIS.

Figure 7 Illustrated example of data capture, query and GIS analysis using intrusive survey data

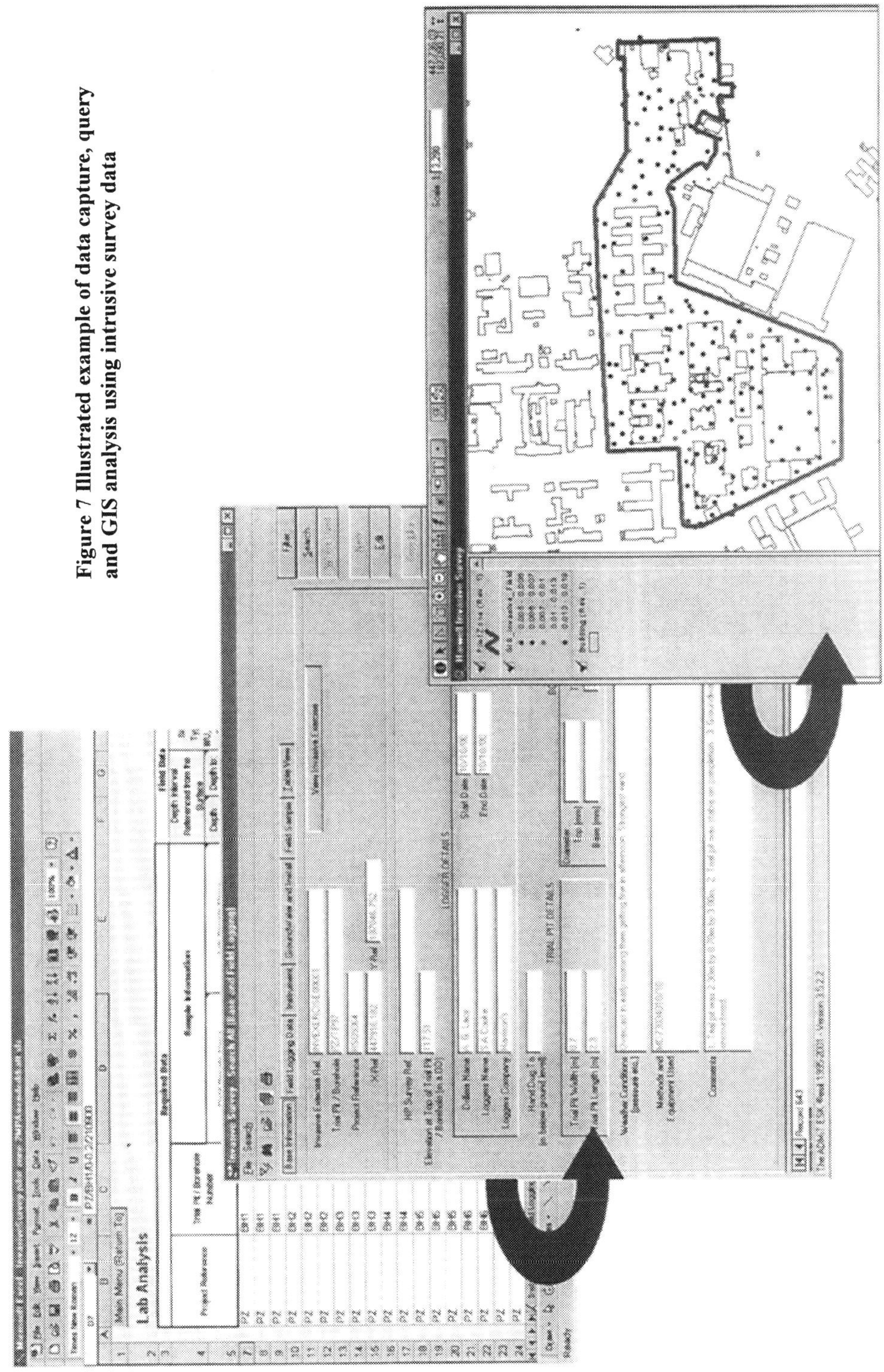

3.3.6 Drain surveys

Decommissioning activities may require drain monitoring to ensure integrity and to ensure that there is no residual internal radioactive contamination. This is carried out using a monitor that incorporates both a video recording device as well as a gamma spectrographic detector. Results can be processed and visualised through a GIS interface. Drain data can also be linked to intrusive survey results – where subsurface contamination might be associated with a breach in the drain system.

3.4 INFORMATION CONTROL

As was outlined earlier in the paper, all information associated with delicensing or long-term environmental management should be controlled and fully auditable. IMAGES tackles this by placing all data in a workflow procedure, controlling input, access, editing and revision (Figure 8).

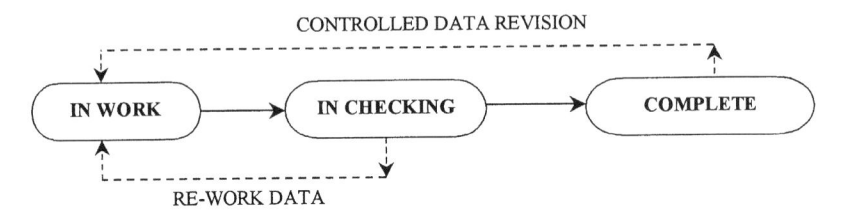

Figure 8 The IMAGES Workflow concept

4. CONCLUSIONS

This paper has outlined the systems developed by UKAEA to capture, interrogate and present land quality assessment data in a controlled and auditable manner for land quality management. Several recent projects have utilised these methods with an overall aim to demonstrate compliance with relevant legislative standards and to be consistent with UKAEA's commitments to environmental restoration.

The development of the IMAGES system, which combines advanced data management with a Geographical Information System, has been an integral part of UKAEA's drive towards the management of land on its sites. Information from this has also been used to form the basis for identifying radioactive and chemical contaminants and their implications to a source-pathway-receptor concept of environmental management. Also, by utilising the spatial data held within GIS, idealised sampling strategies can be formulated for any area of a site and the information fed into a parametric cost estimating system "PRICE" to generate estimates for future land quality surveys.

The focus of IMAGES on data management has enabled UKAEA to record and efficiently analyse all data collected during the land quality assessment for delicensing, from historical documents to information collected using state of the art investigation techniques.

Public Perception

C596/037/2001

Public opinion on nuclear power

B INGHAM
SONE, Supporters of Nuclear Energy

SYNOPSIS

Today's received wisdom is that for nuclear energy to play any role in the future it needs to resolve three issues:

    its alleged lack of competitiveness

    the "problem" of waste storage

    public acceptability.

In practice, the nuclear industry has only one problem: public opinion. Its "lack of competitiveness" and the "problem" of waste storage would instantly disappear if public opinion were not perceived to be averse to it. It would then be welcomed as the ultimate solution to global warming.

There is nothing new in this analysis. Some of us have been offering it for years – ever since climate change became the whisper of the town. What might be new is a willingness not just to accept it but to try to do something about it. If so, we shall be making progress. Supporters of Nuclear Energy (SONE, for short) will have served its purpose, if not run its course.

SONE is a product of electricity privatisation. The moving forces behind it recognised that, with nuclear energy in private hands, it would have no friends unless the market found it irresistible. In that event, it would become as desirable as Marilyn Monroe. But since it was neither the fashion of the hour, nor likely to become one in the foreseeable future, it needed supporters. SONE was formed "to keep the nuclear flame alive", as we rather dramatically put it at its foundation in June 1998.

Put more prosaically, we exist to keep the two words "nuclear power" on the agenda at a time when the Government of the day is desperate to keep it off it. Four times SONE asked for a meeting with the Minister for Energy in the 1997-2001 Blair Government. And four times we were denied. We have grown more determined as well as nastier in the process.

We chose at the outset to have only individual members and deliberately to avoid corporate membership. Our reasoning was that a body composed of individual supporters who wanted to belong to it would not only carry more weight but also be far more flexible in its operations. Its activists answer only to a very high powered membership of some 220. Small we may be, but we don't take "No" for an answer.

We are, in short, a pressure group. That very description of SONE has been known to give even some of our members the vapours, though they are getting over it. We are I think, overcoming the intellectual snobbery from which nuclear energy has suffered from its inception. My first experience of nuclear's early overwhelming sense of infinite superiority was in the Department of Energy where I heard that "those who know [about nuclear energy] don't need persuading and those who don't know don't count".

Such an attitude had no use for communicators, still less for pressure groups and didn't deserve them.

This toffee-nosed approach, combined with a secretive defence background, has all but killed an industry which was originally hailed as beating swords into ploughshares. It left the nuclear power industry at the mercy of the militant "Green" movement which proceeded to demonstrate that it was as unprincipled as it was merciless. No exaggeration was too large for the anti-nuclear nutters, no lie too monstrous, no scare too alarming, no hypocrisy too blatant and no stunt too illegal.

I have read that the sands of Chernobyl are so hot with radioactivity that the snow melts as it hits the ground. A scientist friend who went with me to Chernobyl to investigate has worked out that this is an exaggeration of the order of billions. The BBC speaks of untold thousands of deaths at Chernobyl when the massed ranks of UN scientists have still validated only 45 more than 15 years after the event. Small leukaemia clusters are routinely attributed to areas around nuclear power plants in Britain even though scientists studying the disease put such clusters, if they exist, down to a virus. Plutonium is presented as the most lethal substance known to men yet the nuclear power industry's processes have never killed a single individual whereas coal claimed thousands upon thousands. Plutonium is so lethal that a friend of mine, now 76, has both been injected with it and inhaled it in the interests of medical science and is as wick as a weasel in his highly monitored old age. As for the "Greens" attachment to the outrageously illegal and snoot-cocking stunt, the worse it is the better they think it is. After all they're only there for the cameras.

Dr Goebbels would have been proud of them.

The result is that nuclear energy is now regularly cited at public relations and communications conferences as the classic example of what happens to an industry which abdicates the field to its enemies.

The "Greens" have succeeded in frightening the uninformed nation to death about radiation and radioactivity from nuclear energy even though medical science exposes individuals on average to 144 times more radiation than nuclear power generation, reprocessing and waste management. Yet have you ever heard of a "green" with a broken leg refuse an X-ray? Blithering hypocrites that they are.

They go on and on about the dangers of transporting nuclear waste in steel flasks which not even a train, aimed deliberately at them at 100mph, could damage while an average of 10 people a day are killed on our roads – 3,650 a year.

They have effectively injected not only the population of Britain but of the world with a mortal fear of radiation which we can neither hear, see, smell nor touch. Yet it is a very part of nature and properly harnessed the saviour of so many cancer-stricken lives and the clean, sustainable solution to curbing and eventually overcoming the global warming which could eventually wreak havoc with the world's climate.

The "Greens" have done this with the help of an ignorant, malleable media which, in general as distinct from in particular, never allows the facts to spoil a good story, especially when it is accompanied by dramatic pictures.

In the light of all this, it is easy to say that the nuclear industry deserved what has undoubtedly been thrown at it. But that solves nothing if global warming is a reality and is about to bring momentous changes to our lives.

The first requirement of the nuclear industry and the scientists, technologists and administrators who earn their living from it is to recognise that it has only one problem: public opinion. That is the easy bit. Not many people in the industry are prepared to deny this, though they are much happier arguing until the cows come home, both in private and in public, about the respective merits of different reactors. They seem oblivious to the gift they are thus handing enemies of nuclear.

The hard bit is resolving how to act on this recognition that public opinion is the main obstacle to a nuclear future and having so resolved how to summon up the blood to get on with it.

The fault, dear Brutus lies not in our stars but in ourselves that we are underlings.

Let's not get preoccupied with nuclear's alleged uncompetitiveness with gas. On the latest figures gas has a 0.7p per unit advantage over nuclear – 2.3p compared with 3p per unit generated. But that advantage would be eroded if the price of gas reflected its damage to the environment in the way that nuclear electricity's price has to provide for de-commissioning and long-term environmental protection. A carbon tax could – and probably would – render nuclear competitive, taking account of construction costs and its likely 50-year generating life. The 1997-2001 Blair administration, in its preoccupation to avoid the words "nuclear power" ran away from a carbon tax, and instead gave us the so-called climate change levy which should be referred to the Advertising Standards Authority as the most blatant example yet of gross misrepresentation.

In any case, my 23 years' experience of Government teaches me that economics, far from being the dismal science, is the most creative, malleable, riggable and ingenious science of them all. What would the competitivity of nuclear electricity look like if you set its fairly stable known fuel costs against the price uncertainties of gas which incidentally roughly doubled in price during the year before this talk was prepared? And don't forget insurance costs. In any normal commercial calculation some allowance would have to be made for the price stability and security of nuclear fuel supply compared with the volatility and insecurity of natural gas supplies coming, as they will increasingly in the years ahead, since Britain is no longer self-sufficient, from such areas of instability as Russia, Algeria and the Middle East?

Economics, like computers, depend on the input. Put rubbish in and you get rubbish out. So far only inadequate rubbish has been fed into the calculation of nuclear's competitivity. This is because the public do not yet require a competitive answer.

This leads me to the other conventional problem – but in reality sub-problem – of nuclear; waste management. There is no technological problem that I know of in storing nuclear waste of high, medium or low level radioactivity now or in the foreseeable future or for posterity. Neither our nuclear power stations nor Sellafield are building up dumps which carry any risk of explosion. They are merely returning, in controlled conditions, radioactive material to the earth whence it came – or would be if any Government of any hue had the guts to sanction an underground store of the kind they have in Sweden.

The reason they lack the guts is public opinion which has been fed column miles of "green" guff. Public opinion does not demand a solution.

It is at this point that the nuclear industry falls into a trap. Instead of resolving to take on the "green" hysterics, much given to lying, exaggeration and chicanery, and the political duckers and weavers, it falls in with their non-answer; energy conservation and the so-called benign and renewable sources of energy.

The argument is that we should not unnecessarily set ourselves at odds with the "greens", notwithstanding the fact that their motto is death to nuclear. Instead, we should go along with their pretence that if only we used less energy and where we have to use it we used it more efficiently, and if only we also harnessed the biomass, the wind, the waves, the tides, the sun, the moon and no doubt the stars, plus technologies unborn, undeveloped and certainly unproved, we could replace fossil fuels, do away with nuclear and live happily ever after. Well, I shall not go along with it. I absolutely and utterly refuse to fall in with this monstrous deception of the public. It is utter bunkum and balderdash.

We cannot yet run the fourth largest economy in the world by burning coppiced wood and chicken muck; wrecking our countryside with God knows how many wind farms which produce only intermittent, unreliable electricity; surrounding this sceptred isle with Salter's nodding ducks to produce wave power which, in the immortal words of Tam Dalyell MP, has so far succeeded at a cost of £1.5m in generating enough juice off Islay to power 25 electric fires; exploiting hot rocks in Cornwall; building barrages across untold estuaries without knowing the consequences for the environment; swamping umpteen more valleys under water with every knowledge of the consequences for the environment; adding solar power converters to every south facing roof, as recently advocated by a Liverpool professor who can't count; and processing biomass into car fuel. Nor are we remotely near the hydrogen economy, especially as we shall need nuclear energy to produce it, or the photo-voltaic cell future.

As for energy conservation being part of the solution, I speak as the first to be in charge of the old Department of Energy's energy conservation division. We could save a great deal of energy if we used it as sparingly as our grandparents used it because they couldn't afford it. But we saw a year ago the hostility to the use of taxation to try to damp down demand. Luxuries undreamt of by even our parents are now regarded as necessities. And the Government is absolutely determined we use more and more energy by trying to wire us all up to the internet.

Technology is undoubtedly getting more useful work out of energy – and will continue to do so – but all the evidence and rational analysis of human nature tells me that, outside a severe national emergency, we cannot expect to close any power stations because of energy conservation.

Anybody who thinks alternative sources and energy conservation can fuel and power the future should be carried away to the funny farm and given every encouragement to practice what they preach as a cure for their peculiar delusions.

Of course, we cannot rule out new technologies emerging later in the 21stC. But we cannot fuel and power the present and immediate future on pious hopes, gleams in the eye or even theoretical possibilities. We have to make the best of what we have got while at the same time doing our best to encourage better ways of doing things.

We have no excuses for Alice in Wonderland dreaming when California shows us the consequences of getting wrong.

So where does that leave us? The answer is a nuclear future, provided the nuclear industry can summon up the blood to take on and vanquish its mortal enemies – and, incidentally, the mortal enemies of not just British but global economic development – and win over public opinion.

But there then arises another potential excuse for inertia; namely, the inevitability of a nuclear future when the penny drops. This argument – which, I confess, I have done much to propagate – relies on the essential hypocrisy of man. It teaches that even the most dreaded and violently opposed solutions become instantly desirable when there is perceived to be no other alternative.

I am one of the most fervent adherents of this theory. I have been reinforced in my convictions by the utterly hypocritical recent performances of "Greens" in Scandinavia, Austria and Germany towards nuclear energy. This is not to mention Rotherham Council which signed up to "green" energy with the result that this almost certainly means that when it deliberates late into the night in the Town Hall its weighty considerations will be illuminated by base load nuclear power. Nuclear is, of course, the greenest fuel but I am sure that is not how Rotherham conceives it.

This sort of Fabian theory of the inevitability of gradualness do not, however, cope satisfactorily with practicalities. There are now some signs that Civil Servants, if not the politicians, are getting the wind up. It is not that Government energy policy, in so far as we have one, is a mass of contradictions. It's the arithmetic.

I do not propose to tax you with numbers. Just four simple ones. Today, nuclear generates 25% of Britain's electricity. By 2020 it will, as things stand, generate a mere 3%. Government policy envisages only 10% of that gap being filled by renewables. So where is the missing 12% to come from if we are not to burn more fossil fuels and not replace ageing nuclear power stations as they close?

Stand by for panic.

The result of that panic can only be nuclear, beginning by extending the lives of existing nuclear stations. But panic is not a sensible background for developing a nuclear power programme. Nor does it necessarily reduce the time involved. It takes years – possibly ten – to secure planning consent and build nuclear power stations.

So, all supporters of nuclear energy now need, in the national interest, to shake off their reticence, their inhibitions about speaking out, the excuses for doing what has come naturally for years – that is, doing next to nothing in the face of its enemies – and to campaign to awaken the nation to reality.

That reality is that nuclear power presents the only tried, tested, safe, reliable and economic solution to global warming. SONE exists to hasten the dawning of this reality in Britain.

Are you prepared to join SONE and help with that campaign? Are you prepared to fight for your industry or do you prefer to leave it to others? This battle will not be won by you letting someone else write to their MP, someone else bombard their local and national newspaper editors with pithy exposes of the ignorant but inventive codswallop served up in the average newspaper story and correspondence column; someone else raise hell with the broadcasters when they exhibit their anti-nuclear prejudices, someone else take on the lying "greens" on their own hypocritical patch wherever and however they manifest themselves.

Dammit, why should I exert myself in retirement when I don't earn my living in the nuclear industry?

SONE expects you to do your duty and it is watching whether you do – or whether you leave it to SONE.

Finally, SONE exists to employ among its arguments the moral dimension. If the West means what it says in espousing the cause of Third World development, then we, of all people, should make use of the one advanced means by which we can secure our power through non-polluting generation. There is no morality in our turning our backs on uranium if that means we ultimately deny the developing nations any chance of enjoying our current standard of living.

So, the nuclear industry should concentrate on one essential; winning back the public support it has so wilfully thrown away by neglect. If it does that, all else will fall into place and a sensible future will be secured for mankind.

C596/038/2001

Public acceptance and the development of long-term radioactive waste management policy

C CURTIS
Environment Centre, University of Manchester, UK

ABSRACT

The UK Government will shortly embark on "a detailed and wide-ranging consultation" concerning future policy for the long-term management of solid radioactive wastes. The Radioactive Waste Management Advisory Committee, RWMAC, has argued that the way to formulate a policy that has a good chance of being eventually implemented is by wide public consultation from which the policy is seen to emerge with clearly articulated public views. Such views will legitimise the policy. Suggestions are offered concerning how the process of policy formulation should be conducted and how it should be overseen. Some thoughts on incentives for success are appended.

1. INTRODUCTION:

The United Kingdom (UK) Government has stated (1) that it will be carrying out "a detailed and wide-ranging consultation" on future policy for the long-term management of solid radioactive wastes. At the time of writing (July 2001) it was expected that the consultation paper would be published in September. The Radioactive Waste Management Advisory Committee (RWMAC) understands that the policy consultation will focus primarily on the process for deciding policy rather than specific policy choices.

RWMAC provides (and publishes) advice to Ministers on a range of radioactive waste management issues, including methods for establishing consensus (2,3). It has prepared advice on the form it believes the consultation should take (4) and much of what follows is taken from this advice: it outlines what RWMAC believes to be the best chance of developing policy which will the public support necessary to allow implementation.

Decommissioning is the focus of interest here. A number of major decommissioning projects have been successfully undertaken (or are in process) in the UK. All of these have involved, to greater or lesser extent, public consultations and, no doubt, have employed many of the approaches that RWMAC has been looking into. To this extent, no claim is made for novelty!

My understanding of the decommissioning process is that it is initiated once the decision to terminate the operational cycle has been taken. The initial sequence[a], which effectively involves operational procedures, is broadly:

- Stop reactor *(this is then de-pressurised therefore a catastrophic dispersal accident should be impossible)*
- Control criticality *(this involves full deployment of the control rods such that there is no short-term criticality risk short of extremely unlikely events)*
- Remove fuel *(now zero risk of criticality and the vast bulk of all radioactive materials removed)*

The requirement for active management is greatly reduced from this point on and overall hazard is similarly lessened – particularly once spent fuel has been removed from site. In the UK, this is presently transported to Sellafield. Subsequent operations will also require removal of radioactive waste (mostly intermediate and low level waste) before the site can be returned to the "green field" status that is the theoretical end-point of the decommissioning process.

An obvious concern must be that, without a long-term management plan for all the wastes removed from decommissioned Nuclear Power Plant (NPP) sites, *decommissioning can hardly be claimed to be complete.* The greater part of the hazard is just being shifted from one location to another. This is one of the rather obvious reasons why RWMAC is so keen to see that long-term management policy is developed and put in place.

[a]*Following this initial sequence, it would appear that issues of public concern would mostly be to do with local nuisance – noise, transport, visual amenity and employment loss.*

2. THE NEED FOR A NEW APPROACH

The advice on the policy consultation carries forward previously published RWMAC thinking on this issue (2,3). It covers the process only to the point where policy is decided (in line with the consultation itself). Thereafter, further consultation obviously will be required during the even more contentious stages required for implementation of whatever policy is decided: RWMAC believes that these will also need to be founded firmly on principles of open discussion and consensus-building. The advice (4) was prepared by RWMAC's Consensus Working Group, chaired by Professor Andrew Blowers. It was subsequently approved by RWMAC meeting in plenary.

Stepping back a little, it seems clear that Government policies should be in place for all aspects of management of the UK's radioactive waste. Ideally, liquid and gaseous discharges as well as solid wastes should be considered together within a holistic frame. Such an approach would ensure that the hazards and risks associated with all radioactivity are dealt with in a consistent way.

In practice, discharges and solid wastes have been treated separately. For example, consultations on the UK Strategy for Radioactive Discharges 2001-2020 (5) and Statutory Guidance on Regulation of Radioactive Discharges into the Environment from Nuclear Licensed Sites (6) have recently taken place.

Previous policy for high-level waste (HLW) management had been vitrification followed by 50 years' storage for cooling with no commitment to further action (Cm 2919 (8)). Policy for the long-term management of intermediate level waste (ILW) and some of the more problematic low level wastes (LLW) extended from conditioning, packaging and temporary storage to deep geological disposal. The demise of the Nirex repository programme in 1997[*] inevitably brought this policy into question. A disposal route is available for more easily manageable LLW at Drigg, Cumbria but the facility there has limited capacity. Some further capacity for LLW will need to be found in the medium term. Existing packaging and stores for HLW and ILW will last for tens of years, whereas some of the radioactivity contained within them will take tens of thousands of years to decay to safe levels.

The UK thus has large quantities of solid radioactive waste for which no long-term management strategy is in place. RWMAC considers that to do nothing in respect of formulation of new policy is not an acceptable option. RWMAC thus welcomes the Government's consultation initiative and hopes that this advice will help to move the process on. Given public concerns about radioactivity, the exercise will certainly be difficult, but it must be joined.

RWMAC has also made clear (2,3) its opinion that the approach formerly adopted in the Nirex programme, sometimes described as "decide-announce-defend", must not be repeated. Given that neither the "decide-announce-defend" nor the "do nothing" approaches are appropriate, RWMAC therefore favours a consensus-building approach, involving wide and open discussion of the issues. In the Committee's view, this approach currently offers the best chance of identifying a policy solution with a reasonable prospect of eventual delivery.

But what is meant by consensus-building? It must surely be recognised that unanimity of view is an unrealistic goal. In its 1999 advice (2), RWMAC took the term consensus to be *"the achievement of a sufficient concurrence of view at various stages to legitimise a decision to proceed with a particular course of action"* and sees no reason to modify this definition. It needs to be clearly spelled out at the outset in order to instil realism early in the process. However, the fact that the issues can be seen to have been widely, openly and fairly discussed, and that those with an interest will have had opportunity to present and justify their views, will clearly be important when it comes to the later implementation stages. As will become obvious, any consensus-building approach will inevitably require significant time, effort and financial resource.

---

[*] In March 1997, the Secretary of State for the Environment, then John Gummer, dismissed an appeal by UK Nirex Limited, the nuclear industry and Government-owned company charged with finding a solution for the future management of the UK's ILW, against refusal of planning permission for an underground "rock characterisation facility" (RCF) at Sellafield. This, in effect, caused the collapse of the national programme to develop an underground disposal facility for the long-term management of solid ILW and some low level radioactive waste (LLW).

It should be noted that Government appears to have already committed itself to some form of consensus-building approach in its response to the House of Lords Select Committee on Science and Technology's report on Nuclear Waste Management (1).

Given the unambiguous preference for a holistic and robust approach to radioactive waste management, three further points need making. First, any policy that is adopted for the management of solid wastes will need to be compatible with that being formulated for control of radioactive discharges. There are interdependencies. Second, a sound policy will cover all the waste forms. Obviously it will cover HLW, ILW and LLW but it is important to recognise that there are additional categories of radioactive materials for which there is a need for clearly stated policies. These include spent fuel not destined for reprocessing, plutonium and uranium (both reprocessed and depleted). Third, policy for the management of solid wastes should encompass a range of possible nuclear futures. Scenarios should include an end to nuclear generation when current reactors close, the replacement of existing nuclear reactors with new reactors, and an expansion in nuclear generation to meet climate change and/or energy supply security concerns

3. KEY GUIDING PRINCIPLES

If it is to be effective, the process will need to be founded on clearly stated and appropriately applied guiding principles. These, in RWMAC's view, remain as stated in its earlier advice on the establishment of consensus (2), namely: *provision of adequate time for exploration and resolution of complex technical issues; early involvement of the public and other stakeholders; openness and transparency; a deliberative, accessible process of decision-making; and commitment to appropriate peer review of scientific and all other expert input.*

A further principle, namely *equity*, was also considered. The context is the extent to which both the process for deciding policy and the policy itself are generally perceived and accepted to be fair. Considerations of equity must, taking account of the needs of future generations and, at the implementation stage, inequalities imposed by facility siting. There is a need to be seen to be fair to those directly affected insofar as is reasonably practicable within the broader context of a need to deliver a solution to a national problem. Involvement in decision-making, provision of incentives and compensation have already been explored (and utilised) in a number of countries.

A further key recommendation made in RWMAC's earlier advice (2) is that the process policy development should be overseen by an independent or, at least, balanced-interest body that is widely accepted as being capable of upholding the broader public interest.

4. THE PROCESS REQUIRED FOR POLICY FORMULATION

4.1. General requirements
Radioactive waste management should be viewed as a problem for society as a whole whose solution requires general public support in moving from the definition of the nature of the problem, through identification of the options for its solution, to the choice and, ultimately, implementation of the option(s) selected.

This implies a staged process with involvement of different constituencies. For any such process to be successful, RWMAC believes that it should be underpinned by all of the key guiding principles set out above and including oversight by the independent body. But probably most important overall will be *provision of adequate time for exploration and resolution of what will inevitably be complex issues.* In respect of constituencies, *early involvement of the public and other stakeholder groups* who have an interest in, and may ultimately be directly affected by, the outcome is of paramount importance.

Other general points are relevant here. The key aims of the process adopted will be to identify the policies around which consensus is more likely to form and to enable decisions thereupon to be made in a well-informed and robust manner. The process must be designed carefully to allay suspicion that the outcome has been pre-determined. Clear presentation and weight of argument must be the basis for any decisions made. The policy advice that will be the product of the exercise must be seen to have flowed logically from the process. That the expectation is for progress by consensus rather than unanimity of view must be reiterated and confirmed.

Until now, the process has tended to focus on detailed scrutiny of single options, for example underground disposal of ILW within the Nirex repository programme. In contrast, RWMAC believes that the forthcoming policy formulation process must be based on the simultaneous evaluation all practical options in a consistent manner against a single set of appropriate and agreed criteria. Although RWMAC does not regard "do nothing" to be an acceptable option, it should certainly be included in the evaluation to bring out clearly the risks that such inaction might entail. The evaluation criteria must be agreed and address risks, costs and all other pertinent considerations.

A very necessary part of the process will be provision of information. It must be made clear that radioactive waste is problem shared by society as a whole and that it is in everyone's interest to find an agreed solution and then to get it implemented. Information must be provided concerning the present location of all solid wastes (and where potential wastes are destined) and how they are being managed. The hazard that these wastes pose to people and the environment, now and in the future, must be objectively and quantitatively explained. The legal and technical requirements for protection of people and the environment must be given and explained, for these must be met by each and every option to be considered.

The overall programme must include time and resource for the assembly of educative material. Expert input will be required here and this must be quality assured. All such material will have to be readily accessible to the public and presented in a manner that members of the public can understand; no easy task. The Internet offers a new, inexpensive and increasingly convenient vehicle for the dissemination of large amounts of detailed information.

4.2. The need for a staged process
RWMAC suggests that a process along the following lines should achieve the goal of demonstrating that the developed policy will be seen to have evolved logically from the process:

Stage 1. Initiation This would aim to identify the concerns, values and expectations of public and stakeholders. This knowledge would then inform design of all subsequent stages of the process. Sounding of views requires the early provision of appropriate information. RWMAC has suggested guiding principles for the process but the public and stakeholders may introduce others.

Stage 2. Development of the Framework for Option Evaluation This is essentially a two-part stage:

> 2A. Option Identification, which, through a process of public and stakeholder discussion, would aim to identify all potential options for management of the solid radioactive materials in question. Which of these options are practicable and which should therefore be carried forward for detailed evaluation would then be decided.
>
> 2B. Identification of Criteria for Option Evaluation, to agree, again through a process of public and stakeholder discussion, the criteria against which the practicable options that are carried forward are evaluated.

Stage 3. Assemble Expert Input Once the options and evaluation criteria have been assembled and agreed at stage 2, there will be a need to assemble expert information upon which to base option evaluation. The requirements are not simple. The scientific and technological understanding of core issues must be facilitated, and this then used to assess the technical risks associated with each of the options. Evaluation of each option will equally require quite different expert input: that relating to social and economic issues, for example.

Stage 4. Evaluation of Options This would involve evaluation of practicable options identified under stage 2A against the criteria identified under stage 2B, drawing appropriately on stakeholder dialogue and public deliberation.

Stage 5. Preparation of Policy Recommendations The collective views emerging from option evaluation under Stage 4 would be analysed by the overseeing body. This would then put forward its recommendations as to which option, or options, should be pursued as the basis for policy. This could then be exposed to one further round of public/stakeholder consultation.

The precise manner in which these stages are implemented will require thought. Various combinations of consultative and participative techniques (see below) could be used. The key need is to engage the wider public, stakeholders and experts in the most meaningful way. The Government's forthcoming consultation should explore public and stakeholder opinion to confirm, or otherwise, support for the kind of staged process advocated here,

It is possible that some of the options identified under stage 2A might be considered to be so inappropriate that they should be eliminated before detailed evaluation under stages 3 and 4. Procedures for such preliminary rejection would need to be clearly formulated and be both logical and transparent.

4.3. Constituencies to be involved in the policy formulation process

There are probably only two constituencies that need to be involved in the policy formulation process. The first is vested interest groups: they normally, even invariably, respond to consultations. The second is the much larger constituency of members of the general public, a tiny proportion of whom get involved in consultation exercises. It will be necessary to use appropriate consultation techniques to identify existing public views in stage 1, and also to ensure involvement of representative members of the public in subsequent stages. These techniques should be supplemented by other mechanisms to ensure the widespread provision of information and opportunity for members of the public to provide feedback as the consultation progresses.

The involvement of the wider public is particularly important. The goal should be to emerge from the policy formulation process with clearly articulated public views. Such views will legitimise the choice of policy.

5. OVERSIGHT OF THE PROCESS

The programme discussed above, and particularly the proposed five-stage process for deciding future UK radioactive waste management policy, will constitute a substantial programme that could extend over several years. It is essential, however, that this process should not be allowed to extend indefinitely. There is a need for proper understanding of the requirement, sound management and adequate resourcing. Planned timescales must adequately reflect the needs of the work but must be clearly defined and adhered to.

As well as being managed in an effective manner, this programme will need to provide for *openness and transparency*, another of the key guiding principles identified earlier. Such openness and transparency will be vital for establishing trust which will be needed in order to achieve a broader societal consensus. It is for these reasons that RWMAC has suggested that the programme of work should be overseen by an independent, or at least balanced-interest, body that is widely accepted as being capable of upholding the broader public interest. This must be established by the Government, presumably after taking views on its constitution and remit in the forthcoming consultation. It seems unlikely that a government department would fit the bill.

A number of questions immediately arise, one being whether or not appointments to the overseeing body should be representative or *ad hominem*. The latter might be more efficient, obviating the need for repeated reference back to sponsoring bodies with potentially inflexible positions.

The overseeing body's core remit would be management of the policy formulation process. This should include responsibility for deciding and managing the detail of the programme, including organisation of events and securing information and expert opinion as necessary. Translating the outcome into policy advice for Government will be its ultimate role. Given that it must represent the public interest, the overseeing body must have an appropriate degree of freedom to pursue its work in the manner which it sees most appropriate for this purpose.

The overseeing body's remit must also be clear about the range of radioactive wastes to be covered. It should see identifying areas of agreement and disagreement in respect of the management of these wastes as a key element of its work at all stages of the process.

As noted above, the process will depend from the very first stage on the provision of authoritative and appropriately quality assured information. This implies the need for the overseeing body to be in place to commission and arrange suitable review of such material from the outset.

The body will need to set its approach for deciding when an acceptable level of agreement has been reached (the yardstick of success) and procedures to be adopted if this is not achieved (how to move forward in the event of failure to achieve sufficient agreement).

While the form of reporting required of the overseeing body (eg to the Government, Parliament and the public) must be identified, it would be sensible to allow some flexibility. Provision of full and readily accessible reports of progress and outcome will be a key part of the overseeing body's work: thereby establishing openness and transparency.

These are RWMAC's suggestions for the role and membership of the proposed overseeing body. It is anticipated that such matters will be more fully explored in the Government's forthcoming consultation. For successful completion of the programme it will also be particularly important for the overseeing body to be adequately resourced for the demanding programme of work that it will be required to undertake. In the first instance, the lifetime of the overseeing body should be limited to that of the policy formulation task.

6. CONSULTATION METHODS AND TECHNIQUES

The policy formulation exercise will need to use consultation methods and techniques that provide for a *deliberative and accessible process of decision-making*. There is a need, in particular, to create a dialogue between the scientific and technical community and the wider public in which each contributes to the process. An important element of this dialogue will be that of putting complex scientific and technical information into a form that the public can readily understand. Only in this way will it be possible to carry the policy dialogue through from problem statement, debate of options for solution, to decisions on policy.

RWMAC has surveyed the variety of consultative and participative techniques (telephone surveys, focus groups, consensus conferences and so on) that can be used to gain appreciation of generic issues and to facilitate specific areas of decision making. Some are aimed primarily at specific audiences (for example; the public on one hand or vested interest groups on the other). Some merely assemble pre-existing views: others seek informed opinion through deliberation or seek to go on to reconcile differing opinions. It may be important to ask if some techniques are likely to have a local or national orientation? The differing characteristics of the various consultation techniques are analysed and discussed in annex 3 of the report (3).

In practice, it is this kind of analysis that will help to decide how these techniques are best employed in the policy formulation process.

7. EXPERT INPUT AND QUALITY ASSURANCE

Expert inputs will be required at various stages of the process. Notably, these will provide material and testimony to support public and stakeholder deliberations. In its publication "A Better Quality of Life: a Strategy for Sustainable Development for the United Kingdom" (8), the Government states the science used for policy decision making must come from "sources of information of high calibre". It goes on to say that "where possible, evidence should be reviewed from a wide-ranging set of perspectives".

Insofar as science and technology are concerned, it is important to appreciate that, at any point in time, the knowledge relating to any particular issue lies somewhere in a continuum between "universally-accepted" and "significantly-speculative" (due to lack of information and/or understanding). Science can therefore rarely supply the reassuring "black or white", "yes or no" answers that politicians, lawyers and perhaps many non-scientists would like.

The implication of this for policy formulation is that, given inevitable predictive uncertainties, science is correspondingly unlikely to be able to provide categorical reassurance on risks into the far future. This needs to be suitably stated and recognised. The basis for the choice of policy should be that it can provide a level of assurance of protection from harm that society and individuals perceive to be reasonable. To this end, expert input needs to be integrated to give an appropriately balanced overall view of the scientific, technical and societal impacts of radioactive waste associated with each of the various options potentially available for its future, longer-term management. Benefits and detrimrnts (including relative hazards and risks) need to be clearly identified. It may also be useful to put associated hazard and risk in the context of those deriving from other suitably comparable day-to-day activities.

In order that the science (and the subject matter of all other expert advice) is seen to derive from "sources of information of high calibre", assessment will be required. This will usually include some form of peer review. Criteria for and methods of such assessment have been critically evaluated (9). This need for *appropriate peer review of scientific and other expert input* is another of the key guiding principles mentioned earlier and is also in line with the approach advocated in the Government's Cm 4345 sustainable development policy statement (8).

There are a number of issues that reviews of science (or any other expert discipline) should address. These include examination of assumptions and whether these are in any way sensitive to disciplinary, institutional or political interests. It is important to establish to what extent the science is "universally-accepted" or "significantly-speculative". Information from different parties should be sought to test credibility and independence. It is also particularly important to look for evidence of bias – including bias by omission. Finally, it is helpful to recognise a distinction between assessment of the pre-existing knowledge base and assessment of the risks entailed in utilising aspects of that knowledge base in specific technical applications (9).

At the all-important stage of option evaluation, this process should be of a participative nature, looking at application of the expert opinion in terms of questions that the public and stakeholders deem to be appropriate and assessing its quality and relevance in this light.

Peer review and quality assurance should be applied to all expert input at all stages of the policy formulation process.

The manner and timing of preparation of expert input, including social and ethical perspectives, will be for the overseeing body to decide. On the one hand its preparation will take some time, which provides an argument for an early start. On the other, making such a start before the practicable options and their evaluation criteria have been identified (stage 2 of the five-stage process) might be taken to prejudice the outcome of the process.

Overall, the process needs to be seen to be one of expert input supporting rather than driving, the policy formulation process.

8. SOME FURTHER OBSERVATIONS ON SOCIAL ISSUES

The principle of *equity* will also need to be considered. This is the extent to which the process for deciding policy, the policy itself and its subsequent implementation are accepted to be fair.

While this principle is relevant at all stages, it will become particularly so when specific options and specific sites are under consideration. Risks from radioactive waste are unevenly distributed both spatially and between generations. Taking a long-term perspective to safeguard the interests of future generations is one of the key objectives of the Government's concern for sustainable development policy (8). The precise implications of this for radioactive waste management are unclear but one key conflict emerges:

   a desire not to pass burdens to future generations (where "burdens" may be effort, cost, risks etc);

   a desire to allow future generations to develop better solutions to the waste management problem than are available with current technology and scientific knowledge.

This and other potential conflicts will need to be discussed as part of the policy formulation process. The best way of doing this, in RWMAC's view, is to identify and agree option-evaluation criteria that reflect all aspects of this potential conflict. The precise implications of option adoption in this context are then made clear. Option-evaluation criteria that seek to do this are suggested in annex 2 of the report (3).

In this context also, reversibility, may be seen to be important. This is the concept where radioactive waste once emplaced in some location can, if deemed necessary for whatever reason, be retrieved. Reversibility is therefore also suggested as one of the evaluation criteria. Identification, assessment and comparison of options that have various degrees of reversibility over various time periods are needed. For example, all deep repository concepts have a relatively high degree of reversibility during waste emplacement but a lower degree of reversibility after backfilling and sealing. So-called "phased disposal" concepts involve a care and maintenance period after the end of waste emplacement and before backfilling and sealing, during which time there is also a relatively high degree of reversibility. There will need to be realism concerning the extent to which reversibility is really achievable over different time periods.

Different long-term management scenarios also entail different monitoring requirements, financial costs, health risks to workers, health risks to the public, risks of disruption due to natural processes and to human intervention – accidental or mischievous.. All these aspects need to be addressed when identifying and evaluating options such that comparisons are made on the proverbial "level playing field".

Other social issues will require thoughtful analysis. For instance, should the process of volunteerism, used in some overseas countries to identify potential sites, be a part of policy? Usually associated volunteerism is the question as to whether a local community should be compensated or rewarded for hosting a national facility that may have perceived adverse societal effects.

Also looking to overseas practice, some might raise the question as to whether the community local to a prospective site for a repository or waste store should have a right of veto and, if so, what form the veto should take and at what point should it be able to exercise it.

These societal issues might usefully be exposed to some initial consideration in the policy formulation debate, not least to give an indication of how individual solutions might need to be implemented in the UK. It should not, however, be simply assumed that what is appropriate in one country is equally appropriate in another.

9. FINAL THOUGHTS

I hope that proposals outlined above are a reasonably faithful representation of RWMAC's collective views. What follows, however, is a personal commentary that has not been shared with Committee.

I fully support the view that something along the lines suggested by RWMAC represents the best chance we have of developing policy that can be implemented. At the end of the day, however, there are vociferous groups who seem to be opposed to any development that might deliver safe, long-term management of radioactive waste. We are also blessed with a press that mostly values story line above quality of information or argument. To these obvious difficulties must be added the fact that democratic governments must face re-election every few years. There can be little doubt that new policy development and implementation will be a long and difficult road to follow.

I believe that new policy for long-term management must emerge from public consultations – in other words that the public and stakeholders must discover for themselves that there is indeed a need for such policy. This is where some of the principles referred to above take on special significance.

9.1. Openness and transparency
The military and experimental legacy wastes held in the UK and those presently accumulating from NPP electricity generation must represent a real hazard to health and the

environment. The aim of long-term management must be to convert these wastes to a form which is as stable as possible – one that is chemically unreactive and within which radiotoxic elements are immobile (therefore there is least likelihood of their dispersal within the environment). It must then seek to emplace the wastes in a facility/location that requires minimum (ideally zero) active management in order to prevent dispersal by natural disruptive forces or malicious (especially terrorist) attack. The industry has developed the technologies to achieve the conversion stage (conditioning) and has made progress with it – but much waste remains in a "raw" state. Facilities for long-term management have not been developed due to public rejection of earlier projects and the absence of policy now.

These are the facts as I see them and as such should be openly and transparently detailed in the policy formulation process. *This is the information that justifies the case for urgent development of new policy for better and longer-term management.* It has seemed to me that almost everyone associated the industry (including the regulators) has always played down the hazard – partly to avoid negative publicity and reasonably because they are all working intensively to achieve and maintain safe, day-to-day management. If new and sound *long-term waste management policy* is to be formulated and implemented, the *mid- to long-term hazard* must be acknowledged and detailed within the policy development process. This will require something of a culture change.

The problem is a national one. It is also complex in that solid waste management and radioactive discharges are interlinked – the conditioning of raw wastes to a safer, more passive state may entail small discharges to the environment. These have to be set against the real benefit of reducing overall hazard. *By treating solids and discharges separately, the all-important holistic judgements are made more difficult.* We have the technology (which is largely quite conventional) to make significant gains by converting raw wastes into much more stable (passive) forms. Government, Regulators and Industry have a collective duty to work *together* here and progress can be made ahead of new policy emerging. Again, something of a culture change is called for.

9.2. Quality assurance of all expert testimony

For some years past, the advice of "experts", usually scientists, has been increasingly brought into question. In part this may be a consequence of poor communication skills and partly an impression of arrogance created by a reluctance to join public debate. A great (although perhaps uncomfortable) strength of the proposed formulation process is that expert advice will be quality assured in a formal way. Within the RWMAC proposals, the same assessment procedure will be consistently applied to *all* expert testimony not just scientific; presumably including the evidence supplied by *all* stakeholders. Under these conditions, the playing field may become a little more level.

ACKNOWLEDGEMENTS

Obviously the real substance of this paper is the product of many people – particularly the members of RWMAC's Consensus Working Group chaired by Professor Andrew Blowers. I would like to thank them all for this contribution and also for much thoughtful analysis of quite different topics. Dr L.A. Mitchell gave me specific help with this manuscript; and this is gratefully acknowledged. Full details of RWMAC membership and publications can be found at: http://www.defra.gov.uk/rwmac/index.htm

REFERENCES

1. The Government Response to the House of Lords Select Committee on Science and Technology Report on the Management of Nuclear Waste. Department of the Environment, Transport and the Regions. October 1999

2. The Radioactive Waste Management Advisory Committee's Advice to Ministers on the Establishment of Scientific Consensus on the Interpretation and Significance of the Results of Science Programme's into Radioactive Waste Disposal. Department of the Environment Transport and the Regions. April 1999.

3. The Radioactive Waste Management Advisory Committee's Advice to Ministers on the Process for Formulation of Future Policy for the Long Term Management of UK Solid Radioactive Waste. Department for the Environment, Food and Rural Affairs. In press.

4. Twentieth Annual Report of the Radioactive Waste Management Advisory Committee (Chapter 3, Building Consensus on Future Radioactive Waste Management Policy). Department of the Environment, Transport and the Regions. November 2000.

5. UK Strategy for Radioactive Discharges 2001-2020: Consultation Document. Department of the Environment, Transport and the Regions. June 2000.

6. Statutory Guidance on the Regulation of Radioactive Discharges into the Environment from Licensed Nuclear Sites: a Consultation Paper. Department of the Environment, Transport and the Regions. October 2000.

7. Review of Radioactive Waste Management Policy: Final Conclusions (Cm 2919). Her Majesty's Stationery Office. July 1995.

8. A Better Quality of Life: a Strategy for Sustainable Development for the United Kingdom (Cm 4345). Department of the Environment, Transport and the Regions. May 1999.

9 Sound Science and the Environment. Fisk, D. Science and Public Policy, pp46-49, Spring 1997.

C596/039/2001

Public requirements in nuclear waste management

R E J WESTERN
Friends of the Earth, London, UK

ABSTRACT

The public requirement for sound radioactive waste management comprises two areas - the substance of what should happen to the radionuclides and secondly how decisions are implemented. The risk burden must be minimised. To achieve this further production should be phased out together with further plutonium separation and finally conditioning should be brought in as a matter of urgency.

For the implementation, the four functions of Rule Making, Policy Making, Research and Implementation should be addressed together with the hard question of how committed the personnel of the institutions are to delivering a robust radioactive waste management system.

MAIN TEXT

The nuclear industry produces radioactive wastes which might cause cancer. To try and make sure that the wastes don't cause cancer enormous engineering and financial effort is required to deal with the risks. Furthermore, in addition to the cancer risk there is the risk that the plutonium which is produced as an intrinsic part of the nuclear industry may end up being made into a nuclear weapon - so in a nutshell - cancer and bombs, that's what the public cares about in the context of nuclear waste management and the public requirements in nuclear waste management is to address these concerns.

The real question is: are these public concerns being addressed? In some ways they are, but in others - the most important - they most certainly are not. Public concerns are being addressed in the sense that august bodies have been set up to bear witness to the day to day goings on of the nuclear industry - but they are not in the sense that the routine scrapes and skirmishes show no sign of abating. Furthermore there has never been in the history of the UK nuclear industry an appraisal of the infrastructure of the waste producing aspects of the industry in order to minimise the waste and the weapons risk that is produced.

This paper is being written in mid June to be delivered in October. The Labour Government has just been re-elected and the waste review is expected to be underway in the autumn - so we may have just missed it. The following is a consideration of issues that need to be looked at if we are going to have a hope of getting to grips at all with the problem of radioactive waste management - looking first at the physical picture and then at the institutional picture.

There are two separate parameters that need to be quantified when developing a physical picture of what the radioactive management problem includes. The first is the basic question of amount -
simply how much of the stuff is there and the second is the question of form - what physical and chemical form is it held in.

The question of amount is absolutely vital. The risk of weapons manufacture or cancer depends critically on how much radioactivity needs to be carefully handled. The larger the amount the greater the risk. This means that it is imperative that we close our nuclear power stations as soon as possible and build a future where nuclear electricity plays no role.

The second parameter is less obvious. Although the actual inventory of material is determined in the reactor during the transmutation, activation and fission processes (followed by decay) the handling of the materials is often critical to the risk that it presents. This is most apparent for the fuel rods that are removed from the nuclear reactor after a certain amount of energy has been extracted.

The initial reason for the development of nuclear energy - first in the States during the Second World War - and then over here at Windscale in Cumbria was to turn uranium into plutonium so that the plutonium could be used to make weapons. The actual creation of plutonium was just the first stage though as the plutonium first produced was locked in with the uranium and the fission products and required complex chemical intervention to release it - and all behind thick radioactive shielding. To produce the plutonium large capital intensive infrastructure was required.

When the weapons programme was developed to produce electricity the power was first seen as a bi-product with the prized plutonium as the main output from the new reactors.

Unfortunately this post Second World War thinking has stayed with us and as a result of weighted contracts of the seventies and beyond utilities still send their spent fuel for plutonium separation.

This plutonium separation comes at a very high price. Of course obviously there is the financial penalty and Friends of the Earth have released work demonstrating the enormous degree of savings that would be available if utilities were to switch to the cheaper option of dry storage. More importantly, from the perspective of this talk there is the impact in terms of cancer and weapons risk. The weapons risk is an obvious one. Plutonium separation was developed as a technology to generate separated plutonium for weapons - the more it is continued the greater the stockpile that has to be dealt with.

The wider set of risks go to the heart of the need to think carefully about just how you want a given radionuclide to behave. In the spent fuel rod the radionuclide is likely to be either surrounded by uranium metal or uranium oxide. As an intrinsic part of the plutonium separation process the fuel rod is chopped up and dissolved in hot nitric acid. In terms of the public requirement for the risks to be minimised this step is absolutely disastrous. The advantage of the fuel rod matrix is that, particularly for the oxide, it is relatively passive.

Dissolving the intensely radioactive radionuclides in acid is just is just creating an accident waiting to happen. Enormous levels of scrutiny and engineering effort are constantly required to prevent the dispersal of the liquid solution.

In addition to the liquefication problem there is also the problem of direct release. Radionuclides are directly released to the air and the sea as a direct result of the plutonium separation process and huge amounts of miscellaneous material becomes contaminated as waste. This contamination is extremely problematic. A radionuclide is not a nut or a bolt - it is a chemical element with the whole range of properties that a chemical element can have. An example that makes the issue easy to illustrate is that of carbon. Carbon in sugar will dissolve readily, but carbon in a diamond will effectively not dissolve. Similarly uranium can show just such a range of behaviours depending on how it is linked up. The nuclear industry came very unstuck many years ago at a uranium mine in Brazil when they underestimated uranium solubility by 200 million times because they assumed the wrong chemistry. Similar problems are found for other elements.

Similarly mixing up the diverse range of radionuclides found in a plutonium separation plant with the papers, solvents and other chemicals necessary for the process creates a mixture that is extremely difficult to make accurate predictions for. For the risk predictions for long term disposal the criterion is a risk less that one in a million per year - an error of 4 - 6 orders of magnitude would obviously be problematic.

This gives us two items that must be addressed:

- the early end to nuclear power
- the early phase-out of plutonium separation

The third item applies across the board to wastes that have been produced - particularly in the distant past. As discussed above the chemical and physical form of radionuclides can affect their behaviour. Particularly in the part the industry was set up in such a way that the bulk of the wastes lay untreated for protracted periods of time. The reports of the NII have demonstrated just how dangerous this can be and they have highlighted the need for early conditioning of these raw wastes. So a final item on the wish list is early conditioning. The requirement for early conditioning is particularly important for decommissioning. The critical criterion for conditioning is the requirement for passivity. As long as random events can disperse the radioactivity, or as long as intense institutional effort is required to secure the radionuclides the material cannot be considered passive and presents an unnecessarily high risk.

BNFL's wish to leave reactors in a non-passive state for a protracted period must be rejected and the NII's drive for earlier decommissioning must be welcomed.

Of paramount importance in the context of conditioning is the issue of separated plutonium management. This must be immobilised with fission products to reduce the risk that it is diverted to weapons use. There is no credible economic route for plutonium as a fuel and therefore it must be treated as a waste and handled to minimise the risk that it represents. The development of an immobilisation option must be seen as a priority. Thus the third issue that must be addressed is

- the need for early conditioning and passivity

The next question that arises how the three issues are to be implemented. This comprises two separate questions. The institutional arrangement of the different personnel and implementing bodies - and also the make up of the different bodies.

One point that must made at the out-set - a point that was lost during the eighties is the vital need for the separation of poacher and gamekeeper. The Flowers report of the seventies which highlighted the need for action to be taken pointed to the need to avoid awkward conflicts of interest. This edict was ignored and Nirex the 'Nuclear INdustry Radioactive Waste Executive' was set up owned and run by and for the nuclear industry. It is now vital that Nirex is separated from the nuclear industry. The function of Nirex is a public function, though it should be funded by the nuclear industry, it should be run separately from it as the calls on its priorities may otherwise become clouded.

One of the first functions of a reformulated Nirex would be a radical rethink of just what is the best thing to do with radioactive waste in the long-term. Our previous fixation on disposal may not be the best - other options such as Guardianship should be seriously addressed. Nevertheless Nirex should retain its role as advisor of packaging and conditioning.

In addition to the implementor the other obvious requirement is that of the rule maker. Someone must be there to act as the arbitrator, the benchmark setter. This goes without saying. What is a little less obvious is the need for in-house knowledge in this area. It is all to easy for the Government Agency to farm out the responsibility for guidance to consultants, retaining only rubber stamping capacity themselves. This trend must be reversed.

The final two areas of responsibility apart from rule making and implementing are that of policy development and research. Policy making must sit with Government. In addition to policy development the House of Lords have argued for a Commission to oversee policy development. Unless care is taken this could become heavy handed.. Although RWMAC has not carried much clout - the opposite problem of having a body that has too much power must be avoided. Rule by diktat in this area must be avoided - this is an area where policy must evolve incrementally with a broad range of inputs from a wide range of interested parties. What is lacking in terms of the present stakeholder input is any direct representation of those who will be affected by radioactive waste in the future. There is not a strong voice demanding segregated funds or an independent research body. To redress this I think that we should upgrade our national policy oversight from RWMAC to an institution that has full time membership and a more influential role and wider environmental representation.

The research role is essential. In addition to the on-going requirement for the attempted quantification of the rate that radionuclides would migrate from a repository the alternative option - that of Guardianship, mentioned above, should also be researched. The philosophy of Guardianship is that we can never know enough to irreversibly commit radioactive wastes to a hole underground where it may be subject to uncontrollable leaks. The implications of this approach need to be looked into so that it can be put forward as an alternative option to disposal and the relative merits assessed.

Thus it may be seen that there are four functions for waste management. Rule Making, Research, Policy Making and Implementation. In some areas there needs to be clear separations - particularly for rule making an implementation. Precisely how this is done should be the subject of the forthcoming consultation. What is as important as the precise institutional arrangement is the make-up of the institutional bodies.

The neatest arrangement of deckchairs on the Titantic could not hide the fact that the ship itself was headed in the wrong direction. If the personnel in waste management teams do not really care about what they are doing then any amount of effort in constructing the correct designation of responsibility will be irrelevant.

The industry is at a disadvantage here. In the past nuclear waste has been seen as the definitely unglamourous end of the industry. Although new build has shifted off the agenda this historic artefact still ramifies to the present day. One thing that is definitely needed is an injection of new young into the industry, across all fronts, who see waste management as a contribution to a clean future. That is not to say that we don't already have some good people ready to drive forward a finally agreed policy - although of course the period required for agreeing such a policy does seem extremely protracted.

In conclusion, the public requirement for sound radioactive waste management comprises two areas - the substance of what should happen to the radionuclides and secondly how decisions are implemented. Of course the risk burden should be minimised. To achieve this further production should be phased out together with further plutonium separation and finally conditioning should be brought in as a matter of urgency.

For the implementation, the four functions of Rule Making, Policy Making, Research and Implementation should be addressed together with the hard question of how committed the personnel of the institutions are to delivering a robust radioactive waste management system.

The waste has been piling up for fifty years. Its time we took responsibility for sorting out some of the mess.

Innovation

C596/002/2001

The potential role of high-power lasers in nuclear decommissioning

L LI
Department of Mechanical, Aerospace, and Manufacturing Engineering, UMIST, Manchester, UK

ABSTRACT

This paper presents an overview of current global R & D activities on the use of high power lasers in nuclear decommissioning. The overview covers the use of lasers for decontamination (e.g. paint stripping, metal oxide removal, concrete surface removal), waste sealing and structural material cutting/dismantling. The characteristics and limitations of using lasers (including CO_2, CO, COIL, Nd:YAG and high power diode lasers) in nuclear decommissioning are described and comparisons with other techniques are given. The potential roles of high-power lasers in future nuclear decommissioning activities are discussed.

1. INTRODUCTION

Nuclear decommissioning is a multi-billion dollar global activity. Fig. 1 shows the estimated number of nuclear reactors world-wide that would need decommissioning over the next 30 years, based on a typical 30-year life span of these reactors. From Fig.1, it can be seen that, there is already an urgent need for suitable decommissioning technology to be applied. Over 1000 nuclear reactors would need to be decommissioned within the next 30 years. Each nuclear reactor could generate over 120,000 drums (roughly 24,000 m^3) of solid waste [1]. The cost of waste disposal in UK is estimated at £20,000/m^3. Minimising the waste volume is thus one of the important economical and environmental issues in nuclear decommissioning. Safety of the operating personnel involved in the nuclear decommissioning is another important issue. Therefore, special decommissioning techniques are required to satisfy the above requirements. The fact that any decommissioning equipment in direct contact with radioactive materials will be contaminated and become waste should also be considered when selecting suitable decommissioning techniques.

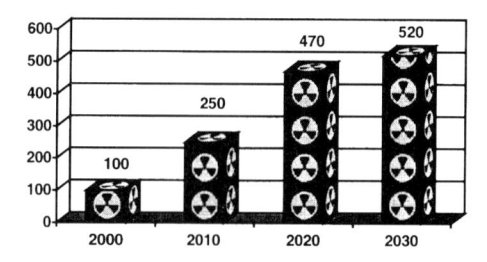

Fig.1. World-wide nuclear reactors to be decommissioned, data from [2,3].

A typical nuclear decommissioning procedure is shown in Fig. 2, from which it can be seen that decontamination and dismantling technologies can play very important roles in the overall nuclear decommissioning process. A viable nuclear decommissioning technology should be able to minimise the waste volume (thus reducing the overall waste disposal/management costs) and be safe and reliable to operate.

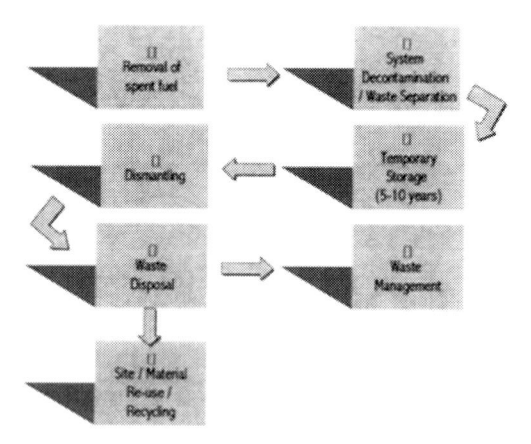

Fig.2. A typical nuclear reactor decommissioning procedure.

High power lasers have been recognised as non-contact and remotely controllable precision tools for cutting, welding and surface cleaning. They are widely used in modern industries. Whether these existing techniques can be applied directly to nuclear decommissioning is still a subject of investigation. There are a number of competing conventional techniques that can be readily available for use in nuclear decommissioning. These include chemical decontamination, dry ice blasting for surface cleaning, diamond saw cutting (for concrete), plasma arc cutting and flame torch cutting (for steels). These processes can deal with complex surfaces, large areas and thick section structural materials. It is necessary to understand the special characteristics and limitations of laser technology as compared to alternative methods, so that its suitability for specific applications in nuclear decommissioning can be identified.

This paper reviews the current global R& D activity on the use of high power laser technology for nuclear installation/structure decontamination and dismantling. Comparison is made between different lasers and with some of the competing techniques. The potential roles of high power lasers, especially high power diode lasers and diode pumped solid state lasers, in nuclear decommissioning are discussed.

2. LASER BASED DECONTAMINATION TECHNIQUES

In a nuclear power plant, radioactive contamination can be found on the inner surfaces of piping, valves, pumps, heat exchangers as well as on the surfaces of large machinery, tanks, vessels, reservoirs, handling equipment, instruments and painted walls/floors. The materials that the low / medium level radioactive contamination is normally associated with can generally be grouped into three categories:

a) Oxides / rust on a metal (mainly stainless steel and carbon steel) substrate. The depth of contamination is normally within the oxide layer that is less than 100 µm.

b) Multiple layer paint (e.g. chlorinated rubber and epoxy paint) on concrete/plaster and metal substrates. Most contamination is within the paint layer which is normally less than 1 mm in thickness.

c) Contaminated concrete. Most of the radioactivity remains within 5 mm from the surface [1,28]

Therefore nuclear decontamination involves the removal of these contaminated layers from various substrates. Damage to the substrate is not a factor to be seriously concerned. The required decontamination factor (DF, the ratio of radioactivity before and after decontamination) needs to be above 100 to meet the work-atmosphere dose – equivalent rate [4]. The decontaminated surface should have radioactivity below 5000 dpm/100 cm^2 for release [5]. Furthermore, secondary waste generation needs to be minimised in the decontamination process. A minimum of 2 m^2/ hour decontamination rate would be desirable to enable the technique practically feasible. Various decontamination techniques have been investigated to meet the above targets. These include the use of lasers, chemical solutions as well as abrasive blasting techniques. These are discussed and compared in the following.

2.1. Oxide layer removal from metal substrates

Large area metal surface decontamination (removal of contaminated metal oxides) can be achieved by immersing the contaminated components in concentrated acids (e.g. oxalic acid, chloric acid and nitric acid) as well as diluted acid solutions (e.g. dilute chloric acid reduction agent of inhibitor-laced chloric acid and vanadium chloride mixtures [4]). Decommissioning factor of 100 can be achieved. The chemical method has an advantage of high efficiency for decontaminating complex geometry components and difficult to access areas such as inner walls of pipes and small holes. The method is, however, not suitable for decontaminating large objects such as tanks/flasks. In addition, despite up to 70% chemical solution recycling rate [4], considerable secondary waste is generated. Furthermore, metallic surfaces can be activated by the chemical process, thus making them vulnerable to recontamination.

Another non-laser based technique that has been investigated is the use of abrasive jet (e.g. water + zirconia powder) to remove metal oxides [4]. Decontamination factor of 100 can be achieved. It can be applied effectively to the exterior of objects and hot spots. Again, this process generates considerable secondary waste and the collection of the waste is difficult due to the high-pressure abrasive jets.

Laser decontamination techniques involve the use of a short pulsed (typically with a few nano-second pulse width) and high peak powered (up to several MW) laser to remove the oxide layer by thermal ablation (vaporisation) and the associated thermal shock effects. Since the thermal penetration depth is proportional to the square root of beam-material interaction time, the use of short pulsed lasers enables minimum energy loss through thermal conduction to the substrate material. Multiple pulses or multiple scans are normally needed to provide high DF. Since vapour/plasma plume generated during laser ablation travels in the direction normal to the surface, to avoid plume interference with the laser beam, laser beam can be fired to the surface at a lower angle (i.e. < 90°). A low-pressure gas jet (a few l/min) coaxial to the laser beam can be used to protect the laser optics and to disperse the laser-generated plume. Helium gas has been found to be able to improve removal efficiency by 50% over other gases when a 90° beam incidence is used [6] since helium gas has higher ionisation potential (24.6 eV) than Ar (15.6 eV), N_2 (15.6 eV) and O_2 (12.1eV). Uniform, rectangular beam geometry, rather than a Gaussian beam with a circular spot, is desirable to achieve uniform ablation and to maximise coverage rate. Therefore, a cylindrical lens, rather than a spherical lens, is preferred in laser ablation cleaning. The removed particles can be collected via high efficiency particulate air (HEPA) filtration. Up to 95% collection rate has been reported [7] and a decontamination factor up to 257 has been demonstrated [7]. Re-deposition of the removed particles can sometimes occur. Multiple passes and the use of a cylindrical lens can minimum this effect. For example, in a case study, with a cylindrical lens, re-deposition rate was < 0.2% compared with 5% with a spherical lens [6]. A typical processing arrangement for laser ablation cleaning is shown in Fig.3.

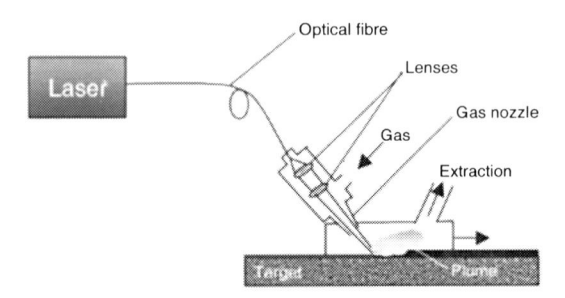

Fig.3. A typical process set up for Nd:YAG laser ablation of contaminated oxide layer from a metal substrate.

Beer-Lambert's law can be used to relate processing parameters and the removal depth for short pulse laser ablation:

$$F_{th} = F \cdot \exp(-\alpha Z_d) \qquad (1)$$

where F_{th} is the threshold fluence for ablation, F is the laser fluence, α is the absorption coefficient of the material ($1/\alpha$ is absorption length), Z_d is the depth of material removal. For multiple pulses, $Z_d \approx nd_s$ where d_s is the removal depth per pulse and n is number of pulses delivered. The removal depth per pulse, d_s, can thus be found from:

$$d_s = \frac{1}{\alpha n}\ln F - \frac{1}{\alpha n}\ln F_{th} \qquad (2)$$

Lasers that have been used for oxide removal include Transversely Excited Atmospheric (TEA) CO_2 lasers, Q-switched Nd:YAG lasers and excimer lasers. A comparison of the effects and characteristics of using different lasers for metal oxide removal is given in Table 1. Generally speaking, a TEA CO_2 laser (far infrared wavelength) is well absorbed by the oxides and much reflected by metallic materials. It is also much cheaper than Nd:YAG and excimer lasers. The disadvantages of the CO_2 laser include the difficulty for beam delivery since its beam cannot be transmitted through normal silica optical fibres due to its far infrared wavelength. Also the long laser wavelength promotes plasma generation which is undesirable in laser ablation since plasma can absorb and deflect the laser beam. Nd:YAG lasers can provide near infrared (1.06 μm at fundamental mode), visible (532 nm wavelength, operating at 2nd harmonic mode) and ultraviolet (operating at 3rd harmonic mode: 355nm) wavelengths which can all be transmitted through optical fibres, although there is a high loss at the UV wavelength and shorter pulse width is causes more physical strain on the optical fibre [57]. Nd:YAG lasers at infrared and visible wavelengths are well absorbed by metals but poorly absorbed by oxides. They generate less plasma than CO_2 lasers thus improved removal efficiency is expected. The UV beam can be absorbed well by both ceramics and metals. Excimer lasers provide high-energy (over 0.5 J per pulse) UV pulses. Compared with other lasers currently available, highest ablation removal efficiency can be generated with excimer lasers (Table1). An excimer laser beam at 308 nm (XeCl) wavelength can also be transmitted through optical fibres, although transmission loss is high (50% [8]). 2 - 6 m^2/hr removal rate has been achieved with excimer laser ablation. The ablation process can be monitored with electrical [8], optical and acoustic methods [9]. When an Nd:YAG laser (used for pre-heating) is combined with an excimer laser (used for ablation), increased ablation rate is observed [1].

Table 1. A comparison of various lasers used in oxide ablation from metal substrates.

Characteristics	TEA CO_2 Laser	Nd:YAG Laser	Excimer Laser
Laser wavelength	10.6μm	1064 nm, 532 nm (2ω), 355 (3ω)	248 nm (KrF) 308 nm (XeCl)
Laser pulse width	120 ns [10]	5-100 ns (Q-switched) 5-100 ps (mode locked) [11,12]	20-70 ns
Typical ablation threshold fluence for oxides on stainless steel.	5 Jcm^{-2} [10]	3-5 Jcm^{-2} [13,14], ns pulse	2.5-5 Jcm^{-2} at 308 nm [8, 13]
Beam absorption characteristics	High for oxides, low for metals	High for metals, low for oxides	High for both metals & oxides
Removal efficiency	10^{-13}-10^{-12} m^3/J est. from [10]	10^{-12}-10^{-11} m^3/J [11,12]	0.5x(10^{-11}–10^{-10}) m^3/J est. from [6]
Relative removal rate (estimated)	0.04 – 0.1 m^2/hr	0.4 -1.2 m^2/hr.	2-6 m^2/hr [8]
Beam delivery	Mirrors	Optical fibre 60-73% efficiency [12]	Optical fibre (308 nm only) – 50% efficiency. Mirror (other wavelengths)
Equipment and running cost	Low	Medium	High

Lasers have also been used to combine with chemical processing method for the removal of metal oxides from metal substrates. Hirahayashi et al reported using an Nd:YAG laser (1.064 µm wavelength) to radiate corrosion products (on stainless steel and mild steels) in chlorine gas to produce metal chloride which is soluble in water and thus can be removed easily [1]. Laser heating was essential to enable the chemical reaction. Decontamination factor up to 361 has been achieved for oxides up to 100 µm thick. The process could introduce additional waste volume if water is used to remove the metal chloride. The same research group also reported the application of gel-decontamination reagent (60 vol % water + 20 vol % liquid glass and 20 vol % of 5mol/L sulfuric acid) onto the surface to be decontaminated followed by low power density 5-10 W cm^{-2} heating using an Nd:YAG laser [1,15]. Fe contaminant was changed to hydroxide through the reaction induced by the laser irradiation. The advantage of the technique is that there is no need to immerse the component in chemical solution. Thus it may be suitable for "hot spot" decontamination.

Yavas et al reported a combination of pulsed laser radiation with basic electrolyte solution (1:1 mixture of 0.15 M boric acid and 0.15 Sodium tetraborate, Ph=8.6) charged with a cathodic potential (threshold > 1.5V) for the removal of metal oxides [14]. A Q-switched Nd:YAG laser (1064 nm wavelength, 14 ns pulse width) at a laser fluence (0.56 J cm^{-2}) well below the ablation threshold of the material was used. The process is based on laser promoted reduction process of H^+ and O_2. This process has the similar disadvantages as the chemical process for nuclear decontamination.

2.2. Paint stripping

In nuclear processing plants, paints are used to form a sacrificial layer on the surfaces of buildings and equipment in low level radioactive environment. Repeated paint coating is used to trap down and fix the particulate contaminants. During decommissioning, these paints have to be removed and disposed of in a controlled manner. A common practice is the use of chemical solutions, such as methylene chloride, phenolics, alkaline and acid activators [16-17]. The use of chemicals not only increases the waste volume, but is also likely to cause disposal difficulties and further contamination of the substrates. Some of the other physical paint stripping methods such as plastic media blasting (PMB) [18-19], sodium bicarbonate media blasting [20], water jet systems [21-22] and ice particle blasting [23], currently used for aircraft and industrial equipment all result in increased waste volumes. Alternative physical paint stripping methods, which do not result in waste volume increase, include dry-ice (solid CO_2) blasting [24] and light based (laser and flash lamp) devices. The dry ice system presents difficulties for waste collection due to local high pressure and particle scattering. Also the excessive CO_2 gas involved in the process makes it mainly suitable for outdoor paint stripping. Light - based systems are considered to be the only true non-contact methods which have the advantages of controllability, flexibility, convenience for waste collection and minimum opportunity for recontamination. Laser paint stripping has been successfully demonstrated on helicopters and military aircraft to facilitate necessary periodic metallurgical inspection. The lasers used are pulsed TEA CO_2 lasers, Q-switched Nd-YAG lasers and excimer lasers. Some recent study also revealed the possibility of using continuous wave diode lasers for paint stripping. The characteristics of various laser paint stripping techniques are described in the following sections.

2.2.1. CO_2 laser paint stripping

Commercial TEA CO_2 lasers up to 2 kW power (up to 7 J/pulse) have been applied for aeroplane paint stripping [25]. TEA-CO_2 laser removal of paint from metal substrate is based on vapour phase thermal ablation or plasma phase thermal/mechanical ablation. In the vapour phase, a thin layer of paint close to the substrate surface cannot be removed due to higher thermal conductivity of the metal substrate [26]. In the plasma phase (higher laser fluence is required), this layer can be removed. However, an increase in residual formation was found [27]. Spraying a small amount of dimethyl formanide on the paint before laser paint stripping was found to enable complete removal of paint without operating at the plasma phase [27]. The sprayed liquid was found to improve beam absorption and introduced additional mechanical shocks. Lower laser ablation threshold is required with the chemical spray. DF can reach 160 with laser alone and DF > 500 when laser is combined with the chemical spray [27]. 17 m^2 / hour removal rate was achieved with a 500 W TEA CO_2 laser [27]. As with oxide layer removal, a line beam is preferred for paint stripping. Study has shown that paint removal efficiency is not affected by the age of the paint [26]. Removing paint from concrete is 3-5 times more efficient than from metal because of thermal conductivity differences between them [28]. Both continuous wave and pulsed laser beams can be used for paint stripping for nuclear decommissioning applications [28,35].

2.2.2. Nd:YAG laser paint stripping

Compared with CO_2 lasers, Nd:YAG lasers have much lower beam absorption by most paint systems. In particular, beam absorption is dependent on the colour of the paint. An Nd:YAG laser at 1064 nm wavelength and 15-30 ns pulse has been used to remove paint from metal substrate [29] and an Nd:YAG laser at 0.3-10 ms pulses has been used to remove chlorinated rubber paint from concrete [30]. Thermal-mechanical process was found to dominate the mechanism of the paint stripping. Ablation threshold around 1 J /cm^2 is needed for paint stripping on metal [29]. Above this threshold the removal efficiency drops to a factor of 4 below the longer pulse or continuous beam removal [29]. For nuclear decontamination applications, Nd:YAG lasers have been used to remove paint and organic material on the surfaces of floor, wall, tank and equipment and hot spots. The process efficiency was found too slow for practical use [4].

2.2.3. Excimer laser paint stripping

Interaction of excimer laser with industrial paints is partially thermal and partially photo-chemical [31]. Since the excimer laser photon energy is higher than many polymer characteristics bond energies, direct breaking of polymer bond without going through the heating process is possible. Epoxy, polyurethane, epoxy/ polyester lacquer on low carbon steel and aluminium substrates have been removed using a 248 nm excimer laser beam at ablation threshold of 0.7 J/cm^2 [31]. Each pulse removes up to 0.7 µm depth. Darker paint was found more efficient to remove. Some products of ablation were black powders, similar to soot deposited near the edges of the treated areas [31].

2.2.4. Diode Laser paint stripping and comparison with other lasers

High power diode lasers (HPDLs) are only commercially available in recent years. They have the highest electrical to optical conversion efficiency compared to other lasers and thus they have the minimum physical sizes. For example, an 8 kW diode laser is the size of home refrigerator, whilst the same powered CO_2 laser would be the size of a lorry and a chemical oxygen-iodine laser (COIL) of a similar power will be the size of two lorries. HPDLs can be found either at 810 nm or 940 nm wavelength. Both can be delivered through optical fibres. Commercial diode lasers of over 8 kW are available and the laser power has been doubled each year over the last 10 years. Furthermore, the price of diode lasers reduces rapidly to the present comparable to that of a CO_2 laser. Thus HPDL could open up new opportunities for nuclear decommissioning applications. Li et al reported complete removal of paints from metal substrate with a diode laser combined with an oxygen jet [56]. Schmidt et al reported the use of diode lasers of 60 W, 120 W and 2500 W operating at continuous mode for the removal of chlorinated rubber paint (approximately 0.6 mm thick) from concrete surfaces [32,33]. O_2 gas was used as assist gas. It was found that the use of O_2 gas enabled the removal rate to increase by up to three times over that with N_2 and argon assist gases. A laser controlled combustion process was identified as the main removal mechanism when O_2 gas was used. There is a linear relationship between the removal rate and laser power [32,33]. Therefore the process is directly scalable. The removed particle sizes are between 1 μm and 2 mm with majority of them falling within 30-60 μm for laser powers below 60 W [34]. Increasing laser power increases the particle sizes indicating a smaller proportion of combustion at higher laser powers. Increasing O_2 gas flow reduces the particle sizes. Compared with other high power lasers, diode laser is not the most efficient one for paint stripping (Table 2). However, it has the advantage of being small in size (thus portable) and the laser beam can be delivered with optical fibres.

Table 2. Comparison of various lasers for chlorinated rubber stripping, data derived from [35].

Optical Sources	Ablation Threshold (J / cm^2)	Removal Efficiency (m^3 / J)	Removal Rate at 100 W for CR paint of 0.6 mm thickness ($m^2 / hr.$)
Arc Lamp	37.4	4×10^{-10}	0.24
CW CO_2 Laser	0.3	10^{-8}	6
Diode Laser	38	6.7×10^{-10}	0.4
Excimer Laser	0.04	8×10^{-10}	0.5
Nd:YAG Laser	4.8	10^{-9}	0.6

2.3. Concrete decontamination

Concrete structures are seen in various places in a nuclear power plant or a nuclear reprocessing plant. For example, waste storage reservoirs are made of concrete. The depth of low/medium level radioactive contamination of concrete has been found to limit to 5 mm below the surface [1,28]. It is therefore necessary to investigate techniques to effectively remove this layer without introducing secondary wastes. A laser would be an ideal candidate, since no additional solid waste would be introduced. Several laser-based methods have been investigated for concrete surface layer removal. These include scabbling (shock cracking), thermal delimitation (glazing) and vaporisation. Furthermore, lasers can be used to tie-down and fix contamination onto a concrete surface to immobilise the particular contaminants. The characteristics of these processes are described in the following sections.

2.3.1. Laser scabbling

By passing a laser beam across the surface of concrete, the top layer can be made to eject violently, without melting or vaporising. This is known as laser scabbling pioneered by the author of this paper [36,37]. This effect is believed to be caused by the rapid dehydration and evaporation of the moisture in the concrete. The process is more effective using large beams, especially when the beam size is larger than that of aggregates (10 –20 mm diameter). Single pass removal depth up to 8 mm has been achieved using a 4 kW Nd:YAG or a 5 kW CO_2 laser having a 80 mm diameter beam spot [38]. A laser power density of 100 – 300 W / cm^2 was found optimum for the process. There is no significant difference in removal characteristics between the two types of lasers. Typical removal rate is 1000 cm^3/ hr.kW [38]. The process was most effective for the first layer concrete removal [1]. This process has the highest material removal rate compared with other laser-based methods as described in the following.

2.3.2. Laser glazing - delamination

In this process, a laser beam is used to melt a layer of concrete. At the same time, heat affected zone (HAZ) is developed below the melt pool. On cooling and by absorbing moisture in air, this layer delaminates from the concrete substrate (Fig.4). This work was initially reported by Li et al using a CO_2 laser and an Nd:YAG laser [36,37] and further investigation was reported by Hirabayashi et al [1] who used an Nd:YAG laser operating at 600 W , 13 mm spot diameter, traversing at 80 mm/min to remove 4-5 mm layer concrete (up to 14 mm) by laser glazing. Multiple passes (1 mm per pass) were used. Over 90% waste can be collected [1].

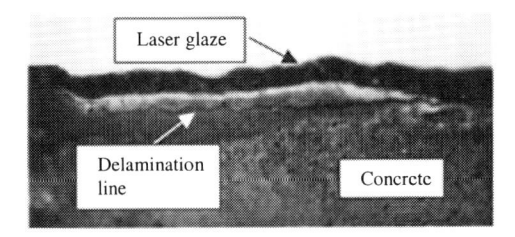

Fig.4. An example of laser glazing delamination of concrete.

2.3.2. Laser vaporisation

A CO_2 laser has been used to vaporise layers of concrete surfaces with a higher degree of depth control than the above two methods [28]. A minimum of 2500 W/cm^2 power density and 5-20 times higher traverse speed than the above two processes are required to avoid significant melting and heat loss. The material removal efficiency is lower than the above two processes.

2.3.3. Dust containment tie down

Between the shut-down and decommissioning of a nuclear power plant there is usually a cooling period of 5 - 10 years. There is a need to temporarily fix the particulate contamination onto the walls so that they are immobilised. Industrial lasers have been used to fix these onto concrete surfaces by glazing [39-42]. It was found that, high power diode lasers could generate a thin (< 1 mm) layer of glaze which does not spall, whilst CO_2 laser glazing of concrete always resulted in spallation. One of the major differences between the diode laser and CO_2 laser glazing of concrete was found to be the absorption length with 177 ± 15 μm for HPDL and 470 ± 20 μm for a CO_2 laser [43]. With a 2.5 kW diode laser, concrete glazing coverage rate can reach 1.94 m^2/hr [44].

3. DISMANTLLING / CUTTING OF STRUCTURAL MATERIALS

Large structural materials in the form of reinforced concrete, stainless steels and carbons steels are seen in nuclear installations. These include reactor pressure vessels (e.g. low-alloy steel plates of over 400 mm in thickness with 6-10 mm stainless steel clad [4]), biological shield walls (e.g. up to 3000 mm thick reinforced concrete with steel bars of up to 50 mm diameters and lined with 10 mm thick steel plate on the inner surface, activated up to 1 m from the inner surface [4]) and reactor internals (50 - 500 mm thick stainless steel [45]). In nuclear decommissioning, it is desirable to reduce waste storage volume. Large structures need to be cut to smaller pieces. Some of these large structures may have to be cut underwater. The requirements for the cutting technology in nuclear structure dismantling include minimum radioactive waste generation/spreading, minimum operating hazards, remote operation and safe to operate. A special feature of the dismantling process is that the quality of the cut is not important. The conventional means of cutting thick-section structural materials include the use of combined arc gouging and propane-oxygen gas flame torch for cutting mild steels with stainless clad (e.g. 420 mm thick carbons steel lined with 6 mm thick stainless steel was cut under water at 3-4 cm/min for pressure vessel dismantling [45]), and the use of large circular saws for cutting reinforced concrete (e.g. cutting 1500 mm thick reinforced concrete wall at 5-10 cm^2/min and concrete without reinforcement at 150 cm^2/min.[45]). The use of a large diameter diamond saw might cause access difficulties and contamination of the tools. A major challenge for laser cutting is to deal with the extraordinary thickness of these structures. The possible advantages of laser cutting include:

- Narrower cut kerf width, thus minimising the generation of radioactive dust and fumes (10 fold reduction in dust and fume emission was found [46])
- Minimum opportunity for the contamination of the operating equipment because of remote operation and long distance beam delivery.
- Power scalable to cope with material thickness requirement.

The characteristics of lasers required for dismantling is different from those for decontamination. A high power, continuous laser beam is normally preferred. Also operating conditions for thick section material cutting are different from those for cutting thin materials. Very long focal length (e.g. over 1 m) is required to maintain the beam intensity over a long distance. Much larger beam diameter and kerf width are required on the top surface compared with cutting thin materials. For under water laser cutting, a local dry zone is provided using high-pressure gas or gas/water jets. A typical set-up for laser cutting of nuclear structures underwater is shown in Fig.5. In the following sections, specific features and limitations of using high-powered lasers for cutting metal and concrete materials are described.

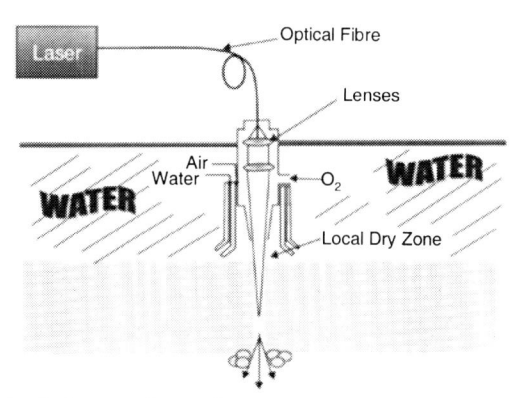

Fig.5. A typical set up for under water laser cutting of nuclear structures.

3.1. *Laser cutting of thick-section metallic structural materials*

Various high power lasers have been used to cut thick-section metallic materials for nuclear decommissioning applications. These include CO_2 lasers, CO lasers, COILs and Nd:YAG lasers. Laser cutting of structural materials is mainly based on the principle of "melt and blow", i.e. the laser beam provides the main thermal energy to melt the materials and a gas jet is used to remove the molten materials. If the gas used is reactive, such as O_2, then depending on the materials to be cut, exothermic reaction between the gas and the cut material may take place at high temperatures. This reaction can generate additional thermal energy into the cut zone. For example, for laser cutting of Fe based materials using an O_2 gas jet the following reaction may take place:

$$Fe + \tfrac{1}{2}O_2 \Rightarrow FeO + 3.43 \text{ kJ/g} \qquad (3)$$

$$3Fe + \tfrac{1}{2}O_2 \Rightarrow Fe_3O_4 + 1.29 \text{ kJ/g} \qquad (4)$$

The cutting efficiency can be increased by up to 200% using reactive gas cutting compared to the use of non-reactive gases. Typical gas pressure is in the range of 2-10 Bar.

3.1.1. *CO_2 laser cutting*

A 5 kW CO_2 laser (focused to 0.1. – 0.2 mm diameter) has been applied to cut fuel channels (e.g. stainless steel tubes of 95 x 5 mm size and zirconium tubes of 3 mm wall thickness and 100 mm diameter) [46]. Each nuclear reactor has about 1600 channels. Stainless steel of 10 mm thickness is cut with this laser at 1.5 m /min rate with N_2 assist gas at 7 bars. Most part of the melted metal particles solidifies just at the outlet from the cutting area and settles down on the fuel channel walls. Laser cutting was found to result in ten-fold decrease in gas and dust contamination compared to traditional methods [46].

In another report, an 8 kW CO_2 laser was used to cut stainless steel of 40 mm thickness at 0.2 m /min cutting speed by incorporating a special high pressure gas nozzle. [47]. This gives a typical of 200 W/mm thickness cutting rate for stainless steels using a CO_2 laser. To cut 300-

mm thick stainless steel reactor internals, a 60 kW CO_2 laser would be needed. A CO_2 laser of such a power rating is currently not commercially available, although theoretically possible. Another disadvantage of the CO_2 lasers is that the beams cannot be delivered through optical fibres. Mirror transmission of the laser beam for 3-D cutting is more complex than using optical fibres.

3.1.2. *CO laser cutting*

CO lasers are not commonly seen for industrial applications. They are far more expansive than CO_2 lasers. CO gas lasers have the advantage of radiating at shorter wavelength (5.2 µm) than CO_2 lasers (10.6 µm) thus the CO laser beams can be better absorbed by metallic materials (Fig.6).

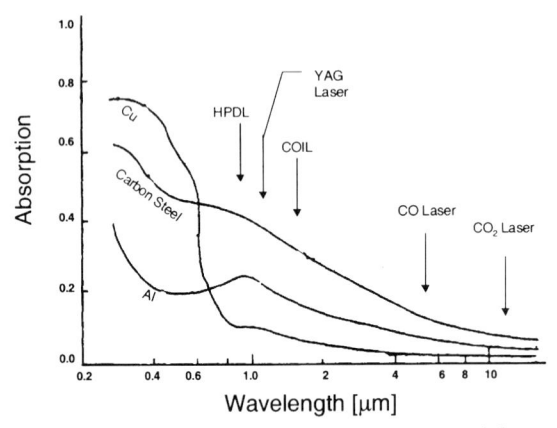

Fig.6. Absorption characteristics of certain metallic materials at various wavelengths.

A 20 kW CO laser was used by NUPEC in Japan to successfully cut stainless steels (a reactor core internal mock up model) of 300 mm in air (using a mixture of 90% O_2 and 10% N_2 gas jet) and 150 mm in water (by providing a local dry-zone with a gas jet: 80% N_2 and 20% O_2) [4]. Dual nozzle configuration was used with a smaller central laser nozzle and a much larger (four times of the laser nozzle) supersonic gas jet nozzle. As the cutting depth increases the proportion of O_2 needs to increase as well. Typical kerf width is 4.5 mm on the front and 8 mm on the back. When the proportion of O_2 increases the kerf width increases as well. For example, for cutting 300 mm thick stainless steel, with 90% O_2, kerf widths of 10 mm on the front and 50 mm on the back were found. Fume production in water cutter was found to be 15% of value for air cutting. Underwater cutting resulted in 50% reduction in debris compared with that in air, due to additional cooling in water [7]. From this work it can be seen that the CO laser cutting efficiency for stainless steels is around 65 W /mm thickness in air and 130 W / mm in water. Compared with CO_2 laser cutting, CO laser is about 2 times more efficient.

3.1.3. *Chemical Oxygen-Iodine Laser (COIL) cutting*

COIL was originally developed for military use. The principal energy in a COIL system does not come from electricity but from chemical reactions. Thus a large power supply unit is not needed. The physical size of a COIL laser is much larger than other high power lasers. The advantages of a COIL system is that, it emits very short near infrared laser wavelength (1.315 µm) which is ideal for optical fibre beam delivery (Fig.7) and also it enables much better beam absorption by metals (Fig.6). A 20 kW COIL was demonstrated in 1995 [48] and a 7.36 kW COIL beam was successfully delivered through a 0.9 mm fibre [48].

For nuclear decommissioning applications, a 5 kW COIL was used to cut 60 mm stainless steel at 100 m /min, under water [49]. Long focal length and large beam size were found beneficial for cutting thick section metals. Around 85 W /mm thickness under water cutting efficiency is seen from this work. This is much more efficient than both CO and CO_2 lasers. COIL cutting speed was found 3-5 times faster than that by CO_2 lasers [3]. To cut 300-mm thick stainless steel reactor internals, under water, a 25-30 kW COIL laser would be needed [3,14]. 50-60 kW COIL laser would be required to cut pressure vessels in nuclear power plants [50]. COIL lasers of this power level are not currently available.

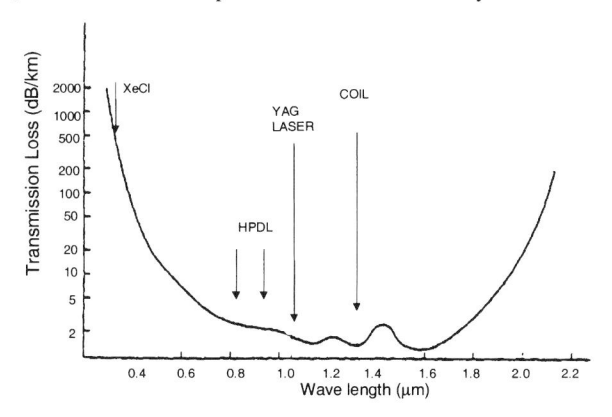

Fig.7. Transmission characteristics of a fused silica fibre.

3.1.4. Nd:YAG laser cutting

Nd:YAG lasers operate at 1.06 µm wavelength which enables the laser to be delivered over long distance by optical fibres. They are normally used in pulsed mode for drilling, welding and precision cutting. For nuclear structural material cutting applications, continuous beam would be preferred. Because of the low energy efficiency of the arc lamp pumped Nd:YAG lasers (0.3-2%), commercial arc-lamp pumped Nd:YAG lasers are below 4 kW. Only in the last few years, diode-pumped Nd:YAG lasers (much higher energy efficiency) are available for powers beyond 4 kW. Application of Nd:YAG laser in nuclear structure dismantling is currently limited to relatively thin materials. For example, a 2 kW Nd:YAG laser was combined with an industrial robot to cut 20 mm thick stainless steel (waste storage tank) at a cutting speed of 200 mm/min [1]. This gives a typical cutting efficiency of 100 W /mm thickness in air, which is 100% better than that of CO_2 laser.

3.1.5. *Cutting parameter relationships*

A simplified equation to relate laser parameters and cutting depth is given [50,51] as:

$$h = \frac{\alpha P}{Vd + \beta} \qquad (5)$$

where h, P, V and d are cut depth, laser power, cutting speed and beam spot diameter respectively. α and β are two constants dependent on the type of laser, assist gas and material properties. For example, $\alpha = 2$ (CO_2 laser) and 5 (YAG laser and COIL). $\beta = 0.1$ (carbon steel cut with O_2 gas) and 0.7 (stainless steel).

Table 3. Comparison of various lasers for cutting.

Items	CO_2 Laser	CO Laser	COIL	Nd:YAG	Diode Laser
Wavelength	10.6	5.2	1.315 µm	1.06 µ	0.94 µm
Operating Efficiency	15%	8%	20 - 27%	0.3-2% (arc lamp pumped) 20% (diode laser pumped)	20-40%
Absorption Characteristics	Very poor for metals Best for non-metals	Good for metals	Good for metals	Good for metals	Very good for metals
Beam delivery	Mirrors	Mirrors	Optical fibre (lowest loss)	Optical Fibre	Optical Fibre
Physical size	Large	Large	Very large	Large (lamp pumped) Medium (diode pumped)	Small
Capital/running Costs	Medium	High	Very high	High	Low
Cutting efficiency (Stainless steel) in air. Twice the value in water	200 W/mm	66 W /mm	50 W /mm	100 W /mm	50 W /mm (estimated)
Commercial available maximum power	30 kW	21 kW	20 kW	8 kW	8 kW

3.1.6. *Comparison of various lasers*

Table 3 compares the characteristics and performances of various lasers for cutting thick-section steels. Since shorter wavelength lasers have much better beam absorption by metallic materials, they are much preferred for cutting metals. Only COIL, Nd:YAG and HPDL can be transmitted through optical fibres. The high power diode lasers and diode pumped Nd:YAG lasers can be much more economical to run than COIL and their available powers are increasing rapidly [55]. The potential use of diode pumped Nd:YAG and diode lasers in nuclear decommissioning would be expected to increase.

3.2. Laser cutting concrete structures

Cutting through biological shield walls made of reinforced concrete up to 3 m thickness is of great challenge for high power lasers. Until now, such an objective has not been achieved yet.

Notwithstanding this, investigations have been carried out to study the feasibility of cutting thinner concrete walls using high power lasers.

Lenk et al reported the use of a 3 kW Nd:YAG laser with optical fibre beam delivery for cutting 70 mm concrete (without reinforced bars) at 25 mm/min cutting speed. A pressure chamber was used to facilitate the cutting process [52].

Yoshizawa et al reported the use of a 1 kW CO_2 laser for cutting 43 mm thick concrete at 20 mm/min cutting speed and a 10 kW CO_2 laser for cutting concrete up to 130 mm thickness at 25 mm/min. The work has identified that laser power density above 10^6 W/cm^2 is needed for cutting concrete and 10^7 W/cm^2 for cutting reinforced concrete. The cutting thickness was largest when laser beam was focused 0-5 mm below the work surface [53]. The work has shown that the variation in assist gas pressure does not have significant effect on cutting depth. Pulsed and continuous lasers perform similarly for cutting concrete. One of the main difficulties is the removal of molten slag during laser cutting. A gas jet was used. Hamasaki and Munehide reported the use of a 5 kW CO_2 laser for cutting 100 mm thick concrete with 10 mm diameter reinforced bars. No results were reported [54]. Kasai reported cutting 180 mm thick reinforced concrete using a 15 kW CO_2 laser at 25 mm/min cutting speed. There is very little difference in cutting speed and depth with and without reinforced bars. The cutting speed is a little higher for the steel bars when an oxygen assist gas is used [58]

From the above reported work it can be seen that, for concrete cutting, CO_2 laser is slightly more efficient than Nd:YAG laser. To cut 3 m concrete, a 300 kW laser would be needed, cutting at 25 mm/min, based on the current laser cutting method.

4. SUMMARY

A review has been presented on the use of high power lasers in nuclear decommissioning. The work shows that, laser processing can offer significant advantages in terms of much reduced secondary waste, scalability and remote controllability. There are also limitations to the laser processing techniques for large area and thick section applications.

For contaminated metal oxide removal, short pulsed (typically nano-second pulse width), high peak power (10^8-10^9 W/cm^2) lasers are preferred. Shorter wavelength lasers down to UV range can provide higher removal efficiency. The potential of using fibre beam delivery for excimer laser beams and diode pumped solid state laser beams would provide practical means for hot spot decontamination. 2-6 m^2 / hr. removal rate can be achieved. Combined UV and IR beam ablation could increase the removal rate further.

For paint stripping, although CO_2 lasers would have the highest removal efficiency (e.g. 17 m^2/hr at 500 W power), high power diode lasers can play a significant role when combined with an O_2 gas jet. Since an HPDL beam can be delivered through optical fibres and much higher power diode laser systems can be made available in the future, HPDL applications in large area paint stripping in nuclear decommissioning could become practical. For paint stripping from metal substrates, complete removal of paint requires much higher laser energy density. Spraying a small amount of dimethyl formamide on the paint before laser paint stripping has been found to enable complete removal of paint on metal substrate.

Concrete decontamination can be carried out by vaporisation, glazing and scabbling. Whilst vaporisation and glazing delamination are more controllable, scabbling is the most efficient method. Typical material removal rate for laser scabbling of concrete is 1000 cm^3/(hr.kW) at a laser power density of 100-300 W/cm^2. The scabbling process is also insensitive to laser wavelengths. Although CO_2 and Nd:YAG lasers have been used, the potential use of an HPDL for concrete scabbling cannot be ruled out. Contamination tie-down on concrete can be best carried out using an HPDL. A coverage of 1.94 m^2 /hr. with a 2.5 kW diode laser has been demonstrated.

For thick-section metal cutting, the maximum depth for stainless steel is 300 mm in air and 150 mm in water for the currently available lasers. Shorter wavelength lasers such as CO and COIL are much more efficient than the use of CO_2 lasers. Laser cutting has been applied to dismantling modelled reactor core internals, storage tanks and fuel channels. By examining the characteristics of HPDL and diode-pumped Nd:YAG lasers, it would be possible to foresee that, their potential applications in thick-section metal cutting will increase.

For concrete cutting, the maximum thickness achievable so far is 180 mm using a 15 kW CO_2 laser. Much research is needed to further develop the laser technology so that it may one day become practical for cutting biological shield walls of 3000 mm.

REFERENCES

1. Takakuni Hirabayashi, Yutaka Kameo and Masato Myodo, "Application of laser to decontamination and decommissioning of nuclear facilities at JAERI", *High Power Lasers in Civil Engineering and Architecture*, Proc. Of SPIE Vol. 3887 (2000) pp.94-103
2. J.Varley, Editor, "1994 World Nuclear Industry Handbook," *Nuclear Engineering International*, 1994
3. W.C.Solomon and D.L.Carroll, "Commercial applications of COIL" *High Power Lasers in Civil Engineering and Architecture*, Proc. of SPIE Vol. 3887 (2000) pp.137-151
4. N.Ogawa, S.Saishu, T.Ishikura, "The development of decommissioning technology for nuclear power plant", *High Power Lasers in Civil Engineering and Architecture*, Proc. of SPIE, Vol. 3887 (2000) pp. 78-93
5. "Surface Radioactivity Guides," in *Radiation Protection for Occupational Workers*, USDOE Order 5480.11, Pub. 21 Dec.1988, Attach. 2, p1.
6. H.M.Pang, R.J.Lipert, Y.M.Hamrick. S.Bayrakal, K. Gaul, B.Davis, D.P.Baldwin and M.C.Edelson, "Laser decontamination: a new strategy for facility decommissioning", *Nuclear and Hazardous Waste Management* in TL. MTG: Spectrum'92, pp. 1335-1341
7. S.Saishu, S.Abe and T.Inoue "Applying laser technology to decommissioning for nuclear power plant" *High Power Lasers in Civil Engineering and Architecture*, Proc. Of SPIE Vol. 3887 (2000) pp.118-127
8. M.L. Sentis, P.Delaporte, W.Marine, O.Uteza, "Cleaning of large area by excimer laser ablation", *High Power Lasers in Civil Engineering and Architecture*, Proc. of SPIE, Vol. 3887 (2000) pp 316-325
9. J.M.Lee and K.G.Watkins, "In-process monitoring techniques for laser cleaning", *Optics and Lasers in Engineering*, Vol.34, Nos 4-6, (2000), pp429-442.
10. S.S.Miljanic, N.N.Stjepanovic, M.S.Trica, "Possibilities of a metal surface radioactive decontamination using the pulsed CO_2 laser" *High Power lasers in Civil Engineering and Architecture*, Proc. of SPIE Vol. 3887 (2000) pp.357-363

11. X. L. Zhou, K. Imasaki, H. Umino, K. Sakagishi, S. Nakai, C. Yamanaka "Laser surface ablation cleaning of nuclear facilities" *High Power lasers in Civil Engineering and Architecture*, Proc. of SPIE Vol. 3887 (2000) pp.326-334
12. H.Furukawa, K.Nishihara, C.Yamanaka, S.Nakai, K.Imasaki, X.L.Zhou, H. Umino, K.Sakagishi, S.Funahashi, "Investigation on laser cleaning for decontaminated surface" *High Power lasers in Civil Engineering and Architecture*, Proc. of SPIE Vol. 3887(2000) pp.128-136
13. C.T. Walters ,"Laser-based cleaning processes for solvent replacement", 42nd Int. SAMPE Symposium (1997) pp. 247-256
14. O. Yavas, R. Oltra,. O. Kerrec, "Enhancement of pulsed laser removal of metal oxides by electrochemical control", *Applied Physics A*,Vol. 63,(1996), pp 321-326
15. Y.Kameo, K.Aoki, T.Gorai, T.Hirabayashi, "Development of laser decontamination technique for metal wastes" *Proc. Of JAIF* (1998) pp. 571-574
16. R.M.Operowsky "Chemical Immersion Piant Stripping", *Metal Finishing*, 94(5A), 1996, pp.380.
17. H.B.Hans "Paint Stripping with Nontoxic Chemicals", *Metal Finishing*, 93(4), april 1995, pp34-38.
18. K.E.Abbott, "Dry media blasting for the removal of paint coatings on aerospace surfaces", *Metal Finishing*, v 94, n7 (Jul 1996), p33-35
19. Wolbach, McDonald, "Paint Stripping Using Plastic Media Blasting" *Journal of Hazardous Materials*, 17(1), Dec. 1987, pp.109-113
20. Gould, "Bicabonate Adds fizz to Aricraft Stripping", *New Scientist*, 144, Oct.1994, p24.
21. Anon, "High Pressure Water Removes Paint Build Up", *Products Finishing*, 6(9), 1996, pp.108-109.
22. D.W.See., "Large Aircraft Robotic Paint Stripping (LARPS) system and the High Pressure Water Process", *AGARD Report* 791, pp11.1-11.19
23. T.Foster, "Paint Removal and surface Cleaning Using Ice Particles", *AGARD Report* 791, pp14.1-14.9
24. Stratford and Scott, "Dry Ice Blasting for Paint Stripping and Surface Preparation", *Metal Finishing*, May 1996, 94(5A), pp395-401.
25. G.Schweizer and L.Werner, "Industrial 2-kW TEA CO2 laser for paint stripping of aircraft", *Proceedings of SPIE – The International Society for Optical Engineering*, v 2502 (1995), p 57-62.
26. C.A. Cottam, D.C.Emmony, A. Cuesta and R.H.Bradley, "XPS monitoring of the removal of an aged polymer coating from a metal substrate by TEA-CO$_2$ laser ablation", *Journal of Materials Science*, v 33, n 13 (Jul 1998) p 3245-3249
27. A.Tsunemi, K.Hagiwara, N.Saito, K.Nagasaka, Y.Miyamoto, O.Suto, H.Tashiro, "Complete removal of paint from metal surface by ablation with a TEA CO$_2$ laser", *Applied Physics A* vol. 63, (1996), pp. 435-439
28. L.Li, W.M.Steen, P.J.Modern and J.T.Spencer, "Laser removal of surface and embedded contamination on/in building structures", in *Laser Materials Processing and Machining, Proceedings of SPIE – The International Society for Optical Engineering*, Vol. 22246 (1994), p 84-95.
29. C.T.Walters, "Short-pulse laser removal of organic coatings", *Proceedings of SPIE*, vol.4065 (2000), p 567-575
30. M.J.J.Schmidt, L.Li, J.T.Spencer, "Ablation of a chlorinated rubber polymer and TiO$_2$ ceramic mixture with a Nd:YAG laser", *Applied Surface Science*, v154 (2000) pp53-59
31. L.M. Galantucci, A. Gravina, "An experimental study of paint-stripping using an excimer laser", *Polymers & Polymer Composites*, Vol. 5 no. 2, (1997), pp. 87-94

32. M.J.J.Schmidt, L.Li and J.T.Spencer, "Characteristics of high power diode laser removal of multi-layer chlorinated rubber coatings from concrete surface", *Optics and Laser Technology*, v31, n2 (1999) p171-180.
33. M.J.J.Schmidt and L.Li, "Removal of embedded contamination in chlorinated rubber coatings using a portable high power diode laser", *Journal of Laser Applications*, Vol. 12, n 4 (2000), pp.134-141.
34. A.A.Peligrad, M.J.J.Schmidt, L.Li and J.T.Spencer, "Ash Characteristics in controlled diode laser pyrolysis of chlorinated rubber", *Optics & Laser Technology*, Vol. 32 (2000) pp.49-57.
35. L.Li, M.J.J.Schmidt and J.T.Spencer, "Comparison of the characteristics of HPDL, CO_2, Nd:YAG and excimer lasers for paint stripping", *Laser Institute of America Proceedings, Laser Materials Processing* (ICALEO'2000), Vol. 89A, 2000, pp40-49
36. L.Li, P.J.Modern and W.M.Steen, "Concrete decontamination by laser surface treatment", *European patent*: 94307937.6 (27 Oct. 1994), *US patent*:08/335357 (3 Nov. 1994), *Japan patent*: JP6-270869 (3 Nov. 1994)
37. L.Li, P.J.Modern and W.M.Steen, "Laser surface modification techniques for potential decommissioning applications" Proceedings of RECOD'94, 4th *International Conference on Nuclear Fuel Reprocessing and Waste Management*, 24th –28th April 1994, London, Vol. III, pp427-440.
38. E.P.Johnston, G.Shannon, W.M.Steen, D.R.Jones and J.Tspencer, "Evaluation of High-powered lasers for a commercial laser concrete scabbling (large scale ablation) system", Laser Institute of America Proceedings (ICALEO' 98), *Laser Materials Processing* Vol. 85A, 1998, pp210-211.
39. L.Li, P.Modern and W.M.Steen, "Laser fixing and sealing of radioactive contamination on concrete substrates", *Proc. of LAMP'92, Laser Advanced Materials Processing* (ISSN 0918-2993), Nagaoka, Japan, June 1992, Vol. 2, pp843-848
40. J.Lawrence and L.Li, "Comparative study of the surface glaze characteristics of concrete treated with CO_2 and high power diode lasers Part I: glaze characteristics", *Materials Science and Engineering A*:, v284, n 1 (2000), p93-102
41. J.Lawrence and L.Li, "Comparative study of the surface glaze characteristics of concrete treated with CO_2 and high power diode lasers Part II: mechanical, chemical and physical properties", *Materials Science and Engineering A*:, v287, n 1 (2000), p25-29
42. J. Lawrence and L.Li, "Influence of shield gases on the surface condition of laser treated concrete", *Applied Surface Science*, v168, n1-4 (Dec. 2000), p25-28
43. J.Lawrence and L.Li, "Determination of absorption length of CO_2 and high-power diode laser radiation for ordinary Portland cement", *Journal of Physics D: Applied Physics*, v33, n8 (2000), p945-947
44. J.Lawrence and L.Li, "Surface glazing of concrete using a 2.5 kW high power diode laser and the effects of large beam geometry", *Optics and Laser Technology*, v31, n8 (1999) p583-591.
45. T. Yamamoto, "Decommissioning situation and research & development for the decommissioning of the commercial nuclear power station in Japan", *International Conf. on Nuclear Engineering*, Vol.5, ASME, (1996), pp. 107-109
46. V.Y.Panchenko, Y.I.Slepokon, V.M.Ryakhin, P.P.Kuznetsov, V.F.Panasyuk and A.M.Zabelin, "Laser Decommissioning of RBMK nuclear Reactors Fuel Channels", Proceedings of SPIE, Vol 3887 (2000*), High-Power Lasers in Civil Engineering and Architecture*. pp.110-115

47. H.Herfurth, M.Deidel, D.Petring, "State of the art laser cutting high speed and thick section applications", *Laser Institute of America Proceedings* (ICALEO' 98), *Laser Materials Processing* Vol. 85A, 1998, pp204-209.
48. W.P.Latham, K.R. Kendrick, B. Quillen, "Applications of the chemical oxygen-iodine laser" *High Power Lasers in Civil Engineering and Architecture*, Proc. Of SPIE Vol. 3887 (2000) pp.170-178
49. Hideki Okado, Takashi Sakurai, Junichi Adachi, Hidehiko Miyao, Kunio Hara "Underwater cutting of stainless steel with the laser transmitted through optical fibre " High Power lasers in Civil Engineering and Architecture, Proc. Of SPIE Vol. 3887 (2000) pp.152-160
50. J.Adachi, N.Takahashi, K.Yasuda, T.Atsuta, "Application of chemical oxygen iodine laser (COIL) for dismantling of nuclear facilities", *Progress in Nuclear Energy*, Vol. 32 No. 3 / 4 , (1998) pp. 517-523
51. W. P. Latham, A.Kar, "A review of the simple model for metal cutting with the chemical oxygen-iodine laser" *High Power Lasers in Civil Engineering and Architecture*, Proc. Of SPIE Vol. 3887 (2000) pp.205-210
52. A.Lenk, G.Wiedemann, E.Beryer, "Concrete Cutting with Nd:YAG laser",Proceedings of SPIE, Vol 3887 (2000), *High-Power Lasers in Civil Engineering and Architecture*. pp.45-48
53. H.Yoshizawa, S.Wignarajah, H.Saito, "Study on laser cutting of concrete", *Transactions of the Japan Welding Society*, v20, n1 (1989), p31-36.
54. M.Hamasaki and M.Katsmura, "cutting a way through the pressure vessel problem", *Nuclear Engineering International*, v 31, n380 (Mar 1986), p44
55. L.Li, "Advances and characteristics of high-power diode laser materials processing", *Optics and Lasers in Engineering*, v34, n.4-6, (Oct 2000), pp231-253
56. L.Li, J.Lawrence and J.Spencer , "Materials processing with a high power diode laser", Proc. of Laser Institute of America: *Laser Materials Processing* (ICALEO'96), v 81E, 1996, pp.38-47
57. M.C.Edelson and H.M.Pang, "A laser-based solution to industrial decontamination problems", Proc. of ICALEO'95, Laser Institute of America, FL, USA. pp768-777.
58. Yoshio Kasai, "Demolition of concrete structures by heating*", Concrete International*, March 1989, pp.33-38.

© Lin Li 2001

C596/021/2001

Innovations in nuclear decontamination techniques

K RILEY, G FAIRHALL, I D HUDSON, and **F BULL**
British Nuclear Fuels plc, Sellafield, UK

INTRODUCTION

This paper describes British Nuclear Fuels, Research & Technology Department's involvement in the development, and application of three commercially available surface preparation techniques for radioactive decontamination.

The techniques being:

- High Pressure Water Jetting (HPWJ) for cleaning internals of pipework on THORP.
- Wet Abrasive Blasting (WAB) and Dry Bead Blasting (DBB) for cleaning the outside of highly active (HA) waste glass product containers produced in the Sellafield Vitrification Plant, commonly referred to as the Windscale Vitrification Plant (WVP).

The development work associated with the high pressure water jet cleaning of the THORP, Receipt & Storage ponds (Building B560) cooling water return line (CWRL) was performed at the BNFL JoRDI[2] test facility, before final deployment on plant.

The Wet Abrasive Blasting project was developed in conjunction with Abrasive Development Ltd, Henley-in-Arden and was also tested at BNFL JoRDI[2].

The Dry Bead Blasting project was performed in conjunction with Guyson Ltd at their manufacturing facility in Settle, Yorkshire.

1.0 B560 Cooling Water Return Line Decontamination

1.1 Introduction

The THORP, B560 fuel receipt and storage ponds at Sellafield are used to store AGR and PWR fuel prior to reprocessing. The pond water is maintained at a temperature of 25-30°C by heat exchangers and a cooling water tower system. After cooling, the water is returned to the pond via a cooling water return line (CWRL). Over a period of time radioactive particulate has plated out in this line, and the point reached where radiation dose levels inhibit pond skip handing operations
The CWRL is a $\varnothing 300/400$ stainless steel line running the length of the ponds (~160 m) with two outlets, one in each pond. The line is located on the ponds west wall, suspended above the water level. (Figures 1 & 2).

Research and Technology carried out two reviews:

1. To determine the mechanism and form of the contamination within the CWRL.

2. To determine the appropriate technique for the decontamination of the CWRL.

The outcome from the reviews identified that the contaminant species was predominantly Cobalt[60] in the form of an iron crud particulate. Chemicals could not be used to remove this material because of constraints imposed by liquid effluent disposal routes. The only feasible option was to use high pressure water jetting (HPWJ) techniques. A commercial water jetting system was identified which is used to clean drains, ducts etc. this employs a centralised jetting head, which is self-rotating and self-propelled. (Figure 3)

1.2 Trials

To ensure that the HPWJ process was suitable, a development testing programme was started with R&T's Decontamination & Decommissioning Group to:

- Remove a short spool piece from the centre of the CWRL. The contamination on the inside was to be sampled and characterised. The spool piece was subsequently decontaminated in BNFL's Ultra High Pressure Water Jetting (UHPWJ) facility, using 8 l/min of water @ 2,500 bar. A new 'Y' shaped spool piece was re-fitted into the centre of the CWRL to allow access for the jetting head.

- An inactive rig was constructed incorporating the complex features of the CWRL i.e. 'Z' joggle, change in diameter, side branches and flanged valve. The jetting equipment was deployed down the rig to check its progression / removal, and the ease of which the simulated contaminated pipework could be cleaned.

1.3 Results of Trials

UHPWJ removed the contamination from the inside of the spool piece, so effectively, that a decision was made to use a lower pressure with a greater volume of water jetting system, i.e. HPWJ 90 l/min @ 1,000 bar. The spool piece was later disposed of as Low Level Waste (LLW).

SPOOL PIECE ACTIVITY LEVELS		
Readings	*Before decontamination*	*After decontamination with HPWJ*
On contact	2.5 mSv/ hr / βγ 2.5 mSv/ hr / γ	2.0 μSv/ hr / βγ 1.0 μSv/ hr / γ
Swabs	5.0 mSv/ hr / βγ 1.2 mSv/ hr / γ 5 – 8 cps DP3 α	<15 cps DP3 β <2 cps DP3 α

Deployment trials highlighted that the originally proposed jetting head centraliser was inadequate, and a new design centraliser was built and successfully tested. Changing to a lower viscosity damping oil also increased the speed of rotation of the jetting head. The jetting system successfully cleaned the simulated contamination from the rig at a rate of 1 m/min.

1.4 Decontamination of CWRL

Due to the length and configuration of the CWRL, the decontamination was carried out in 2 sections, the first section of line in Pond 2 ~85m long, followed by the pond 1 section of line ~75m long.

1.4.1 Decontamination operations: Pond 2 section of CWRL

- A scaffold platform was erected to provide a large working area.
- A 'T' piece was fitted to the CWRL to allow access into the line and still maintain a liquid effluent outlet into the pond. This also incorporated a jetting hose washdown water spray.
- Backout / spray eliminator plates were fitted to the access point on the 'T' piece, after inserting the jetting head into the line.
- A diesel driven HP pump was located outside the building in a clean area, and over seen by an operator with a radio link to the work-face.
- The jetting hose was reeled off a powered drum to allow the jetting equipment to move down the CWRL.
- Control of the jetting operations was maintained by an operator using a foot operated safety dump valve, located near to the line entrance.

- Traverse speed of the jetting head down the CWRL was determine by the activity in the liquid effluent flowing back from the cleaning process.
- Effluent from the jetting operations was collected under the pond water level from the outlet from the CWRL using a collection box / pump arrangement. The effluent was directed to the pond purge system 'Funda filters' via a plastic flexible line routed around the pond East wall.

1.4.2 Decontamination operations: Pond 1 section of CWRL

The decontamination operations for the pond 1 section of the CWRL were the same as the pond 2 section except for the following:
- The jetting head was loaded into the recently fitted 'Y' piece.
- Backout / spray eliminator plates were fitted to the access point on the 'Y' piece.
- Liquid effluent from the jetting process drained into the pond via the existing pond 1 outlet.

RCWL PIPEWORK ACTIVITY LEVELS		
Readings	*Before decontamination*	*After decontamination with HPWJ*
Highest reading in pond hall	7 mSv/ hr / γ	25 µSv/ hr / γ
RCWL overhead at AGR	5 mSv/ hr / γ	6 –25 µSv/ hr / γ
Highest reading at RCWL outlet	3 mSv/ hr / γ	10 –18 µSv/ hr / γ

1.5 Problems encountered

- The high pressure water jetting flexible rubber coated hose picked up more contamination from the inside of the line during jetting operations than was expected. This will be remedied by the use of a smooth PVC coated hose in the future.
- The 'Bounden' effect caused hose couplings to leak whilst under pressure. A single length of hose with minimum joints will be used in future.

1.6 Conclusions

- High pressure water jetting proved to be a successful decontamination process in the cleaning of THORP's pond cooling water return line to background radiation levels. This was achieved in a single pass.
- Cleaning rates of approximately 1m/min where achieved from the decontamination equipment.
- After cleaning there was no apparent evidence of wear, scoring or abrasion of the CWRL pipework bore.
- Approximately 10 months after completion of the decontamination the CWRL remains still at pond hall background level, no radiation build-up detected.
- The collection box and transfer system proved to be a trouble free, and an effective method of transportation of the jetting process effluent to the 'Funda filters' for treatment.
- The two permanent access points installed into the CWRL as part of the decontamination process will allow periodic cleaning of the line when required, thus lower dose up-take in the future.
- The radiation dose to the staff engaged in the decontamination was negligible, due to the radiological controls imposed on the rate of advancement of the jetting head within the CWRL, and thus the dilution effect on the effluent flowing from the pipework.

2.0 Development of Blast Cleaning for WVP glass product containers

2.1 Introduction

The aim of this Blasting Development project was to demonstrate whether commercially available blasting techniques, the Wet Abrasive Blasting (WAB) and the Dry Bead Blasting (DBB) processes could be adapted to become a valid radioactive decontamination technique for waste product containers.

Decontamination of Windscale Vitrification Plant (WVP) highly active (HA) glass waste product containers, prior to export back to the country of origin, is a very important issue for BNFL. Previous R&T trials and the experience of the French Nuclear Industry have indicated that Wet Abrasive Blasting process is capable of meeting the requirements of the 'guaranteed parameters', as defined in the Customers Contract.

For this reason a Wet Abrasive Blasting machine was designed, manufactured and tested on the cleaning of inactive product containers. Results from these trials indicated a potential problem in the treatment of liquid effluent from the machine in downstream effluent treatment plants. After deliberation a Dry Bead Blasting (DBB) machine was manufactured and tested against the benchmark of the WAB machine.

The specification for the two blasting machines were prepared by R&T, Sellafield in conjunction with the Vitrification Engineering Review (VER) group at BNFL Risley. The specification ensured that the design incorporated features, wherever practicable, suitable for use in nuclear environment.

Main objectives of the trials were to:

- Optimise the blasting processes.
- Demonstrate the blasting process as a valid cleaning/decontamination technique.
- Collect information to assist design of a Blasting System for active use.

Trials were conducted using:

- Non-active Vitrified Product Containers arising from a 'pilot plant' or during commissioning of the WVP Line 3.
- A dummy container with facility to attach specially prepared panels.
- A dummy heated container, (in the case of the WAB process only).

2.2 Scope of Trials

The R&T scope of work was to determine the cleaning effectiveness of the blasting processes after completion of machine set up testing, and parameter optimisation. It was also required to record the effect on the following physical parameters for the container:

- hardness
- surface roughness
- emissivity
- surface inclusions/embedment

These parameters were investigated to inform the Customer if there are any changes to the values of these parameters with the exception of emissivity, which is included in the list of "guaranteed parameters".

There was also a plant requirement to determine the following information for each blasting process:

- the container coverage,
- the cleaning cycle period,
- the media life, and
- the media and waste characteristics

2.3 The Wet Abrasive Blasting Machine

The Wet Abrasive Blasting equipment used throughout these trials (Figure 4) was based on the ***Vaqua Wet Blasting*** principle. This is the bombardment of components with a high volume recirculated flow of solid particles in water, with the aid of compressed air. The solid media, such as glass beads, Pink Alumina or stainless steel grit, removes all the contamination, but there is always a cushion of water between the component surface and the media solid particles. This cushion is claimed by the system manufacturers to prevent the impregnation, surface damage and excessive breakdown of the blasting media, whilst removing the fixed or smearable contamination.

The blasting machine consisted of a blast cabinet containing a rotary turntable, a water/media mix (known as a slurry), a sump, two blasting gun deployment systems (each carrying two blasting guns complete with air injection system), and a slurry pumping system.

The slurry pump pulls the media/water mix from the cabinet sump and pushes it in a constant high volume flow to the process-blasting gun. Before the gun, a proportion of the water is diverted down a by pass line to provide slurry agitation, within the cabinet sump. A controlled flow of compressed air is injected into the blasting gun to accelerate the slurry, which is then discharged onto the container.

The container is rotated in front of the blasting guns via a set of drive rollers, electric motor and a gearbox arrangement. Two blasting guns are rigidly fixed to an electrically operated vertically rising deployment mast, two other guns are attached to a horizontal deployment mast to cover the top and base of the container. The number of guns running at any one time, their orientation, deployment feed rates, container rotational speed are all dependent upon the profile of the container being cleaned.

2.4 The Dry Bead Blasting Machine

The machine (Figure 5) comprises of a booth housing a container rotation drive assembly, a media transfer elevator unit and two media blast wheels. The container rotates about the vertical axes. The two blast wheels are positioned one above the other in a vertical plane along the height of the container. The blast wheels spin about their respective fixed horizontal axis. Media is transferred from the sump at the bottom of the machine to a hopper at the top, by an elevator system. Media is fed by gravity from the hopper to the blast wheels, via two pipes. An extract system was provided to carry away the damaged media particles and the dust generated during the blasting operations.

The container rotate, the blast wheels, the booth lid and the booth sump arrangement is designed to give full container surface coverage. A separate orifice plate for each wheel controls the rate at which the media is fed to the blast wheel.

Media is damaged during the blasting process and is extracted by the machine ventilation system, thus reducing the quantity of media charge in the machine.
It is essential that a minimum quantity of media be maintained in the machine for efficient operation of the blast wheels. This is approximately 60kg for this machine.

For this machine, there was no automatic media charge facility provided. The operator observed the media level at the feed point to the blast wheel. The operator would add the media in the machine manually as and when required.

2.5 Blasting Trials

Typical shoulder and mid-height panels used for demonstrating cleaning performance of the blasting machines and the results after blasting are shown in (Figure 6). The special surface preparation included simulation of:

- Melted glass product
- Burnt on plastic/rubber from hoses/cables
- Oxidation due to the heat from the pour glass.

Samples were taken from the blasted containers and were analysed for the following parameters: hardness, surface roughness, emissivity, and surface inclusions/embedment.

Samples of fresh media, media resulting from blast operations, dust/debris collected in the ventilation extract box/cyclone and dust from the filter elements were obtained for characterisation of media in the machine.

Scanning Electron Microscope (SEM) pictures and photographs of pre-blast and blasted surfaces were obtained from selected samples.

Photographs for a pre-blast container used for a vitrified glass pour in the Pilot Plant, are given for comparison against a blasted container in figure 6.

2.6 Comparison of DBB and WAB Processes and Equipment

Parameter	DBB	WAB	Comments
Process and Equipment			
Process	Dry	Wet	
Media			
Type	SS Shot (Recommended: Grade 30)	Glass Abrasive (Recommended: Vaquashene Grade 400/600)	
Initial Load	65kg	35kg	
Machine Booth Dimension	Approx. footprint: 1.2m x 1.5m and 2.7m high	3.5m high x 3.5.m W x 3.5m deep (Including Waste removal – 9m deep)	
Weight	unknown	4000kg (excl. water and media)	
Blast Equipment	Two wheels in a vertical plane along the height of the container	Four guns: One vertical manipulator for barrel carrying two guns and two horizontal manipulators, one for top and one for bottom of the container each carrying one gun.	
Cleaning Process	Dry shots blast with spinning wheels.	Water/abrasive slurry blast at 3.2 – 3.3bar with air assist at 80psi	
Media Transport	28 bucket elevator	Slurry pump.	
Performance			
Container Coverage	100%	100%	
Cleaning Cycle Time at recommended optimum set-up parameters	Oxidised surface – 20-30mins Fixed glass removal – 45-60mins	11min 33-44mins	
Surface Hardness (Vickers)	Shoulder – 238-261 Mid-height – 210-247 Lid Weld (sub-surface approx. 0.05 –0.1mm) – 177-230	221-239 268-304 145-237	Not blasted - 188-202
Roughness (μm)	1.8 – 2.8	1.8 – 2.9	Not blasted - 0.4 – 0.7

Parameter	DBB	WAB	Comments
Emissivity (200 degC) **Thermometer type – OHF** 35/20/18L (1 μm – 12 μm)	Neck/shoulder – 0.34 – 0.37 Container mid-height. – 0.39 – 0.41	0.46 – 0.51 0.47 – 0.49	Not blasted - 0.29 - 0.31
Surface Inclusions/embedment	Not quantifiable	Not quantifiable	Not Blasted - N/A
Surface Contamination	Currently no provision for cleaning settled dust	Clean water wash facility.	
Stock removal	Not quantified but indications are that it is negligible.	Approx. 1μm per blast at recommended parameters	DBB process seems to be much less aggressive than the WAB.
Waste Arising			
Damaged media Top-up Rate	0.3 – 0.4kg/hr Based over 250hrs run.	No comparable data available as tests were carried out for batch run ONLY.	
Carryover into:	Based on 5hr run	Over a long period	DBB Ventilation system not designed for this equipment.
Filter box	43gm/hr	N/A	WAB system has a 50mm cyclone to take out water and solid particles.
Filter element	39gm/hr (Filter type not known)	Not measurable (HEPA filter used)	

2.7 Conclusions

2.7.1 Wet Abrasive Blasting machine

The Wet Abrasive Blasting Process can clean the product containers that will have surface roughness and emissivity parameters very close to those required by the 'guaranteed parameters', for example:

- Four guns, two for the vertical and one each for the top, and the bottom horizontal oscillations in conjunction with the container rotation, proved successful for full product container surface cleaning.
- The process removes oxidation stains in a single pass. Three to four passes are required to remove melted plastic/rubber.
- The blasting operation removes container stock material at a rate of approximately 1micron per pass, within the optimised blasting process parameters.
- The process is very aggressive, and can cause significant damage if an object is subjected to the blast for a long duration.

- At the start of a fresh charge, the abrasive media is not very effective for the first one or two full height passes.
- Pink Alumina 54/70 blast media is suitable for the equivalent of 37 full height container passes, compared to only 12 for Vaquashene 400-600 Glass Bead media.
- The slurry has to be agitated continuously during the blasting process, as it easily de-waters and hence could cause blockages in process lines etc, the machine requires to be flushed before shutdown.

2.7.2 Dry Bead Blasting machine

The following conclusions are made for the Dry Bead Blasting process. Essentially, they are limited to the parameter requirements for the Customer 'guaranteed parameters', or are required for informing the Customer if the current values are likely to change due to the revised container processing operation.

- The process provides 100% container coverage.
- The process does leave a fine dust layer on the container surface and coats surfaces within the machine. This will require removal.
- Surface hardness for blasted surface is higher than that for the oxidised surface resulting from glass pour.
- Emissivity is in the range 0.34 to 0.41.
- Surface roughness is in the range 1.9 to 2.8 µm.
- There does not seem to be any evidence of surface inclusions/embedment from this process.

Acknowledgements

B570 R&S Technical plant support team, Sellafield.
VPS Project team, Risley.
D. Bainbridge WVP, Sellafield.
G. Mistry R&T, Sellafield.
BNFL, Jordi facility

References

R&T[1]   BNFL, Research and Technology organisation.

JoRDI[2]   A joint research and development partnership between BNFL R&T and INBIS Ltd.

Figure 1. THORP Receipt and Storage Pond Hall

Figure 2. Schematic of CWRL Decontamination equipment

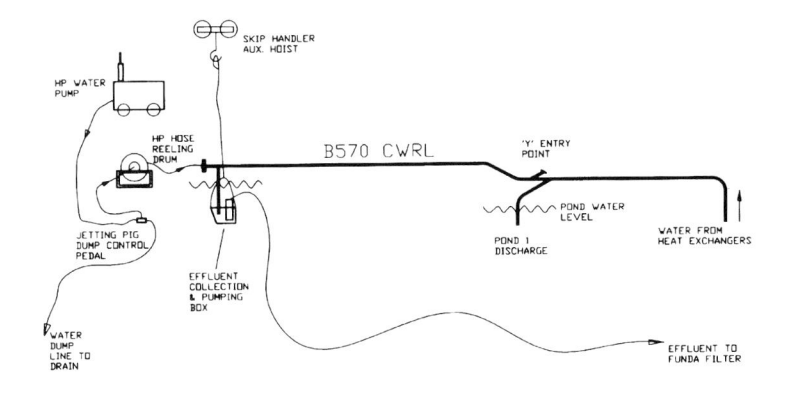

Figure 3. Jetting head with centraliser

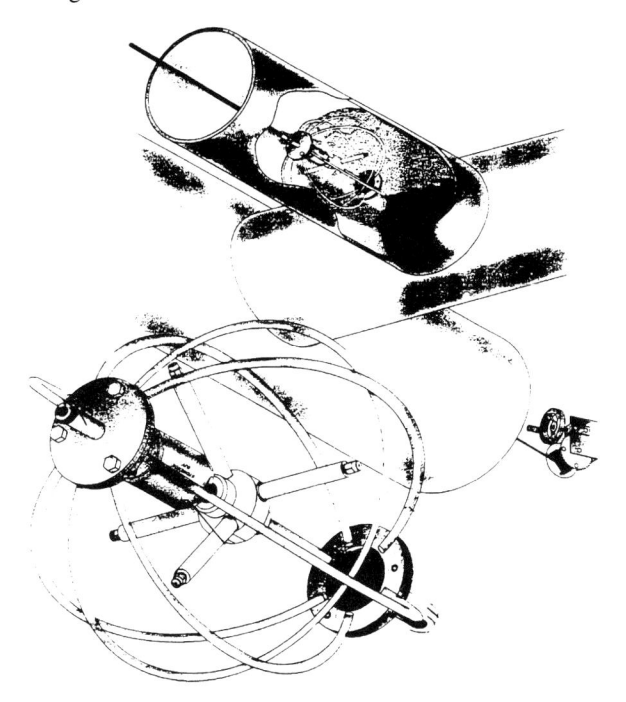

Figure 4. Wet Abrasive Blasting machine

Figure 5. Dry Bead Blasting machine

Figure 6: WVP Glass Product containers

Top Left - Oxidised prior to blasting
Top Right - After Dry Bead Blasting & Bottom - After Wet Abrasive Blasting

© British Nuclear Fuels plc 2001

C596/001/2001

The Dounreay PFR liquid metals disposal project

B BURNETT
Waste Management and Decommissioning, UKAEA Southern Sites and Magnox RDU

Synopsis

A major Decommissioning task at Dounreay is the extraction and disposal of some 1500 tonnes of Sodium and Sodium /Potassium liquid metals which remain from the Prototype Fast Reactor programme. This work started by NNC under contract from UKAEA in 1995 and is well advanced with liquid metal destruction in progress. This paper is an update on the status on this challenging project. A brief overview of the project will be given both in terms of the technical issues and the project structure that has so successfully enabled UKAEA to work closely with contractors. Finally the lessons learned are discussed in the context of how such demanding tasks can be successfully completed in a safe, cost effective manner whilst managing the project risks.

1   INTRODUCTION

The UKAEA Prototype Fast Reactor (PFR) at Dounreay used liquid sodium to remove the heat from the very high power density core. It also used another liquid metal, sodium-potassium eutectic (NaK) in the decay heat removal system. When PFR was shut down in 1994 the reactor vessel was left holding 900 Tonnes of slightly contaminated sodium. A further 600 tonnes of liquid sodium and smaller volumes of NaK are stored in other vessels.

This material will be removed from the reactor vessel and other vessels by means of specialist equipment, pre-treated and then reacted with aqueous solution of 10 molar sodium hydroxide. During operation the strength of the solution is maintained by the addition of water and sodium and the removal of a proportionate flow of hydroxide. After neutralisation with hydrochloric acid to produce a saline solution, the product will be filtered and processed to reduce the caesium content then passed to the low level effluent treatment plant for eventual discharge to the sea.

In 1995 UKAEA awarded NNC the contract to built and operate the Liquid Metal Disposal Plant. Framatome's successful 'NOAH' process had already been adapted to dispose of 37 tonnes of coolant from the Rapsodie experimental fast reactor in France and was scaled up to form the central plant design. With Framatome and AEA Technology as design consultants for some fluid processes, NNC designed the control and instrumentation and the specialist reactor equipment. The plant is part of a decommissioning process that will complete operation with three years of start-up so it is a cost sensitive application requiring proven technology to avoid the risk of delays in development. The sodium destruction process being only the second of its type and the first on this scale, at the time of the award of the contract, uses a high density of instrumentation and automated equipment to facilitate remote operation and automated control of the process.

This paper describes the process of extraction of the primary sodium from the PFR reactor and its subsequent processing, destruction and neutralisation for disposal. The latest project progress position is also described.

2   THE LIQUID METAL DISPOSAL PROJECT – PLANT OVERVIEW

The original design of PFR did not include provision for removal or disposal of the liquid metal after shutdown of the plant. After evaluating the risks and costs of various options, UKAEA decided to invite fixed price bids for design, construction and operation of a plant for disposal of 1500 tonnes of liquid metal and reactor based equipment to facilitate its extraction.

The process design was split into seven main areas with design subcontracts placed both with Framatome ((France) and with AEA Technology. The design included several novel process and instrumentation techniques particularly in the aspect of sodium extraction.

Following the liquid metal disposal process through from end to end, plant can be considered in five stages. Figure 1 is an overview of the process.

1)   Liquid metal supply to extract the sodium from the reactor and other vessels
2)   Liquid metal circuits – to store and prepare the sodium and NaK
3)   NOAH process – to react the liquid metal with sodium hydroxide
4)   Neutralisation process – to produce saline effluent
5)   Filtration, ion-exchange and sampling – to control the aqueous effluent quality

The Sodium Disposal Plant incorporates its own mechanical services such as the ventilation systems, bulk acid plant and gas supplies and it also integrates with existing plant.

The liquid metal supply comprising both old and new systems, feeds the Sodium Disposal Plant. It comprises several large new plant items – drills, heaters and an extract pump all of which extend to the lower regions of the PFR reactor up to 18m from the reactor top entry locations.

After shutdown the reactor sodium was kept in a molten state by the recirculating pumps. Before the level in the reactor vessel could be reduced below the inlet of the circulating pumps, alternative methods of heating had to be engineered. Two new heaters were provided to cope with falling levels during disposal:

1)   A skid-mounting heating plant was developed to provide 220 kW of heat into circulating NaK through a heating loop in the reactor using an electromagnetic induction pump.

2) A large electric multistage immersion heater was built to provide 180kW of direct sodium heating.

A sodium pump driven by nitrogen is immersed in the reactor vessel and is used to lift the liquid metal more than 15 metres out of the vessel and deliver it through a shielded pipeline to the disposal plant.

Drill and punching machines have been developed to penetrate the core-support strongback top plate and a primary sodium pump discharge line at the base of the reactor to release trapped sodium and equalise the levels of liquid sodium in the inner and outer pools.

After delivery to the Sodium Disposal Plant at 200°C the sodium is weighed in batches of up to 16 tonnes in a buffer tank. A small quantity of NaK is added in for disposal in the process up to a maximum of a 1% concentration.

The liquid metal is then transferred by applying nitrogen gas pressure to a batch tank where it is cooled by an air jacket to 120°C to precipitate out contaminated sludge before filtering. One day's worth of processing capacity is stored in a further tank (the day tank).

The sodium passed through the sodium conditioning circuits can be diverted through a sodium sampling station, to allow up to five samples per transfer to be extracted for analysis.

An electromagnetic induction pump is used to lift sodium from the day tank into a header tank ready for injection into the sodium reaction vessel.

Framatome provided design details of their patented 'NOAH' sodium/aqueous reaction system. This was based on the small sodium disposal plant they had built for decommissioning the old Rapsodie reactor in France. The plant has been scaled up to around three tonnes per day required for the Dounreay Plant.

On leaving the header tank a novel design of metering pump uses hydraulic fluid impulses coupled through two diaphragm stages to inject sodium into the 'NOAH' reaction vessel. The rate of injection can be continuously varied between two and 138 litres per hour. More than 25 individually controlled heating zones in this part of the plant alone ensure that the optimum conditions for injection into the reaction vessel are met. By cooling some lines with air jackets, the sodium can be redirected into a recirculation mode and the pumps tested out without operating the reaction vessel.

The disposal process is a continuous one that works by injecting a fine dispersion of molten sodium into the 'NOAH' vessel in opposition to a stream of aqueous sodium hydroxide. The reaction takes place at the point where the two streams meet in the nitrogen filled space of a reaction vessel evolving hydrogen gas. The controlled start-up, operation and shut down of the sodium/water reaction process takes place in a number of defined control states with requirements that must be met before moving on to the next state, as described below.

During operation hydrogen is continuously discharged from the 'NOAH' vessel through a conditioning system. It is then diluted with air to well below the lower explosive limit and discharged through a stack equipped with tritium and particulate contamination sampling equipment.

The sodium hydroxide is continuously diluted with water and the excess hydroxide is transferred to the neutralisation plant for conversion to pH neutral solution by a batch reaction with hydrochloric acid. The solution is filtered by a 'rapid pulse filter' which builds up a thick cake of precipitate for dry disposal. The clear liquid is passed through a Caesium removal process, sampled and then transferred to the site's new low level effluent treatment plant.

A mass balance of the destruction process is shown in Figure 2.

Access to the plant during operation is very limited. To reduce the radiation exposure of persons operating the plant it is required to be remotely operable and consequently it has a very comprehensive set of instrumentation. CCTV cameras are provided to help view parts of the process including one camera (with a microphone to monitor the reaction sound) looking through a window in the 'NOAH' reaction vessel.

The 'NOAH' reaction process uses demineralised water from a dedicated plant for dilution of the sodium hydroxide to sustain a constant molarity. The resulting gaseous discharges from the reaction process and plant ventilation systems are continuously monitored by a dedicated stack sampling system including real time particulate radioactivity monitoring elemental tritium and tritiated water vapour sampling.

The neutralisation and discharge processes also incorporate sample glove boxes. During neutralisation, and discharge pH monitoring is continuously carried out by recirculation of a representative sample through a pH monitoring station.

Sampling of the liquid effluent are extracted for further analysis after the Caesium removal process.

3 SPECIFIC ENGINERING CHALLENGES

3.1 Extraction of Reactor Sodium

One of the first and possibly most significant engineering challenges of the Liquid Metals Disposal Project was the method by which to extract the liquid sodium from the Prototype Fast Reactor Vessel. The original design of PFR did not include provision for removal of the liquid metal after shutdown of the plant. The design of the reactor vessel dictated that the sodium must be extracted out of the top of the reactor vessel.

A sodium extract pump was designed for this purpose. The pump is driven by nitrogen, using nitrogen pulses to lift the liquid metal up the mast of the pump. A vortex diode is used to trap the liquid metal as it is lifted out of the reactor to prevent it from draining back out of the pump nozzle.

The extract pump is installed into the reactor vessel to a total depth of 18 metres and is used to lift the liquid metal more than 15 metres out of the vessel and deliver it through a shielded pipeline to the sodium disposal plant. The pump is installed through the Vickers plug penetration following removal of the existing Vickers plug. This allows the pump assembly to be lowered into the reactor outer sodium pool.

Once installed in the reactor and ready for operation, the pump assembly incorporates a 25mm nozzle on a retracted section of the flexible pipework which is deployed down to the base of the reactor below the core support structure allowing the maximum possible quantity of sodium possible to be extracted.

The extract pump incorporates an array of heaters and temperature monitoring up the main mast assembly to prevent the pump from becoming blocked with freezing sodium as the reactor level is lowered and more of the mast becomes exposed to the nitrogen cover gas.

3.2 Primary Sodium Heating

After shutdown, the reactor sodium was kept in a molten state by operation of the recirculating primary sodium pumps. However, during extraction of the reactor sodium for disposal the sodium level will quite quickly fall below the pump inlet level. To compensate for this two diverse methods of heating have been developed to maintain the liquid metal temperature to much lower sodium levels.

1) At the time of the original filling of the PFR vessel a heating loop was installed in the base of the reactor. This was filled with sodium-potassium eutectic (NaK) so that it could be kept liquid at moderate temperatures outside the reactor. Tests on this loop have shown that it was still intact and a new skid-mounted heating plant and associated extension to the NaK heating loop pipework has been developed to provide 220 kW of heat input into the NaK and circulate it through the existing heating loop in the reactor using an electromagnetic induction pump.

    The NaK heating skid incorporates an automatic fire suppression system which uses automatic isolation dampers and a nitrogen injection in the skid in the event of a NaK leak and resulting fire.

2) A large electric multistage immersion heater has been built to heat the reactor liquid metal directly by insertion into the reactor outer sodium pool.

    The immersion heater is 18m long with six heater banks spread over the bottom 3m of the heater mast. The heater is installed into the reactor through the Reactor Transfer Port (RTP) and extends through the reactor rotor, through a vacant bucket port.

    This required the removal of the bucket exchange machine shock absorber from the base of the reactor, before the heater could be installed.

    The electrical immersion heater does not extend to the base of the reactor vessel due to the restrictions of the RTP shock absorber socket. This prevents the heater from being able to heat the liquid metal below the shock absorber level. For this reason the immersion heater will be used in preference to the NaK loop heater during the bulk sodium extraction to minimise the demand on the NaK loop heater. The NaK loop heater will then be used to heat the residual liquid metal once the immersion heater is no longer effective.

3.3 Reduction of Trapped Reactor Sodium Volumes

In addition to the bulk sodium extraction, a significant quantity of liquid metal remains or is held up within the reactor vessel as the liquid metal level is lowered. To reduce the trapped volumes of liquid metal specialist equipment was developed to common together sections of the reactor vessel.

A special drill has been developed to penetrate the core-support strongback top plate and pleated insulation at the base of the reactor to release trapped sodium and equalise the levels of liquid sodium in the inner and outer pools, allowing the inner pool to drain.

This machine has to operate in a depth of 14 metres in liquid sodium and is to be installed in place of a sector 10 neutron shield rod, removed from the inner pool of the reactor.

A pipe-piercing machine was also developed to facilitate the draining of trapped sodium volumes within the reactor from the Diagrid and Primary Sodium Pump discharge lines. It will rupture sodium pipework within the reactor by operation of a hydraulic punch.

The pipe piercing machine on its strongback support is installed through one of the Primary Sodium Pump Valve channels. Once in position the pipe piercing machine punch is deployed on a short train and must be mechanically steered down the correct sodium outlet pipe, to ensure the lowest sodium discharge line is ruptured, before punching can take place.

The pipe piercing machine can only be deployed once the reactor sodium level has been lowered sufficiently for the Primary Sodium Pump Valve outlet channels to become exposed to allow the manual steering of the punch and train to be carried out.

A diagrammatic representation of the reactor based equipment is shown in Figure 3.

4. UNIQUE ENGINEERING SOLUTIONS

4.1 Hydrogen Management

Hydrogen management is one of the key aspects of the Sodium Disposal Plant, with a hydrogen explosion constituting one of the most severe potential hazards of the plant. At the maximum sodium destruction flow rate of 138 l/h, hydrogen off gas is generated at a rate of up to 70 Nm^3/h. There is also a component of Tritium in the off-gas from the primary sodium destruction.

The hydrogen off-gas is controlled by keeping the hydrogen concentration levels outside of the explosive concentration region, well within the upper and lower safe concentrations. The sodium reaction vessel and hydrogen conditioning line are initially purged and padded with nitrogen to levels of less than 0.1% oxygen. Under these conditions the destruction process can be started and the hydrogen generated is allowed to displace the nitrogen padding gas within the reaction vessel and conditioning line to build up to a concentration level of ~99.5% at a pressure of ~150 mbar.

Hydrogen generated is passed through a sodium hydroxide scrubber used to reduce the caustic content of the off-gas. The gas is then dehumidified by means of a mist eliminator, condenser and reheater and filtered at 50°C using HEPA filtration.

Following filtration the hydrogen is discharged into the building ventilation extract system using purpose designed hydrogen diffuser rings, in a section of high air velocity ductwork.

The diffuser rings are used to inject the hydrogen off-gas into the ventilation air flow as an array of small, non-overlapping jets, to minimise the distance over which an explosive concentration of hydrogen exists. Even in the event of a jet of hydrogen being ignited this will not cause ignition of adjacent jets as the jet plumes only overlap at a hydrogen concentration below the lower explosive limit and will self extinguish due to the high discharge velocity. The hydrogen diffuser rings have been tested under these conditions for hydrogen flows of up to 80 Nm^3/h.

Hydrogen carryover into the neutralisation process is minimised by the use of a breakpot on the aqueous overflow line from the reaction vessel. The breakpot is a small nitrogen purged vessel, the level in which is controlled to maintain a liquid barrier between the reaction vessel and neutralisation process in addition to a float trap after the overflow from the vessel.

4.2 Plant Control

Control of the sodium disposal process is carried out using a SCADA (Supervisory Control and Data Acquisition) Distributed Control System.

This system was built and configured from proprietary modules supplied by Rotork. It uses six Process Control Stations (PCS). These are connected using a dual redundant Ethernet and through Network Interface Modules (NIMs) to three Operator Work Stations (OWS). Two of the PCSs are in turn connected to Input/Output modules (Scatterpods) in plant cubicles for remote signal connections to the PFR Reactor Hall and SDP Bulk Acid Plant. Figure 4 is an overview of the DCS architecture.

The software configured for the DCS consist of two 'types' of control sequences.

1. Supervisory sequences monitor the state of the plant against the desired operating state. If discrepancies occur, i.e. parameters out of the desired range or valves in incorrect positions, the supervisory sequence will shutdown a process, raise an alarm or will take action to correct the problem, e.g. perform a changeover to standby equipment.

2. Operating sequences that carry out the process steps required for normal operation of the plant. These sequences generally run through once, at the request of either the operator or a supervisory sequence, and then terminate.

A manual mode of operation is also provided in which both types of control sequences are disabled.

The distributed control system provides both supervisory and automatic control functions for all plant systems via a series of the automated sequences to carry out routine operations such as sodium transfers, sodium cooling and sodium hydroxide transfers, recirculation, neutralisation and filtering for the aqueous circuits.

The NOAH process is a continuous process and utilises a different style of control philosophy. The NOAH plant has seven defined plant states based on the configuration of over 30 key items of equipment such as pumps, valves and chillers. Three of these states are start-up/shutdown states which are sequentially selected from the distributed control system.

Following successful progression through the start-up states the next two operational plant states are selected by means of a hardwired switch which also conditions some parameters of the safety system, according to the mode of plant operation. The distributed control system monitors each state change and warns or inhibits the operator in the event of any discrepancies.

4.3 The Safety and Protection System

Two NOAH plant states are trip states driven by the safety system. The first trip State is the Security State, which initiated an immediate shutdown of the reaction process and purges the reaction vessel and hydrogen lines with nitrogen to diffuse and dilute the hydrogen. The second and ultimate trip state is the Safety State, which is again the immediate shutdown of the reaction process, but with the hydrogen vent system isolated from the hydrogen diffusion rings to contain any remaining hydrogen.

In the event of the reduction in flow or the loss of the ventilation system to the reaction vessel cell and hydrogen diffuser rings, the NOAH plant not only trips to the Safety State and isolates the cells from the ventilation system duct.

In the event of a NOAH plant trip, the trip state is determined by the parameter that triggered the operation of the safety and protection system. Each input to the safety and protection system for the NOAH plant is categorised to trigger either a Safety or a Security state trip. The sodium conditioning and neutralisation plants also incorporate hardwired Safety trips.

The Safety and Protection System is an entirely hardwired system, completely independent of the Distributed Control System. The System utilises dual redundant trip circuits monitoring key plant parameters for each plant area.

The state of the SPS and its inputs are fed to the DCS to provide alarms and assist with fault diagnostics.

5 PROJECT STRUCTURE AND MANAGEMENT ISSUES

The contract was originally let by UKAEA on a turnkey basis to NNC with major subcontractors AEA(t) and Framatome assisting on design aspects and extensive use of local subcontractors for the site work. The special process equipment skids and the innovative drill and punch components being designed and supplied by NNC's special purpose equipment business at Risley. The extent of supply included all the traditional aspects of design, supply, construction and commissioning, but also included the operational phases.

These arrangements served the project well through the design supply and construction phases of the project. There was substantial commitment to the success of the project by all sides aided by open and transparent discussions on the interfaces and risk areas.

When the setting to work activities started and the pre-commissioning safety case development began it became apparent that there was potential for duplication of effort and lines of management as NNC sought to manage the project obligations and UKAEA its proper licensee role. This project has significant complexity as it must be carried out within the existing facilities with numerous physical, safety and administrative interfaces.

In response to this, together with the client, we established a Test and Commissioning panel chaired and led by the client which approved procedures, test team appointments reports etc.

This had the benefit of a clear line of safety management, efficient co-ordination of interfaces and a fast decision making process necessary for the effective progress in commissioning of novel complex plant.

The success of this arrangement and other factors led to the negotiation of the handover of the plant for active commissioning and operations to UKAEA with NNC retaining technical responsibility for the performance of the plant. An NNC site team remains in place to provide technical assistance to UKAEA. This arrangement removes management duplication and avoids contractual conflict that may have arisen through the operational phase.

The manner in which NNC and its subcontractors have worked on this complex challenging project in close co-operation with UKAEA says much for the commonality of purpose of both organisations and the professionalism and team working of the individuals. This points the way forward for tackling such projects for the future.

It is possible to preserve commercial disciplines inherent in client/contractor relationships whist managing risk in the most appropriate manner if organisations can work together and recognise the long term benefits to the industry. Each challenge has to be reviewed on its merits and we can envisage that partnering, alliances and key supplier agreements being the way forward.

Decommissioning Projects inevitably have risk factors. It is not a matter of who takes the risk, but more establishing who is best equipped to do so.

5.1 Current Progress Position

The SDP plant has been fully set to work, all safety systems tested, all subsystems run and tested to demonstrate their functionality. The plant is fully proven by the processing some 29 tonnes of clean sodium. This has generated 290 tonnes of sodium chloride solution, 15,000 m^3 of Hydrogen. This cumulated recently in a five day proving run at full output during which the plant was demonstrated to recover from plant trips back to full disposal flow rate.

The plant is being prepared for operation with active material. The strategy for this phase of commissioning is to progressively challenge the plant with activity to demonstrate the radiological exposure levels from the process plant are in accordance with the safety case requirements.

With respect to Liquid metals supplies systems all items have been installed or ready for deployment and the first plant connections made. Testing is in progress on the heating and pump systems in preparation for the first Sodium transfers.

Figure 1

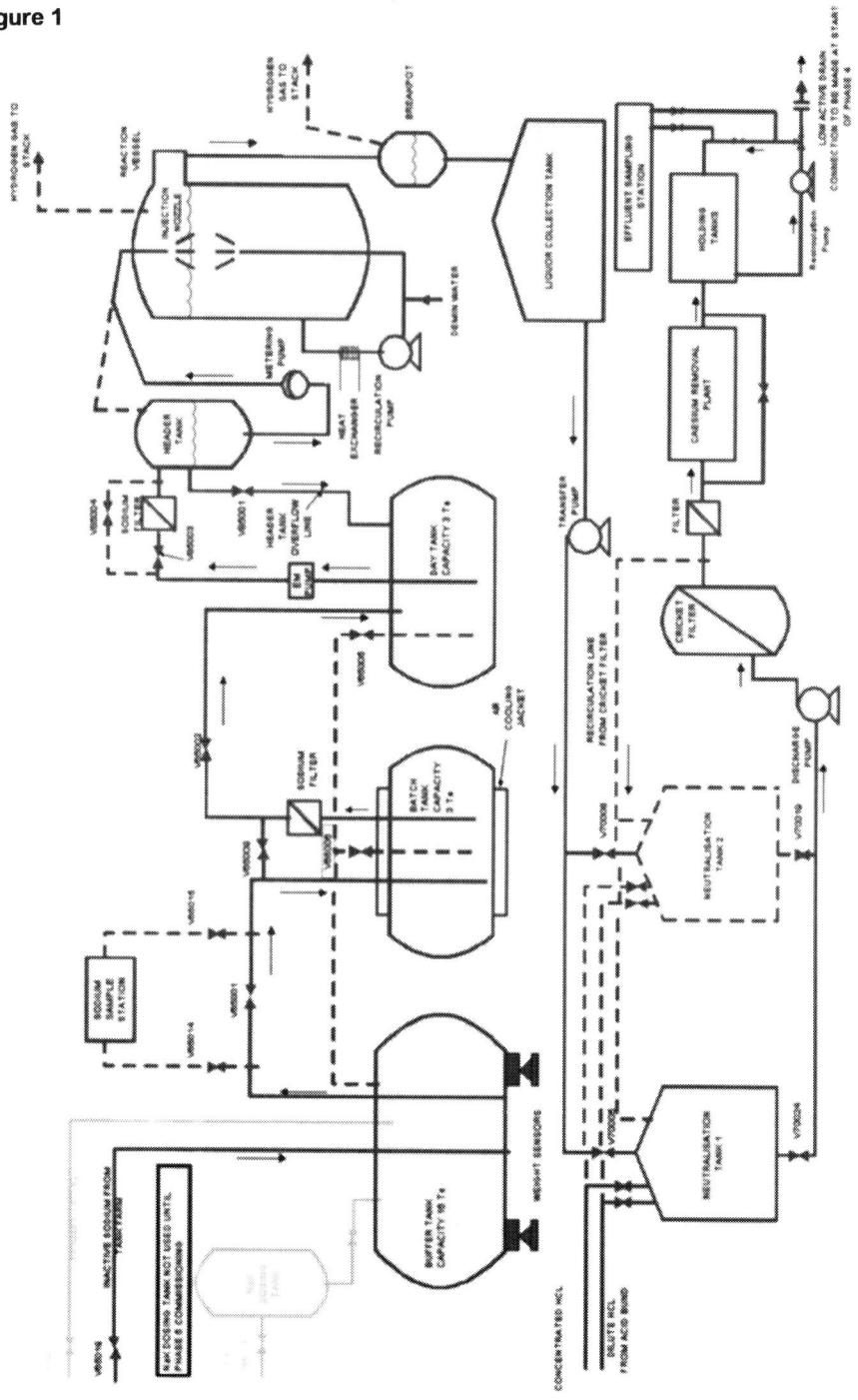

Figure 2

SDP Overall Mass Balance

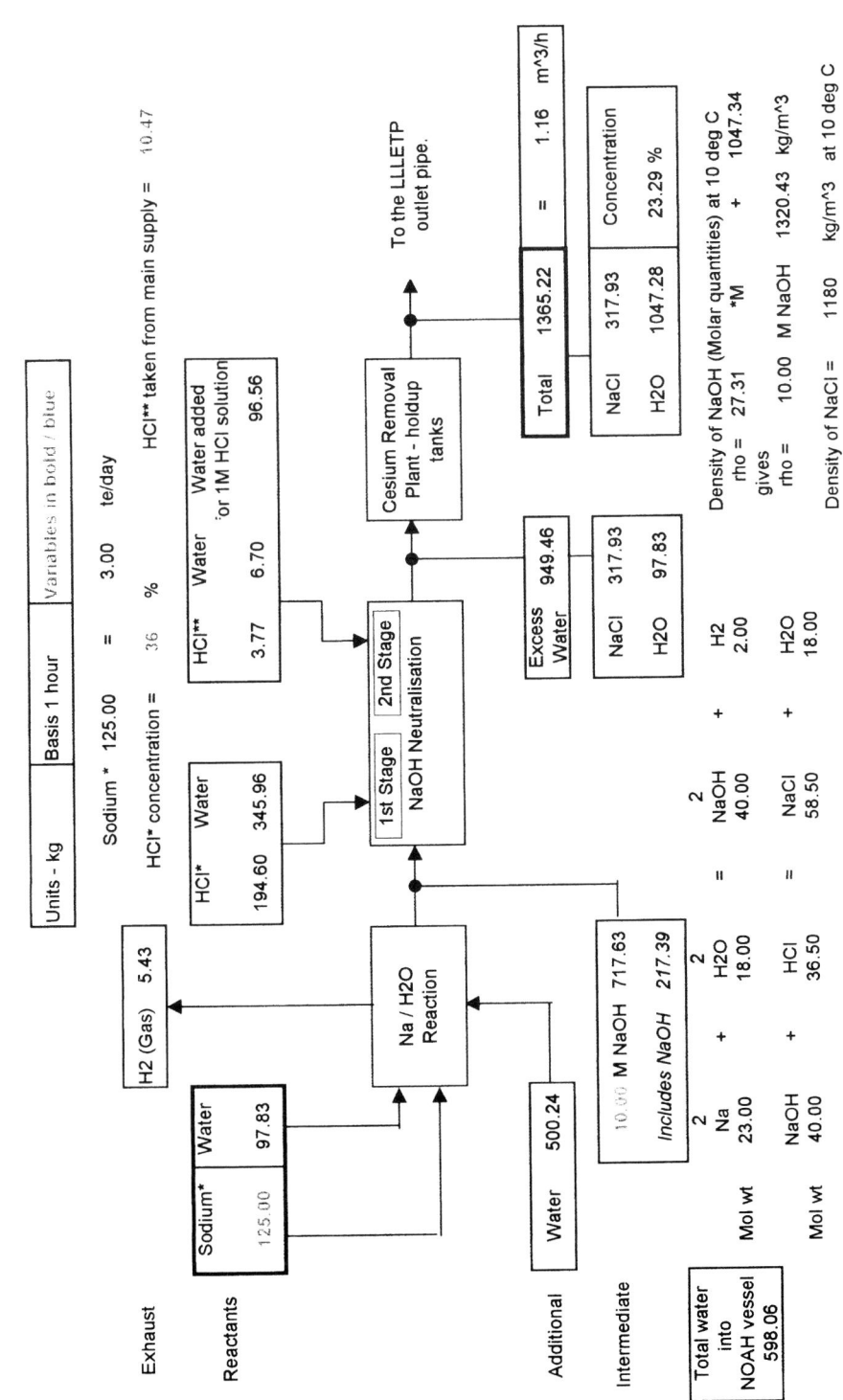

Units - kg	Basis 1 hour	Variables in bold / blue

Sodium * 125.00 = 3.00 te/day
HCl* concentration = 36 %
HCl** taken from main supply = 10.47

Exhaust
H2 (Gas) 5.43

Reactants
Sodium* 125.00 | Water 97.83

HCl* 194.60 | Water 345.96

HCl** 3.77 | Water (for 1M HCl solution) 6.70 / 96.56

Na / H2O Reaction

1st Stage NaOH Neutralisation

2nd Stage

Cesium Removal Plant - holdup tanks

→ To the LLETP outlet pipe.

Excess Water 949.46
NaCl 317.93
H2O 97.83

Total 1365.22 = 1.16 m^3/h
NaCl 317.93 | Concentration
H2O 1047.28 | 23.29 %

Additional
Water 500.24

Intermediate
10.00 M NaOH 717.63 | H2O 18.00
Includes NaOH 217.39

Mol wt 2 Na 23.00 + 2 H2O 18.00 = 2 NaOH 40.00 + H2 2.00

Mol wt NaOH 40.00 + HCl 36.50 = NaCl 58.50 + H2O 18.00

Total water into NOAH vessel 598.06

Density of NaOH (Molar quantities) at 10 deg C
rho = 27.31 *M + 1047.34
gives
rho = 10.00 M NaOH 1320.43 kg/m^3 at 10 deg C

Density of NaCl = 1180 kg/m^3

C596/001/2001

Figure 3

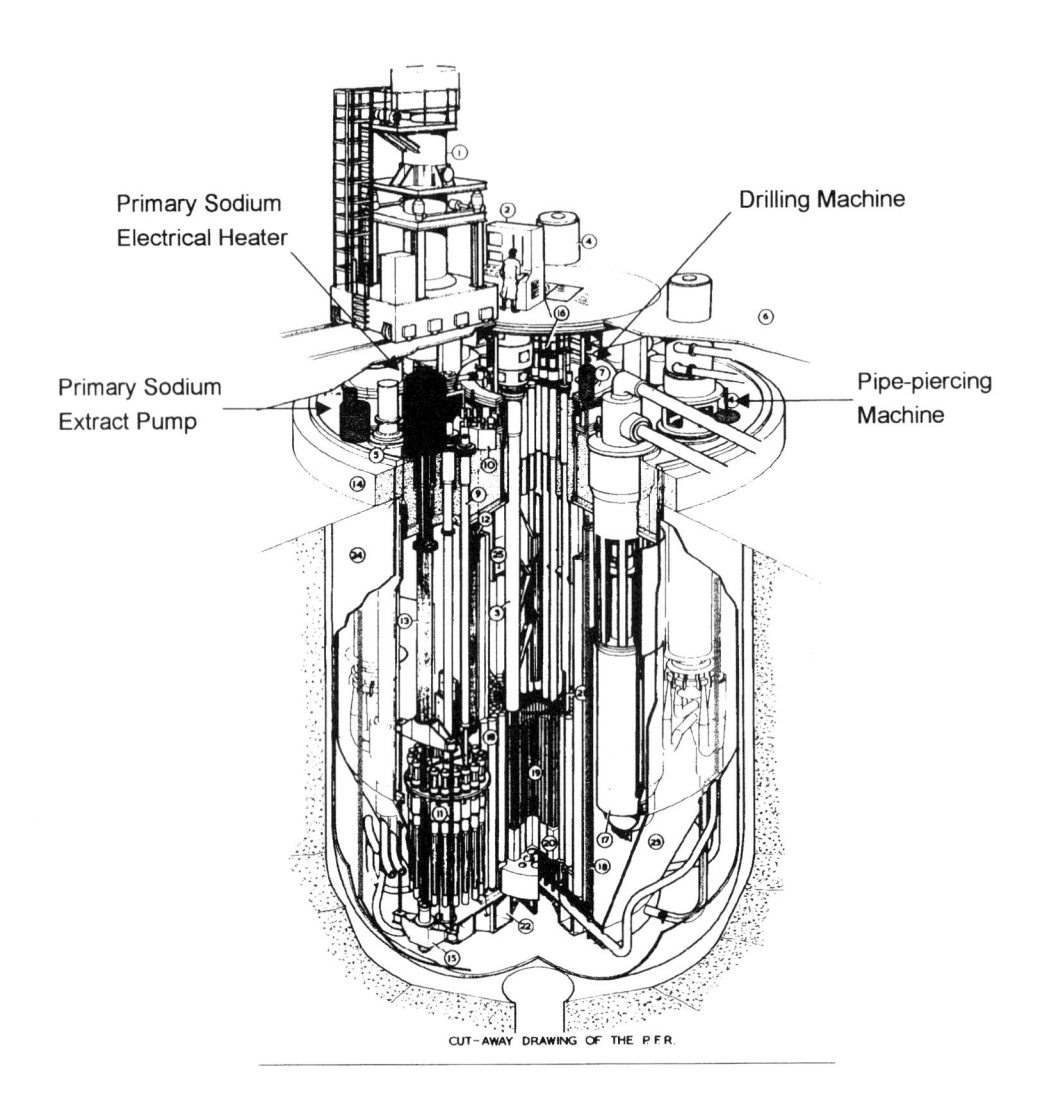

CUT-AWAY DRAWING OF THE P.F.R.

Primary Sodium Electrical Heater
Drilling Machine
Primary Sodium Extract Pump
Pipe-piercing Machine

Figure 4

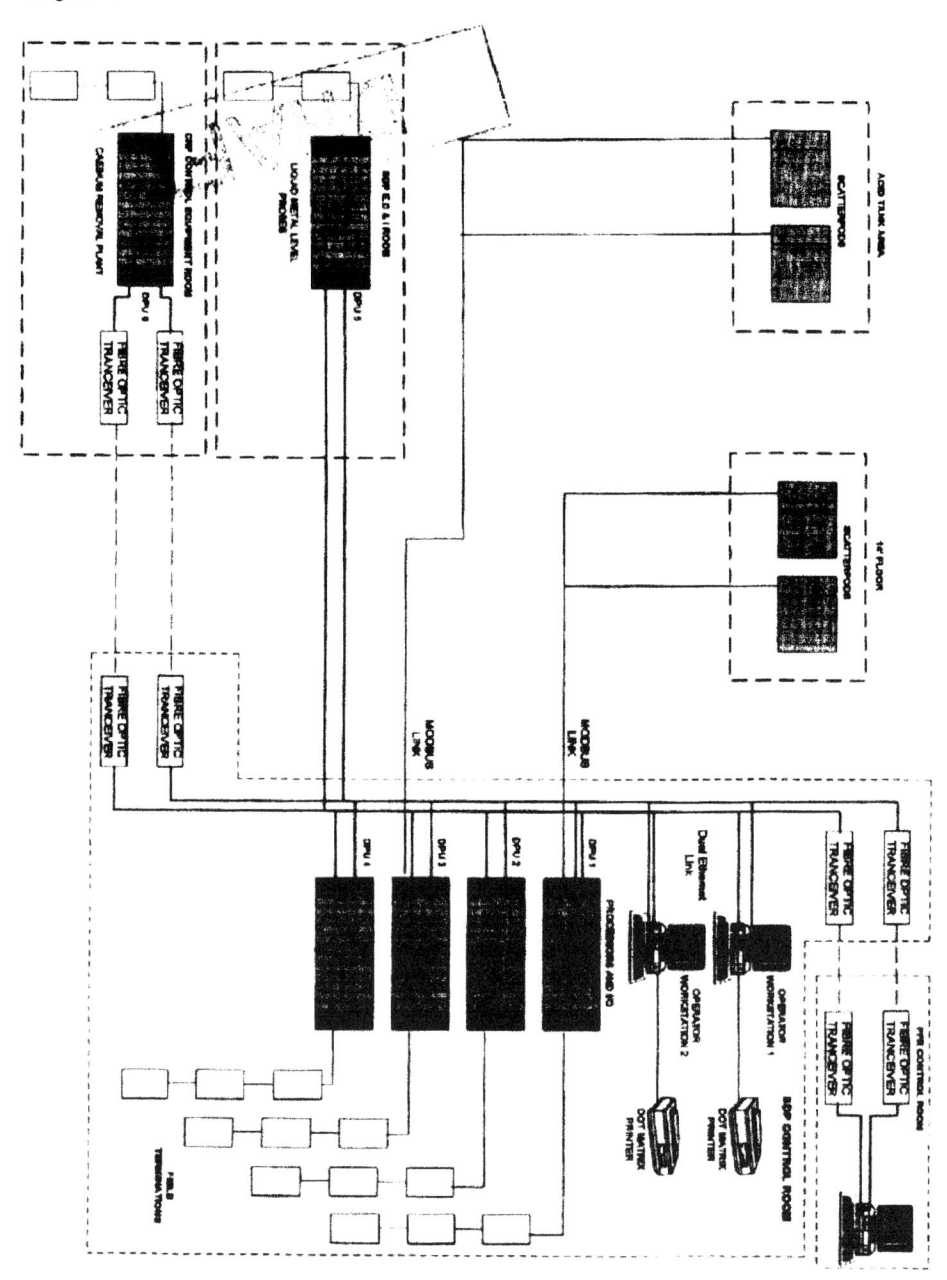

Figure 5

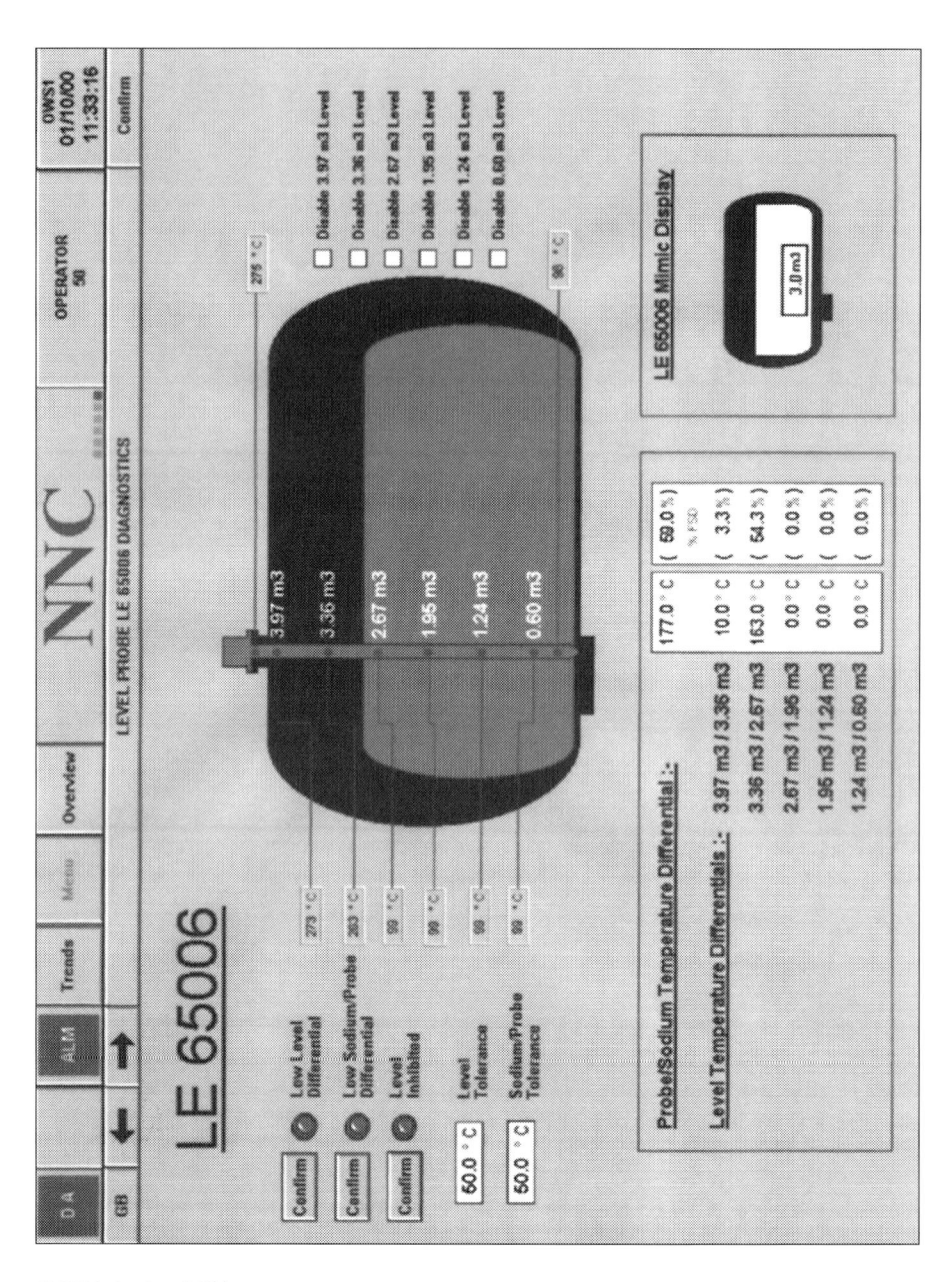

C596/012/2001

The RASP – safe size reduction of herogeneous objects

S BOSSART
USDOE/NETL
S ROSENBERGER
NUKEM Nuclear Technologies, South Carolina, USA
H-U ARNOLD
NUKEM Nuklear GmbH (USA/Germany)
M J SANDERS
NUKEM Nuclear Limited, Dorchester, UK

ABSTRACT

A need was identified for the in situ size reduction of large components such as plutonium contaminated glove box lines and tanks as part of the decommissioning programme. A development programme carried out in Germany resulted in the production of a new cutting method that could be remotely deployed for the in situ segmentation of large components of this type.

The test programme showed the method could successfully cut a wide range of materials and in various geometries, thus demonstrating a tremendous flexibility for tackling challenging segmentation tasks. NUKEM has selected **RASP** as acronym for this technology **R**emotely-**O**perated **A**dvanced **S**egmentation **P**rocess or RASP.

RASP is a remotely operated, dry, mechanical cutting technique that uses the basic principle of operating a hard metal abrasive wire in a reciprocating motion. The length of the wire can be readily changed to suit the size of component

The technology was demonstrated in May 2000 at the test facilities of Florida International University Hemispheric Centre for Environmental Technology and successfully performed segmentation tasks for glove box and tank segmentation.

A NEW CUTTING SYSTEM FOR THE IN SITU SIZE REDUCTION OF LARGE COMPONENTS

Background

In 1998, NUKEM started to provide engineering support for the decommissioning of several fuel fabrication facilities in Germany, including a MOX fuel fabrication facility. During the evaluation of the existing manual and remotely operated cutting systems for glove box size reduction, it became increasingly apparent that new cutting technology was necessary to improve worker safety and maintain or advance the technical segmentation capabilities.

NUKEM set out to address the unique problems associated with the D&D of such facilities by providing safe, innovative, cost-effective solutions. The facilities contained a large number of glove boxes, including plutonium-contaminated glove boxes in the MOX facility. To facilitate their removal, many of these boxes need to be cut up in situ because of their large size and the size of the installed equipment they contain.

A review of the standard manual glove box cutting methods highlighted a number of significant shortcomings including:

- Operative safety could be compromised in many situations e.g. when large openings are cut to gain manual access to internal structures that are not easily accessible
- High energy cutting tools generate high temperatures and can spread contamination over wide areas
- The sharp edges generated can damage the protective suits
- Hand held tools typically have limitations on geometry and/ or materials to cut and can be tiring for operatives to deploy
- Close proximity of the operatives to the tools and contamination increases the risk for an accident and increases dose uptake from gamma emitting isotopes
- The full protective suits that are typically used for this work limits the operative's visibility and reduces efficiency
- During D&D activities, operative are responsible for their personal safety while operating cutting tools in a constantly changing environment.

The evaluation of remote in situ segmentation technologies for glove boxes using – for example robotic vehicles with commercial tool attachments - showed other significant deficiencies and areas of concern. These included the following:

- High energy cutting tools such as grinders, plasma cutters spread contamination
- Remote technologies lack the flexibility required to adapt fast to the different challenges e.g. working in confined spaces
- Remotely handled segmentation tools still require permanent and sometimes multiple operator attention
- Remote systems cannot be deployed in all situations due to their size and footprint
- Although the use of remote tools substantially improves safety for the operators, their use can increase cost and take longer.

The development of new technology development was therefore focussed on improving safety for operatives and developing an inherently safe and flexible technology, which can be used for in situ segmentation of any large component, but particularly glove boxes of various sizes and configurations. A prime goal was to remotely cut the large components/glove boxes without removing the internal equipment, thus improving contamination control. Other important criteria included:

- Minimal space requirements
- Fast set up times for use in confined spaces or high dose areas
- The ability to cut in any direction
- The ability to cut virtually any material and geometry
- The minimization of secondary waste
- Low cutting temperature so flammable materials can be cut

It was concluded that a dry low energy mechanical cutting principle, which did not require cooling, would meet these criteria.

The development program produced a low energy tool that uses the principles of a reciprocating saw together with a novel operating system that applies the principles of fuzzy logic. RASP technology can be used for virtually any cutting task and is among the safest and most versatile segmentation technologies for the in situ processing of nuclear components. The technology deployment program for use of the RASP in Germany was presented and discussed in (1).

The Remotely-Operated Advanced Segmentation Process

RASP, the **R**emotely-**O**perated **A**dvanced **S**egmentation **P**rocess is a remotely operated, dry, mechanical cutting technique that uses the basic principle of a reciprocating saw. It is designed specifically for cutting contaminated metal components more safely, at lower cost and with minimal contamination spread. Tremendous advantages in risk reduction are offered when compared to manual size reduction of glove boxes and other artefacts with conventional methods. The RASP system uses commercially available hard metal abrasive wire that operates in a reciprocating motion. The length of the wire can be readily changed to suit the size of component to be cut. The RASP technology is patented in Europe and patents are pending in the USA.

The reciprocating wire is attached to the two reciprocating tools (compressed air-driven actuators) on opposite sides of the component to be cut. The tools travel along a modular frame arrangement.

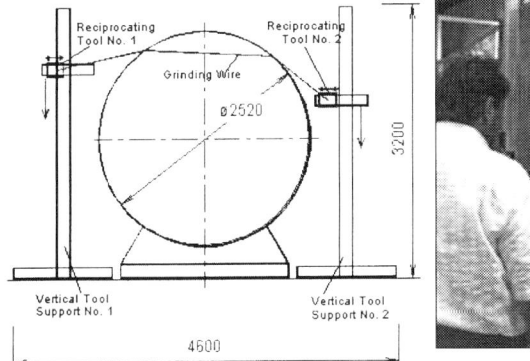

Figure 1: Typical Arrangement of Segmentation Module for Cutting a Tank

Figure 2: RASP Module during Cold Testing on a Glove Box

Figure 1 illustrates a typical for cutting a tank and other geometries such as glove boxes can be placed between tool supports as shown in Figure 2. The tools are indexed along a rigid frame which supports and guides the tools. The stroke length of the tools is governed by available space and if necessary, the tools can be positioned very close to the component being cut.

Newly developed proprietary software is used to control the cutting process, and this makes decisions based on fuzzy logic. The pressure of all systems is constantly monitored and the software adjusts the cutting process in real time. The mechanical resistance of the wire during cutting is the most important parameter, triggering variation of the cutting pressure and angle by the control system and this avoids overstraining and jamming of the wire.

Secondary wastes are minimised because no coolants or lubricants are used. RASP can cut through a wide range of materials, both soft and hard, including copper, tin, aluminum, graphite, plastics, hardened metal ball bearings, ceramics, glass, pespex and other organic materials; cutting speed is the only limiting factor. Figure 3 shows an example of a cut through a mixture of materials - a plastic coated laboratory glass pipe containing metal components fixed with foam. Objects as large as three metres in diameter have been successfully segmented during the technology development.

Figure 3: Foamed Glass Pipe with Several Metal Internals

The standard RASP unit consists of two components – the cutting module and the control station. The control station can be conveniently located outside of the controlled area, which can be a significant advantage. Operators need to spend <5% of their time attending to one segmentation module and several units can be monitored from one control room with a single team of operatives supporting a set of several independently operating modules – a very efficient arrangement. The process is extremely flexible and adaptable and can be configured for use in most situations that are likely to be encountered.

Contamination Control

RASP technology greatly simplifies contamination control because small particles are not ejected from area of the cut and they do not burn to generate radioactive aerosols. The technology also minimises the spread of loose contamination. The limited stroke length enables a Perspex shroud to be used to collect fines with a vacuum system if required.

A gel belt can also be used to effectively trap the fines and seal the already cut area ensuring virtually complete containment until the segments are separated. The gel belt consists of a gel-filled flexible tube or belt and is attached directly along the planned cut line. Other contamination control measures such tie-down coatings and foam filling can be used.

RESULTS OF TECHNOLOGY DEMONSTRATION AT FLORIDA INTERNATIONAL UNIVERSITY

The RASP system was demonstrated in May 2000 at the test facilities of Florida International University (FIU-HCET). The objective was to demonstrate the cutting technology for size reducing glove boxes and tanks as per the specifications of FIU-HCET scope of work and tests were carried out inside an enclosure to simulate the constraints likely to be encountered in a real decommissioning situation.

Horizontal and vertical cuts on a glove box and a kynar-lined tank were successfully demonstrated. Both components were made out of stainless steel with lead cladding. Also included in the tests were Perspex windows and various internals that were embedded in foam. The following table shows some data that was collected during the demonstration (2). More detailed performance information can be found in the subsequent reports issued by FIU (www.hcet.fiu.edu/tap/draftcopy/rasp.htm):

Table I: Cutting Rates of the Demonstration at FIU

Item	Thickness	Production rates
Glove Box	¼ in (6mm) stainless steel + in (3mm) lead shielding	Foamed cut through glove rings: 3.26 in (83mm) /hr Foamed with internal tools: 1.37 in (35mm) /hr Foamed avoid glove rings, cut through perspex window: 4.31 in (110mm)/hr Average production rate: 3.65 in (93mm)/hr
Kynar Tank	Body: ¼ in (6mm) stainless steel + in (3mm) lead shielding, cap flanges ¾ in (19mm)	Horizontal cut: 2.43 in (61mm)/hr Vertical cut: 1.02 in (26mm)/hr

The tests were carried out with a gel belt in place and the airborne level of lead contamination was measured using the NIOSH 7300 sampling method. Levels were below the lower limit of detection. No hearing protection for the workers was required since the noise levels recorded were between 74 dB(A) and 78 dB(A) significantly less than 84 dB(A) that mandates such protection.

The demonstration highlighted the following advantages (2):

- The RASP system is easy to operate
- Remote operations reduced operator fatigue – RASP works all day and doesn't get tired
- Operator exposure to hazardous conditions was minimised
- Secondary waste generation is relatively low and includes used grind wires, dry gel belts, dust and small amounts of airborne particulate.
- RASP technology is quiet

- The cutting temperature of the work piece was below 22°C at 25mm distance from cutting area
- RASP can cut through internal components fixed within organic foam (e.g. screwdrivers, pipe, pliers, electrical conduit, etc.)
- RASP can cut through stainless steel, lead shielding, foam, and Perpex
- Cutting activities are safe for both operators and environment
- Spread of contamination is minimal.

The lifetime of the wire can be in excess of 50 hours but in the trials with the glove box and tank at FIU, the wire replacement frequency averaged about 7 hours or one per shift. Stress released while cutting through flanges and glove rings resulted in pinching of the wire leading to breakage. The manual replacement of such a wire in the wire holder tools takes the operator only a few seconds per side. Automated replacement systems are possible.

SUMMARY

The minimisation of personnel risk and exposure to airborne contamination and radiation gives RASP very important advantages over existing cutting methods. RASP is a low energy, low temperature universal cutting process and can be used safely and successfully for cutting a wide range of items. Examples include glove boxes and other metal and non-metal components used in the nuclear industry such as plywood boxes, ventilation ducting, heat exchangers, scrubbers and storage tanks with internal structures and/or coatings.

The low energy cutting principle minimises the spread of contamination and does not result in significant temperature rise. It cuts through virtually any construction materials in any combination without the production of sharp edges and deployment is not geometry-dependent. The RASP has demonstrated that it is possible to segment glove boxes with a single tool and that segmentation can be performed with complex glove box internals.

The system is simple and easy to set up. It operates remotely with minimal manual intervention. This will result in a substantial reduction in exposure of personnel to radiation. The independent demonstration has verified the system's high flexibility for operating in various orientations and in confined conditions. Secondary waste was minimal and the absence of lead emissions indicated contamination control was excellent.

The ability to cut a wide range of materials and equipment such as gloveboxes complete with internal equipment minimises personnel exposure to hazards, the time spent in the direct vicinity of the glove box and results in faster progress. Filling the void spaces with organic foam prior to cutting fixes the position of internal items and equipment and ties-down contamination, thus minimising its spread of during handling of the cut pieces.

In summary, the RASP technology has proved to be a safe, economic and viable process for cutting up contaminated components such as glove boxes into manageable sized pieces, even where space it tight. This technology minimises the spread of contamination, exposure of personnel to radiation, generation of secondary waste and risk.

REFERENCES

1. F. W. LEDEBRINK and H.-U. ARNOLD, "Dismantling of Large Plutonium Contaminated Glove Boxes," 1999 Waste Management Conference, Tucson, Arizona, March 1999
2. C. APONTE, "Technology Assessment Program (TAP) Decontamination Technology Assessment Summary – Remotely-Operated Advanced Segmentation Process – RASP," Florida International University, June 2000

C596/013/2001

Copper cable recycling system – value gained from the recycling of copper from decommissioning projects

R MESERVEY and C CONNER
Idaho National Engineeirng and Environmental Laboratory
S ROSENBERGER
NUKEM Nuclear Technologies, South Carolina, USA
M J SANDERS
NUKEM Nuclear Limited, Dorchester, UK

ABSTRACT

The United States Department of Energy (DOE) continually seeks safer and more cost-effective technologies for the deactivation and decommissioning (D&D) of nuclear facilities. The Deactivation and Decommissioning Focus Area (DDFA) and the DOE's Office of Science and Technology (OST) sponsor Large Scale Demonstration and Deployment Projects (LSDDPs).

The expanding programme of D&D activities can generate hundreds of tons of electrical cables per facility consuming valuable resources such as wasted disposal site capacity and wasted value of the recycled copper. Increasing environmental concerns as well as economical pressures are key drivers for developing a way for recycling of the uncontaminated copper.

In November of 1999, the NUKEM Copper Cable Recycling System (CCRS) was demonstrated at the Idaho National Engineering and Environmental Laboratory (INEEL) as part of the LSDDP program. CCRS allows recovering and recycling the uncontaminated copper contained within the cables that have been contaminated externally.

A total of 13.5 tons of non-contaminated and surrogate contaminated cables in a wide variety of sizes were successfully processed during the INEEL technology demonstration. The assessment has demonstrated the mobility and flexibility of this new process.

The NUKEM CCRS was originally developed in Germany for use during the D&D of nuclear facilities. To date, the CCRS has successfully processed more than 200 metric tons of contaminated cables in Germany resulting in virtually 100% free release of copper in accordance with German standards.

THE SOURCES OF CONTAMINATED CABLES AND THE BENEFITS OF COPPER RECYCLING

A large quantity and a wide variety of contaminated cables are generated during the decommissioning of redundant nuclear facilities and smaller but significant quantities are generated during maintenance, upgrade and repair. Cable is typically copper and insulated with rubber, PVC, or polyethylene and may be rigid or flexible depending on the diameter.

It is unusual for the cable to be contaminated throughout its length and standard practice has been to either decontaminate it for unrestricted release, which can be very labour-intensive or packaging the material for disposal as radioactive waste; both methods incurring a substantial cost.

As national decommissioning programmes ramp up, a large increase in the volume of contaminated cable arisings will occur and a safe and efficient alternative processing method will be needed to save time and money.

An extensive review of the technologies that are available worldwide clearly indicated that a different approach was needed to meet the nuclear safety and handling needs because copper cable recycling processes used in the non-nuclear industries were not suitable use with contaminated material.

NUKEM set down a number of strategic goals for the new technology. These were to:

- Separate cable into contaminated and non-contaminated materials
- Maintain possible cross contamination of the non-contaminated copper below free release limits
- Accept all commercially used and different cable insulation types such as rubber, PVC, or polyethylene
- Be able to process cable sizes ranging from telephone size wire to multi conductor heavy gauge power cable
- Allow efficient assay of the processed materials
- Enhance handling to prepare contaminated materials for disposal
- Apply ALARA principles for safe handling and processing
- Design the system to be transportable.

This resulted in the successful development of the Copper Cable Recycling System (CCRS) and its deployment at several nuclear decommissioning projects in Germany. The ability was demonstrated to separate the feed stream into contaminated and non-contaminated end products, achieve a substantial volume reduction, granulate the cables into homogeneous materials for better assay and obtain a high purity of individual material fractions for increased value in recycling of the copper. The process is patented.

Recovery of the uncontaminated copper inside electric cables is often economically attractive. Benefits can include the resale of clean copper to the scrap metal market, eliminating the cost of disposing of clean copper as low-level radioactive waste, avoiding the cost of sending clean materials in disposal facilities for low-level radioactive waste (LLW) and unnecessarily depleting available space for disposal of LLW.

THE COPPER CABLE RECYCLING SYSTEM DEMONSTRATION PROJECT AT INEEL

The USDOE plans to decommission about 5,000 surplus facilities in due course. Many of these contain miles of copper cable that are either uncontaminated or contaminated only the external sheathing. (1)

The field demonstration of the NUKEM Copper Cable Recycling Technology is part of a larger series of demonstrations carried out under the Large Scale Demonstration and Deployment

Project funded by the Department of Energy Office of Science and Technology. The main purpose of the Demonstration and Deployment Project is to demonstrate innovative technologies on a large-scale basis in conjunction with Decontamination and Decommissioning (D&D) activities. The new technologies are compared against DOE's baselines and/or industrial baseline technologies. The results of the demonstration are summarised below and have been described in more elsewhere. (2).

The current baseline technology for the disposal of contaminated cables at INEEL is packaging in waste storage boxes for disposal at the Remote Waste Management Complex. The objective of the demonstration was to provide sufficient technical and economical benchmark information against this baseline.

Demonstration Goals and Objectives
The primary objective of the LSDDP is to identify existing technologies that might provide better solutions to DOE D&D challenges but are unproven in DOE applications. The second objective is to quantify the benefits by a side-by-side comparison of the innovative and baseline technologies. Direct comparison provides an opportunity to compare the new technology against the baseline and assess the potential benefits, which include:

- Lower cost
- Less exposure
- Improved safety
- Easier to use
- Faster

The demonstration project scope was to demonstrate the NUKEM Copper Cable Recycling technology and compare it with the existing baseline technology. The demonstration enabled the effectiveness of the NUKEM Copper Cable Recycling Technology to be assessed for providing separate output streams for uncontaminated copper, contaminated insulation and contaminated dust products. The NUKEM process was evaluated in an Innovative Technology Summary Report (2) for efficiency, reliability and potential for cost and schedule savings compared to the current baseline of direct disposal at the INEEL.

Description of the Technology Demonstration
The NUKEM Copper Cable Recycling System was transported to INEEL and set up to demonstrate the recycling of non-radioactively contaminated cable and cable to which a surrogate contaminant was applied.

The Copper Cable Recycling System is housed in two containers the size of standard size shipping containers. The system can process a wide variety of cables ranging from telephone wire with diameter of 5mm or less up to 65mm diameter with the full range of insulation materials. Set up time is 2 days or less. The CCRS produces three distinct waste streams:

- Clean copper
- Potentially contaminated insulation
- Contaminated dust.

A total of 13.5 tons of cable containing various sizes of copper conductors was available at the demonstration site for the demonstration. The total was divided into two parts to allow individual demonstrations using non-contaminated cable and demonstrations using surrogate contaminated cable.

System Setup and Operation

The demonstration project was carried out as if it was using radiologically contaminated material including generation of appropriate documents, personal protection equipment (PPE), monitoring, environmental and air controls and all associated personnel were assumed to be radiation worker trained to an appropriate level. Pre-job and post-job discussions were held with staff involved to note observations, concerns, and opinions of operators, sampling personnel, industrial and safety personnel, and other support personnel. The NUKEM Copper Cable Recycling Technology Operational Procedure was reviewed by all personnel involved in the set-up and operation of the equipment. Personnel were briefed with the procedure and followed all safety guidelines established in the INEEL Jobs Safety Analysis (JSA).

The key components of the CCRS consist of:

- a conveyor to feed the cable scrap into the system
- a pre-shredder
- a multi-blade shredder that cuts the cables into 3 to 5 mm nodules.
- an air separation table that separates the insulation/dust material from the copper.

Figure 1 shows the simplified process diagram.

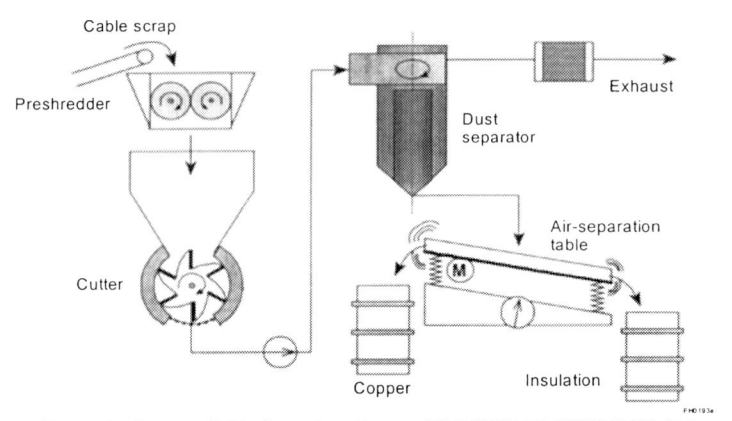

Figure 1: Copper Cable Recycling System PROCESS FLOW DIAGRAM

The cables were placed on the conveyor either as loose coils or as individual pieces, larger cables being cut into approximate 1 m lengths to make handling easier and provide a constant feed. The operators observed an ammeter and cable feed rate was controlled to keep within the normal current operating range.

The shredder granulates the copper wire, the filler/strengthening fibres and the insulation material covering the copper wire and during the process, most of the contamination is removed from the insulation by the powerful shearing action. An air-separation chamber separates the processed cable into contaminated and non-contaminated material fractions. Figure 2 shows pure copper and insulation as a result of processing. The non-contaminated copper wire is recovered from the inside of the insulation/covering, the less dense insulation and dust granules floating on a layer of air above the sieve, while the heavier copper fraction is separated by control of the sieve. Approximately 13.5 tons of copper cable was processed in the demonstration and data was collected throughout. Granulated copper and 'contaminated' cable insulation and dust materials were collected in 200 litre drums or plywood waste boxes. A large quantity of virtually pure uncontaminated copper was collected, which demonstrated that contaminated insulation material could be successfully separated from the uncontaminated copper wire, which can then be recycled or reused.

Figure 2: Before and After Processing

The insulation is collected separately as grainy, non-sticky material that flows easily and although most of the insulation leaving the CCRS is uncontaminated and can be free released, the remaining contaminated fraction is good for filling the void spaces in waste containers, thus allowing virtually free disposal and full utilisation of disposal space. If the free release of insulation must be maximised, the cable sheath can be wiped prior to processing in order to remove loose surface contamination. Tests conducted in Germany have demonstrated an approximately 20% increase of free releasable insulation if wiped prior to processing.

A three-stage process to prevent the release of airborne contamination filters contaminated dust generated by the shredding process. Re-binding of contaminated dust to the insulation material, is prevented by a special design feature of the NUKEM Copper Cable Recycling Technology. The dust filters and the off-gas filter are encapsulated and monitored for particulate build-up and cleaned or replaced as required. The dust was collected in standard waste boxes and is a fluffy material containing most of the contamination and with careful handling, it can be used as void filler or can be compacted. The CCRS operated very well over a period of approximately five days and twenty-one 0.3m x 1.2m x 2.4m waste boxes containing 12,250kg of copper cable were processed during the demonstration. Also generated were 9.5 individual 210 litre drums of clean copper (7,800kg), five 0.3m x 1.2m x 2.4m waste boxes of granular insulation and four 0.3m x 1.2m x 2.4m waste boxes of dust. The unit was capable of processing up to a maximum of six tons of insulated copper cable per day. An average throughput during the entire demonstration was 384kg per hour and this included the time required for setting up the equipment at the beginning of the day and shutting down and cleanup at the end of the day. The overall run-time for the demonstration was 32 hours.

The demonstration provided sufficient data to develop a cost benefit analysis for fair and independent comparison of the potential benefits of the NUKEM Copper Cable Recycling Technology over baseline technologies at end-user sites. These benefits can be re-evaluated by the end-user based on the changing value of copper and changing costs of alternative disposal methods.

Economical Benchmark
Based on the data collected during the demonstration, it is estimated that the innovative CCRS can save $350,000 if processing 50 ton of cables designated for the INEEL Remote Waste Management Complex (RWMC) landfill for disposal (see Figure 3) (2). Normally, such a quantity could be processed in less than 15 single shift processing days. With increasing project size these cost savings will significantly increase over the baseline.

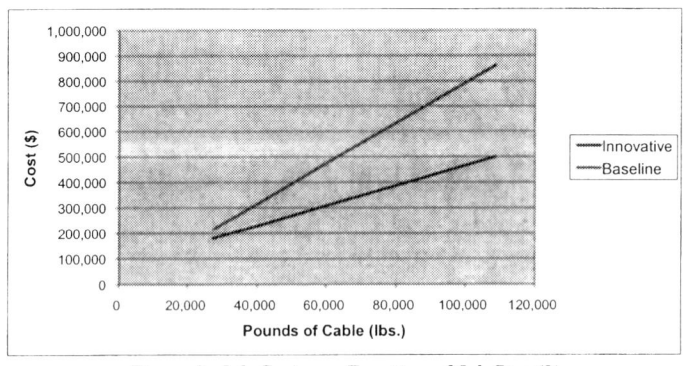

Figure 3: Job Cost as a Function of Job Size (2)

Large decommissioning projects of commercial power plants and DOE facility decommissioning will generate several hundred tons of contaminated cables and the cost savings will be clearly visible over the baseline technology of direct disposal. The handling, packaging, transportation and disposal costs for such projects can be significantly reduced using the NUKEM CCRS.

Lessons Learned During the Demonstration
Each time the CCRS is shut down and restarted, a total of approximately 2 hours is required to optimise the system and therefore it is more cost effective to run continuously for as long as possible; double shifts or around the clock operation will allow even more cost-effective operation than demonstrated.

A large portion of the CCRS costs is mobilisation and demobilisation, which remains relatively constant irrespective of the quantity of cable processed. Consequently, the CCRS compares differently with the baseline technology for different size jobs. For tasks involving more than 11,300kg of cable, the CCRS will be more cost effective than the baseline technology at INEEL.

CCRS PROCESS EXPERIENCE IN GERMANY

The CCRS has been used successfully to separate the contaminated insulation/dust from the clean copper in over 200 tons of cable in several D&D projects performed in Germany including of material from the decommissioning of a fuel manufacturing plant (primarily alpha contaminated) and commercial power plants (primarily gamma contaminated) and the process is therefore considered proven and mature.

Use of the CCRS at various German commercial power plants resulted in the unrestricted release (by German standards) of 74% of the initial cable scrap mass (100% copper, 65% of insulation). A total of 8% of the insulation was disposed at a regular landfill utilising the unrestricted release limits. Only 18% of the cables processed (insulation and dust) remained and had to be treated as radioactive waste. Figure 4 summarizes the typical ratios after processing 97 ton of material.

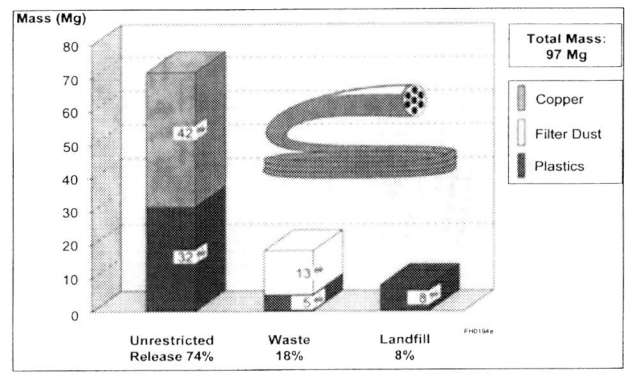

Figure 4: Processing Results for Contaminated Cables in Germany

This very efficient separation of contaminated materials resulted in significant cost savings for the decommissioning project (3). The copper was recycled and the cost savings due to reduced disposal costs and revenue from this sale of copper contributes to the reduction of decommissioning costs for the facility.

IS THERE A FUTURE FOR COPPER RECYCLING?

As worldwide, an increasing number of nuclear facilities reach the end of their 40-year design life, increasingly large volumes of decommissioning waste will be produced. Direct disposal without minimisation and recycle will deplete valuable waste site capacity and will result in increased decommissioning costs. The CCRS demonstration project has demonstrated that there are already solutions available that not only supporting limited budgets but also enable valuable material to be recycled for reuse. The CCRS has its immediate application at DOE sites where facility D&D is planned or underway. The mobile system is able to support relatively small and large copper recycling projects as the demand develops on the individual sites thus eliminating extra handling, storage, container and transportation costs. The CCRS has demonstrated the enormous potential to significantly reduce costs for many D&D projects throughout the United States.

FOOTNOTES

* References herein to any specific commercial product, process, or service by trade name, trademark, manufacturer, or otherwise, does not necessarily constitute or imply its endorsement, recommendation, or favouring by the U.S. Government, any agency thereof, or any company affiliated with the Idaho National Engineering and Environmental Laboratory.

REFERENCES:

1. BOSSART, MESERVEY, ROSENBERGER, JOBSON, "Copper Cable Recycling System Reuse of Copper from Decommissioning Projects," Paper Presented at Spectrum 2000, Chattanooga, Tennessee, September 2000
2. "Innovative Technology Summary Report, Copper Cable Recycling Technology," OST Reference #2958, June 2000
3. BERGMANN, DUEMPELMANN, ROSENBERGER, "High Cost Savings due to the Separation of Clean from Contaminated Concrete and Cables," Paper Presented at the Waste Management Conference 2000, Tucson, Arizona, March 2000

Learning from Other Industries

C596/024/2001

Market forces facing the management of environmental liabilities

R FOUQUET
Centre for Energy Policy and Technology, Imperial College of Science, Technology, and Medicine, London, UK

ABSTRACT

Growing concern for environmental quality is leading to substantial liabilities. In some industries, these liabilities may be large in comparison to the company's turnover and far in the future. This paper presents an economist's perspective on the environmental liability manager's role. It focusses on the incentives and constraints, which result from the external market for environmental quality and the internal market for funds. Taking these forces into account will be crucial for the effective management of these growing environmental liabilities.

1. INTRODUCTION

The growing concern for environmental quality is putting increasing pressure on polluters to take responsibility and pay for their actions. The scale and timing of environmental activities, either to avoid or remediate damage, varies according to the polluters' activities. For certain industries, such as those of energy production, the expenditure required to cover such liabilities can be substantial, suggesting that the consequences of incorrectly estimating them can be costly to individual companies. Also, because some of the activities will occur far in the future, associated variables introduce great uncertainty about these values. Considering the importance of environmental liabilities for both the energy and environmental markets, it is surprising so little research has been pursued.

The purpose of this paper is to present an economist's perspective on some of the issues facing environmental liability management. In particular, it will introduce the concept of the market for environmental quality, and associated markets, which are crucial for determining the context and the level of liabilities a firm will face (Section 2). Then, it will outline the liability manager's objective (Section 3) and the incentives and constraints existing within a company to achieving this objective (Sections 4-9). These ideas will be used to discuss possible future issues for environmental liability management (Section 10).

2. THE MARKET FOR ENVIRONMENTAL QUALITY

Services provided by environmental resources generate a demand for environmental quality. The public as a consequence is likely to be willing to pay to avoid damage to the environment and the services it provides (Pearson 1994). This suggests the potential for a market for environmental quality - as shown in Figure 1, where the social optimal level (i.e. society's maximum level of well-being) is identified by the meeting of the demand and supply curves.

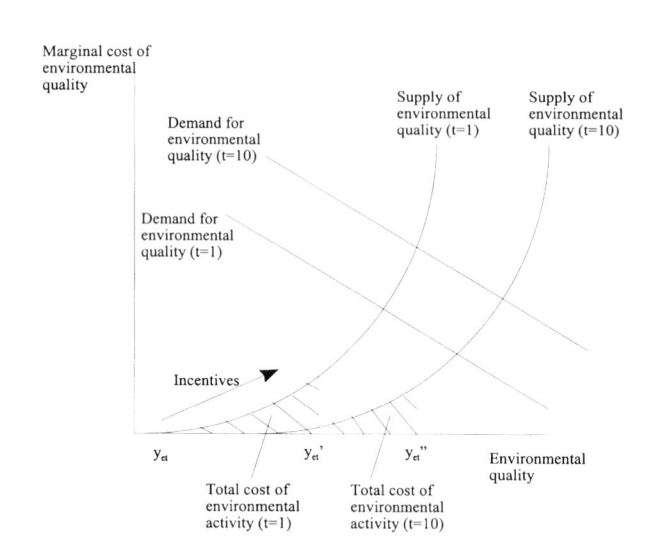

Figure 1. The market for environmental quality

The value of environmental services generates a willingness to pay for environmental quality, which can depend on, for example, income, the perceived cost of improving the environment, preferences, information, education and awareness, and political opportunities. This demand fails to be met directly through market mechanisms. That is, suppliers of environmental quality, mainly the polluters[1] (who reduce its quality), have few market incentives to avoid pollution and, therefore, choose to pollute as long as the net costs are positive. This failure of the market to lead to the social optimal generates the need for environmental regulation. Suppliers of environmental regulation, the politicians and their civil servants, respond to political and other incentives, that

---

[1]

The term 'environmental pollution' is used in its broadest sense. That is, any activity leading to an ecological response is considered pollution. A more legal definition would differentiate between an incident, which occurs within the company's boundaries and, therefore, not (apparently) harming others or causing externalities, and pollution, which directly causes external effects. The main reason for not making the distinction is that a cost is likely to be incurred for both incidents and legally-defined pollution, since both, at least in the long run, need to be cleaned up either through the rehabilitation/decommissioning of sites (in the case of incidents) or through remediation activities (in the case of pollution). So, both incidents and legally-defined pollution will be considered as sources of pollution.

the public and pressure groups provide, by introducing (or delaying - if discouraged) environmental legislation. Such legislation generally takes the form of environmental standards, technological requirements or market-based instruments (such as taxes, subsidies and tradable permits) and, in the case of energy-related environmental quality, leads producers and consumers in the markets for energy, energy technology and environmental pollution permits to incur costs associated with environmentally damaging activities. The resulting adjustments in these markets shifts the supply curve for environmental quality, thus, reflecting to some extent the public requests for environmental services. Shifts in the demand and supply curve for environmental quality will in the long run lead to changes in the levels of environmental quality and the costs of achieving them. These interactions are key determinants of a company's environmental costs and provide the wider framework within which the environmental liability manager acts.

3. THE OBJECTIVES OF THE ENVIRONMENTAL LIABILITY MANAGER

Faced with the prospect of environmental liabilities, a strategy must be developed for dealing with them. The key argument developed here is that, irrespective of resource allocation, the management 'team' will be responsible for an internal (ie. company-wide) market for funds to cover environmental liabilities. Every year, environmental activities will need funds to cover the costs, thus creating a demand. The manager needs to ensure the funds are available, making it a supplier. It, therefore, collects information required to estimate the cost-flow. There is also a role for generating (or producing) the funds, for example, by 'taxing' energy use and of enabling them to grow in the most 'risk-return'-appropriate way. The overarching objective of the manager, therefore, will be to ensure the correct flow and stock of funds - that is, *to minimise through time the discounted costs of error between the demand for and the supply of funds*. We will examine these responsibilities (information collection; fund demand analysis and management; production, management and supply of fund; and public information provision) below.

4. THE COSTS OF FLOW AND STOCK ERRORS

Having suggested that the manager's main objective is to minimise the cost of error between demand and supply, it is necessary to consider what are the costs associated with flow and stock errors. Flow errors, here, refers to a company not having the correct funds available to pay for a liability in a particular year; stock errors describe the same problem over the entirety of the liability (i.e. several years). Looking at flows first, there are asymmetries between the cost of excess supply and demand in any particular year. If excess funds are made available, however, they cannot be invested elsewhere. There is generally, an opportunity cost of not investing the funds. This cost is equal to the return that would have been earned on the excess funds, or the difference between higher returns from not being able to touch them and lower ones from making them available. The return could be equal to the interest rate or higher, if invested in certain higher risk securities.

Insufficient supply will generate different costs. Here, it is assumed that in a particular year insufficient funds were made available, but that within the stock of funds there is sufficient funds to cover these costs - perhaps they are tied up in securities that cannot be traded. There is a lack of liquidity. Traditional economic theory argues that it does not matter where funds come from, provided the investment will eventually generate (discounted) funds - the value of the company is unchanged by the source of funding. Recent analysis indicates that it does matter - borrowing from external sources reduces the value of the company. For example, nobody wants to lend to a company with a large debt burden, possibly because it may use some of the money to pay back past debts or because it is heading towards bankruptcy. This increases the risk to the lender and, therefore, the cost of borrowing money. Thus, there are costs associated with borrowing funds externally (Froot et al 1994).

While it is a little ambiguous, it is probable that insufficient supply will be more costly than excess supply. So, the sign of the error matters. The timing of flow errors is also important. Errors now are more costly than ones in the future - because the future is discounted and because mistakes may indicate a need to reevaluate expectations, which are costly (as is discussed in the next part). Also, the size of errors are non-linear in their cost function. Small excess supply mistakes in one year, say, will probably be balanced out by minor excess demand in the future; they are not likely to influence stock errors. Large errors are less likely to be balanced out in the future; so, they lead to both an increase in flow costs and in the risk of stock errors. Thus, the sign, timing and size of errors are influential in the cost of flow errors.

The cost of stock errors are potentially more important. If the stock of funds for environmental activities turn out to be excessive, then naturally the manager is delighted. Nevertheless, there is a cost to the company as a whole. In a similar way to flow costs of excess supply, there is an opportunity cost associated with having set aside provisions for environmental liabilities, when they could have been invested in the company elsewhere and probably with higher returns.

Insufficient provisions may be more dramatic. In the case of environmental liabilities, there tends not to be an actual direct return on investments. This means that contrary to flow errors, the investment will not generate future funds. There is generally, however, an avoided cost related to the fines incurred from not meeting certain environmental standards. The firm needs to detract funds from other investments that are likely to lead to direct returns. So, substantial costs could be incurred.

The sign of errors will be important to the costs incurred and, again, excess demand is likely to be dearer than excess supply. Because error in the stock of provisions effectively appears in the final year (which probably only occurs when the company stops operating), there is no time dimension to the cost of stock errors. The cost of flow errors will, however, reduce the total funds available for covering environmental (rather than financial) activities, increasing the risk of stock errors, and must be taken into account when balancing out stocks of demand and supply. The timing of the completion of all environmental activities may severely affect the value the manager (and possibly the company) places on the costs of errors; in some cases, such as in the nuclear industry, the completion of activities could be over one hundred years into the future, therefore, the manager will be long gone and may place little value on the cost of errors.

5. DEMAND FOR FUNDS: ESTIMATING ENVIRONMENTAL LIABILITIES

The ability to estimate the future costs depends on the complexity of the site, the remediation technology, uncertainty over the cleanup standards, the number and financial viability of other companies involved in the clean-up activities, insurance coverage, the protracted period over which negotiations with authorities, other companies and insurers occurs, and the availability of records to show the company's impact on the environment (Barth and McNichols 1994).

A problem associated with liability management is the difficulty and cost of estimating the expenditure involved. When deciding on the calculation exercise, numerous factors need to be taken into account. First, much of the calculation cost is the result of paying for labour (or human capital) services; these include hiring accountants and actuaries, and consulting engineers, chemists and environmental assessors. In the future, these costs are likely to rise in real terms. So, the timing of estimation exercises is important. Furthermore, to some extent, many of the activities are uncertain because it is not clear what remediation or decommissioning will be required and at what rate; they depend on, say, future legislation, which are likely to be tighter, although to what extent would be difficult to anticipate. Third, to improve the accuracy of estimates, the team will need to enter into each aspect of the activities in more detail, increasing the costs of calculation. In some cases, particularly for capital and decommissioning expenditure, the costs of calculating can be equivalent to 5-10% of the actual estimates and, therefore, non-negligible. As a result, estimates cannot be done on an annual basis, but must be done selectively. Thus, overall, the liability manager will have to make a trade-off between the cost of reducing the uncertainty and the cost of facing uncertainty (and potentially of errors between demand and supply of funds), both in the depth of analysis and the timing of the exercises.

So, faced with such uncertainties, the management team has to make assumptions about the level of investment, expenditure and, therefore, liabilities for the task of pursuing environmental prevention or clean-up and decommissioning activities. These can be represented by the (production) relationship between the amount of environmental or decommissioning activity, y_{et} in Figure 1. The total amount of environmental 'production' for a particular company is the sum of activities in each year. Because, in general, legislation imposes the standards of environmental prevention or clean-up and decommissioning, the total level of environmental production is fixed. This means that the company cannot adjust the level of environmental output to maximise profits. Its only flexibility is in altering the different amounts of inputs used to minimise costs, where the costs or expenditure are the product of the quantity of inputs used and the unit price of each input - taking account of the fact that prices of inputs can change through time. So, annual environmental liabilities is represented in Figure 1 as the area under the supply curve between y_{et} and y_{et}'. In addition, legislation determines by when activities have to have been completed. The difficulty for the company and manager in seeking to minimise costs and to estimate the total costs (in order to manage the liabilities), therefore, is the uncertainty about amount of total environmental activity and decommissioning required (ie Y_e), about the year in which the activity needs to be completed, and about the future prices of each of the inputs.

While the main responsibility of the liability manager will be to gather information and estimate the demand flow and stock of funds, he/she can also seek to reduce demand. Because the manager will have an overview of the costs of environmental activities, he/she might discover inefficient resource allocation and possibly ways of reducing expenditure or altering the organisational structure.

6. PRODUCTION OF FUNDS

The environmental expenditure required in the long run can be a large sum, even discounted back to the present. This will be especially true for companies with substantial decommissioning costs. As a result, a decision has to be made about whether the company can cover costs out of annual expenditure (known as 'PAYG' - pay as you go) or whether funds need to be set aside (called 'fully funded assets' amongst pension fund managers), or some combination of the two. PAYG schemes generally have lower set-up, running administrative costs. On the other hand, PAYG means that current revenue needs to cover current liabilities, thus, failing to reflect divergences between revenue and liability over time. This would be especially important in companies that might experience declining activity or anticipating rising liabilities. Furthermore, fully-funded schemes tend to be legally protected even in cases of bankruptcy. Consequently, PAYG schemes tend to be useful for small, highly predictable and near term needs, and less appropriate for bigger, uncertain and long-term liabilities. It should also be noted that PAYG is not consistent with current UK accounting practices in UK GAAP and FRS12 (Blake 1995). Even if the manager is not solely responsible for finding sources, he/she will be important in advising the company from where funds should be generated.

The next decision is about the source of the funding. In most cases, it would come from the company, either as a cost (most probably) or out of profits. In other circumstances, government may provide contributions. This will occur if there is a social value associated with ensuring that the company manages to pay for the environmental activities.

Since most of the funds are going to come from the company, there may be tensions within it about setting aside money. While internal policy and accounting arrangements do provide guidelines for selection of fund uses, each department within a company needs funds for the numerous projects, and (to some extent) internal politics may influence the size of the budgets. The choice for the company and its central funding authority (such as the finance department or the Executive) is between investing funds in current projects, presumably with rates of return, or setting aside money for future activities; clearly, the temptation is to discount any future projects. If the department responsible for environmental liability management does not have sufficient power and has not agreed the level of funds to be segregated, it may not manage to accumulate sufficient funds. This may become a serious problem later. This will involve the costly generation of new funds later on if an insufficient quantity was initially set aside.

7. SUPPLY OF FUNDS: PROVISIONS FOR ENVIRONMENTAL LIABILITIES

The fund manager, presented with company assets, faces the task of investing them to ensure that there is the appropriate flow and stock of funds to cover the required environmental activities. His/her aim has been defined as achieving maximum investment return which is consistent with the security of the fund. The manager must take account of the risk-reward profile, the time horizon, any mandate restrictions (for example, limiting investment of certain securities for political, ethical, etc., reasons), and the resources. These factors will be key to determining the initial structure of the portfolio, the adjustments made to it, and its performance. Because of the nature of environmental activities, there is a need to maximise the value of the assets to meet specific but uncertain liabilities. The need to ensure a specific level of funds are available through time is likely to increase the fund manager's aversion to risk and influence his/her strategy.

Because of fluctuations in economic variables, including interest rates, exchange rates or commodity prices, in certain years the company will generate lower revenue than others. If the company was planning substantial investment, this lack of cash flow can cause concern for managers, especially since borrowing funds from external sources while waiting for higher revenue years is seen to reduce the value of the company. Similarly, changes in economic variables may alter the need for investments, creating a different internal demand for funds.

Hedging (or risk management) is needed as a strategy of ensuring that funds are available when needed rather than when the cash flow is there. The main way of reducing the risk of shortfalls in the supply of funds in any year is to invest in financial services that will balance out potential drops in cash flow from fluctuations in economic variables. And, since the key role of the liabilities management team is to allow the internal market for funds to run smoothly and minimise the costs of error between demand and supply, it needs to supply the demanded amount every year. Thus, risk management is - not so much about ensuring that the total future liabilities will be covered as - about ensuring that annual (or flow of) liabilities are covered every year.

In addition to managing risk associated with supplying funds in the correct year, there is the importance of managing the risk of having the incorrect total funds. The manager should have an awareness of the risks involved. This will have been done as part of the information collection process. The information will indicate the range of total liabilities estimates and the key factors that will influence the sum.

Since the costs of insufficient funds are likely to be greater than the costs of excess funds, management may prefer to assume a slightly more conservative (i.e. higher) total liability estimate, deciding the target level of assets to accumulate. In addition, some form of hedging (in relation to stock rather than flow) may be undertaken on the part of the fund manager. If it is clear that certain factors (such as technological development or environmental legislation) will be key to influencing the total liabilities, the manager may be choose to invest in companies that will balance out potential changes. So, the level of assets (i.e. total supply) will rise when the level of liabilities rise (ie total demand). The difficulty will be in ensuring that supply rises sufficiently to cover changes in demand. Particularly in the case of legislation, which can have dramatic impacts on liabilities, the return from investment in counter-balancing securities may not be enough.

8. ISSUES RELATED TO THE ENVIRONMENTAL LIABILITY FUNDS MARKET

The argument laid out above is that there is an internal market for funds. Some brief comments could be made about the nature of the market and its implications. First, there is no competition - there is one supplier (the fund manager). In large companies, there will be numerous environmental activities to be completed all demanding funds. The market structure could be described, therefore, as one of monopoly, facing numerous consumers. Markets with monopolies are notoriously inefficient in their allocation of resources. It suggests a role for regulator, with the power to encourage efficiency and penalise poor allocation. If all the functions (information collection, demand, production, supply, etc.) discussed above are separated, this would be the role of the liability manager.

Second, more generally, since there are likely many collectors of information, many demanders for funds, many producers of funds, many suppliers of funds, each with his/her own objectives and incentives, the liability manager will need to coordinate all these objectives to promote the overarching one of minimising the costs of error. In particular, the demanders have no costs associated with borrowing the funds and the suppliers have no compensation for providing funds. Such a situation does not signal to the consumers the scarcity of resources and to producers the value of resources, which is a powerful way of using individual behaviour to achieve the overarching objectives.

Third, since much of the issues take place over a long time scale, markets tend to discount future activities at high rates. It is unclear at what rate such an internal market would discount rates; again, many agents may not be at the rate that would collectively (ie for the good of the company) be chosen. Fourth, markets are eternally in evolution - with changes in institutional structure, the centres of power and the incentives. The regulator must learn to anticipate the long term evolution to drive down the costs of error from imbalances between supply and demand.

Finally, it is not clear what the implications of the outcome of this internal market will be for the energy and environmental markets. As suppliers of both energy products and environmental quality, the efficiency of the internal fund market will influence the cost of energy and achieving environmental standards. There may also be distribution effects to consider.

9. INFLUENCES ON ENVIRONMENTAL LIABILITY MANAGEMENT

It is interesting to consider briefly how different factors influence the environmental liability management behaviour. These should be useful in understanding how different industries manage their liabilities.

Here, the importance of liabilities relative to turnover (or other measures of economic activity) on liability management is being considered. Do changes in the relative stock of liabilities affect the flow of liabilities, the cost of flow or stock errors, or the risk of flow or stock errors? It would seem that a large relative stock of liabilities will increase the cost of stock errors to the company. Also, it would seem that companies with large relative environmental liabilities will increase the level of resource allocation (particularly, financial expenditure and priority within the organisational structure of the company) for their management, as well as the strategy chosen. So, high stocks may mean that there is a better management team, which may reduce the risk of stock and flow errors. There may also be economies of scale related to the management of liabilities, reducing further the risks. On the other hand, if there are many different types of environmental activities within the company, the management of all these agents within the fund market will be increasing difficult and complex as the size of liabilities grows.

In a similar way, we can ask whether changes in the flow of expenditures affects the stock, the costs of flow or stock errors, or the risk of flow or stock errors? Delays in expenditure reduce cost of errors, since activities further into the future are discounted. Delays also reduce the risk of stock errors, as management have more time to adjust provisions to suit requirements and may gather more information about the cost function. There may be certain path dependency issues that may lock the company in to the use of certain technologies or ideas; that is, the timing at which investments have to be made may have an influence on strategic decisions made, which will have repercussions on future costs and risks of errors.

An important development in the market for environmental quality is the probable increase in the demand and, thus, in regulation. As has been discussed earlier, this is likely to increase both the stock of liabilities and bring forward the flow of liabilities. So, there is likely to be an increase in the need for environmental liability managers with increasing pressures and better teams. This could lead to a more efficient abatement of environmental pollution, thus, reducing the cost of achieving standards of environmental quality. This reduction in cost may spurn on a desire for higher standards.

10. FUTURE TRENDS IN ENVIRONMENTAL LIABILITIES

This section considers some of the trends in environmental markets that may influence the environmental liability manager, focussing mainly on demand-side forces. Total activity and timing are determined mostly by future legislation. Since the evolution of legislation is even more difficult to anticipate than economic factors, little can be predicted. One probable outcome, as discussed earlier, is that the demand for environmental quality will increase, putting pressure on politicians to impose higher standards. Thus, it is likely that the overall level of future environmental and decommissioning activity will be required to rise, which will increase the number of inputs employed and probably total expenditure (See Figure 1).

The future requirements for assessing in which year activities must be completed is ambiguous. Increasing environmental demands are likely to bring forward the date. On the other hand, there are tendencies for policy makers to seek to put off the date. This is partly because other players (e.g. industry lobbies) will seek to delay the timing of activities, as they will tend to reduce the costs, and because a better understanding of problems and technological improvements will make it easier and cheaper to do the job later.

There are likely to be substantial differences in future input price trends. Labour costs will probably rise. Real wages have been rising for the last two hundred years (apart for isolated periods, such as the 1930s). There is no reason to expect the pattern to stop, especially as productivity is likely to continue improving. The only declining force is the probable reduction in income taxes, as governments increasingly impose charges on bads (like pollution) rather than goods (e.g. labour).

Capital costs will probably tend to fall. As more environmental and decommissioning projects are taken-on, more capital will be manufactured; this should enable substantial economies of scale to be achieved and, therefore, a reduction in the unit cost of each machine. These reductions will be accompanied by an improvement in the technology used; each machine will achieve a higher level of environmental services. This also means that if a company delays the investment in environmental technology it can achieve a similar level of service at lower cost. On the other hand, if new, higher standards are imposed, new technologies will need to be developed to achieve these standards, which will raise the costs. So, we are likely to see a marginal capital costs of abatement curve with step jumps, but at each step following a declining trend. If, however, capital costs do decline, there will be a tendency for managers to substitute capital for labour. As capital takes a greater share of the work, this will put more downward pressure on total expenditure.

While there have been periods of upward pressure on energy costs, historically the trend in individual fuels is declining. And, while the real average energy price may have risen over the last hundred years, this hides a switch towards fuels more effective at delivering energy services. This trend is likely to continue. So, although governments are increasingly charging the full social costs of using fuels and, therefore, energy prices will tend to rise, the energy expenditure will most probably fall.

Needless to say, it is very difficult to estimate future long term expenditure in site remediation, rehabilitation and decommissioning. The previous analysis argues that three key uncertainties are the level of activity required, the date at which all work must completed, and the expenditure (i.e. use multiplied by unit cost) on inputs. The level of activity required will probably rise, the date may be brought forward or delayed, and real unit costs of inputs are likely to fall. This means that there are pressures driving up and pushing down total costs.

11. CONCLUSION

This paper sought to present an economist's perspective on environmental liability management. The manager's overarching objective is to be responsible for the market for internal funds for environmental activities and, therefore, to minimise the costs of stock and flow errors between demand and supply. The demand for funds is influenced by legislation (determining level and timing of activities), costs of inputs and the efficient use of those inputs. Abilities to collect information and estimate these demands will be crucial. On the supply-side, within this company-wide market, there are several demanders for funds, several producers of funds, several suppliers of funds. Each agent either requiring, generating or distributing funds will have his/her own objectives and incentives. If the liability manager is the sole agent in each case, there may be some ability to coordinate all these objectives to promote the overarching one of minimising the costs of error. If he/she is not the sole agent involved in this market (as is likely), then it is possible that the market outcome (ie the flow and stock of funds and the costs of error between demand and supply) will not satisfy the liability manager's overarching objective. The manager, therefore, needs to act more as a market regulator, trying to understand the objectives and incentives of each agent involved and use those to minimise the costs of error.

With the growth in concern and legislation on environmental quality, companies' associated liabilities are increasing. For some firms, such as those in the nuclear industry but many others as well, they are large and far in the future. They will be a burden on the value of the company. The costs of mismanagement and benefits of effective management of these liabilities will grow. This paper argues that effective management must start with taking account of the market forces that influence the demand and supply of related funds.

Relating respectively to this overall market, the demand and the supply of these environmental liability funds, a few issues to consider are:
- what indicators might be used to measure liability management performance?
- how will government deal with retroactive liabilities? Is there a role for government to set up a national liabilities agency?
- will liability managers have greater control of the factors influencing the production and supply of funds?

ACKNOWLEDGEMENTS

I thank British Nuclear Fuels (BNFL) plc for funding this research and Robin Sellers, David Warner, Steve Rixon, Bill Wilkinson, Peter Pearson, Gordon Mackerron, Dennis Anderson and an anonymous referee for valuable comments throughout the project.

REFERENCES

Barth M. and M.F. McNichols (1994) 'Estimation and market valuation of environmental liabilities relating to Superfund sites.' *Journal of Accounting Research* 32 (supplement) 177-209.

Blake, D. (1995) *Pension Schemes and Pension Funds in the United Kingdom.* Clarendon Press. Oxford.

Froot, K.A., D.S. Scharfstein and J.C. Stein (1994) 'A framework for risk management.' *Harvard Business Review* (November/Decmeber) 91-102.

Pearson P.J.G.(1994) 'Energy, externalities and environmental quality: will development cure the ills it creates?' *Energy Studies Review* 6(3): 199-216.

C596/007/2001

Health and safety aspects of decommissioning offshore facilities

M LUNT
Offshore Division, Health and Safety Executive (HSE), London, UK

ABSTRACT

Safety considerations during decommissioning need to address the identification, evaluation and consequences of "soft" issues arising from working relationships and contractual drivers, as well as more commonly discussed "hard" issues (such as techniques for cutting, lifting and transportation of components). Risk is greatly reduced if all parties involved are fully aligned and aware of their responsibilities / accountabilities.

The severity of many decommissioning hazards is also related to the efficiency and duration of offshore activities, particularly for weather sensitive operations. Comprehensive onshore proving of equipment prior to deployment in the field along with comprehensive contingency planning will assist in minimising delays, optimising schedule retrieval should problems be encountered and reducing the exposure of personnel to risks.

The paper presents the regulatory framework governing offshore decommissioning, discusses safety aspects associated with the current regime and outlines lessons learnt worldwide by oil companies / contractors when performing decommissioning operations.

1 INTRODUCTION: THE REGULATORY FRAMEWORK

The oil and gas exploration and production industry is highly regulated. An extensive worldwide regulatory framework governs the removal and disposal of offshore installations. The framework includes global conventions and guidelines, regional conventions and national laws.

1.1 International

The UK's international obligations on the decommissioning of offshore installations have their origins in the United Nations Convention on the Law of the Sea 1982 (UNCLOS). The Convention, which came into force in 1994 and was ratified by the UK in 1997, states that "any installations or structures which are abandoned or disused shall be removed to ensure safety of navigation, taking into account any generally accepted international standards established in this regard by the competent international organisation". The competent international organisation for this purpose is the International Maritime Organization (IMO) which in 1989 adopted the IMO Guidelines setting out the minimum global standards for removal of offshore installations. The IMO Guidelines concentrate on removal, and particularly navigational safety, but contain no reference to dumping or disposal.

1.2 Regional

In 1992, a new convention (the Convention on the Protection of the Marine Environment of the North East Atlantic: the "OSPAR" Convention) was agreed. This regional convention, which applies to specific sea areas including the North Sea and parts of the Arctic Ocean, replaces the 1972 Oslo Convention on the Protection of the Marine Environment by Dumping from Ships and Aircraft and the 1974 Paris Convention on Prevention of Marine Pollution from Land-based Sources. It came into force in 1998.

At the First Ministerial Meeting of the Contracting Parties to the OSPAR Convention in July 1998, a new regime for the decommissioning of disused offshore installations was established. Ministers adopted a binding decision (Decision 98/3) prohibiting the disposal of offshore installations at sea. The key points of Decision 98/3, which took effect from 9 February 1999, are:

- All dumping of platforms at sites remote from E&P activities is banned.

- All toppling of platforms in-situ is banned.

- The topsides of all installations must be returned to shore.

- All installations with a jacket weight less than 10,000 tonnes must be completely removed for re-use, recycling or final disposal on land.

- There is a facility for seeking derogation from the main rule for the "footings" of large steel jackets weighing more than 10,000 tonnes (but only where removal is prohibited due to safety, environmental, technical and economic considerations).

- There is a facility for seeking derogation for concrete gravity based and concrete floating installations, as well as concrete anchor-bases.

- All installations emplaced after 9 February 1999 must be completely removed.

- The provisions of Decision 98/3 do not apply to pipelines.

Before any decisions on derogation, Contracting Parties are required to have carried out a detailed comparative assessment of the position and consulted with the other Contracting

Parties, which could involve the holding of a special consultative meeting to address opposing views. In exceptional and unforeseen circumstances an offshore installation may be disposed of at sea or left in place as a derogation from the main rule. However, this option will only be considered if it can be demonstrated that, due to structural damage, deterioration or some other cause presenting equivalent difficulties, there are significant reasons why disposal is preferable to re-use, recycling or disposal on land. Any permit issued by a Contracting Party allowing a derogation must specify the terms and conditions of the derogation, including any necessary subsequent monitoring of the parts left at sea, details of their owner, and specifying the person liable for meeting claims for future damage caused by those parts.

The next OSPAR Ministerial Meeting is planned to take place in 2003, when there is provision for Decision 98/3 to be reviewed in the light of experience and technical developments. However, it seems unlikely that any large steel structures will be decommissioned in the UKCS or the derogation facility within the Decision implemented before 2003.

1.3   National
Legislation has been enacted in the UK to ensure that the operator of an installation has to apply to the relevant government authorities for a specific licence to dispose of any installation.

The abandonment of offshore installations and pipelines is controlled through the Petroleum Act 1998 for which the DTI has responsibility. Before an installation or a pipeline can be abandoned, the owner must obtain DTI approval. The DTI has developed and issued Guidance Notes for Industry covering the "Decommissioning of Offshore Installations and Pipelines under the Petroleum Act 1998". In the Guidance, it is recommended that discussions between the owner / operator and the DTI should commence well ahead of the forecast cessation of production. In the case of a large field with multiple facilities, these discussions may commence 3 years or more in advance. Following this preliminary discussion period, the operator must submit a Decommissioning Programme to the DTI. It may take 9 months from submission of the first draft of the Programme until approval of the final draft is obtained from the Secretary of State (or 18 months for a potential OSPAR derogation case).

In most cases the general rule under OSPAR Decision 98/3 will apply and the Decommissioning Programme will provide for re-use, recycling or disposal on land. A statement is required within the Programme to demonstrate how the principles of waste hierarchy are to be met and to show the extent to which the installation (including the topsides and materials contained within the platform) will be re-used, recycled or scrapped. An Environmental Impact Assessment is also to be included within the Programme.

For more complex cases, and in particular where the owner / operator is seeking a derogation to Decision 98/3, a detailed comparative assessment of the options is required and this will also form part of the Decommissioning Programme.

In addition to the above legislation, following the Piper Alpha disaster in 1988, the central recommendations of Lord Cullen's Report on the Public Inquiry into the disaster were implemented in HSE regulations. Risks from all offshore work activities must be assessed under the Health and Safety at Work Act (HSWA) 1974 and the Offshore Safety Act 1992. Employers must make suitable assessment of risks. The relevant regulations are as follows:

- Offshore Installations (Safety Case) Regulations 1992: SCR (including requirements for an Abandonment Safety Case).

- Offshore Installations and Pipeline Works (Management and Administration) Regulations 1995: MAR (relating to management of offshore installations - fixed and mobile, but not subsea).

- Offshore Installations and Wells (Design and Construction etc.) Regulations 1996: DCR (relating to integrity of installations).

- The Pipeline Safety Regulations 1996: PSR (including decommissioning of pipelines).

Mindful of the obligations of HSWA 1974, the principal legislation governing safe decommissioning offshore is SCR 1992 Regulation 7 which requires the submission of an Abandonment Safety Case (ASC). The ASC must be submitted at least six months before decommissioning can begin and demonstrate that:

- Management systems are adequate
- Audit arrangements are adequate
- Major hazards have been identified
- Risks have been evaluated and measures designed to reduce risk to as low as reasonably practicable (ALARP).

The sequence of decommissioning and abandonment events must be clearly explained in the ASC. The abandonment programme should follow a logical sequence, taking account of the progressive reduction in the availability of plant and equipment on the installation. Major hazard risks should have been identified and assessed, including hazards that may arise at a later stage of the decommissioning process. Consideration should also be given to maintenance and verification during the time lag between the end of operations and the start of the abandonment process.

Well abandonment policies will have been described in the Operational Safety Case. Well operators must ensure the safe physical condition of wells at all stages of the cycle, from design and commissioning through to abandonment.

Descriptions should confirm the extent and availability of safety systems during the abandonment process. The operational status of safety-related plant, equipment and systems should be summarised. Sufficient detail should be provided to show that the management systems can be implemented.

In addition to SCR, other offshore legislation has a bearing on decommissioning. PFEER (Prevention of Fire and Explosion and Emergency Response) 1995 includes requirements for duty holders to protect their personnel from the effects of fire and explosion, and to have effective evacuation, escape and rescue (EER) arrangements. The adequacy of such arrangements will require assessment before decommissioning can begin.

Regulation 16 of DCR requires that "the duty holder shall ensure that an installation is decommissioned and dismantled in such a way that, so far as is reasonably practicable, it will possess sufficient integrity to enable such decommissioning and dismantlement to be carried out safely". DCR 1996 also require new designs to take into account the safety of eventual decommissioning and dismantling, ensuring that structural integrity is maintained during such operations.

2   DECOMMISSIONING ISSUES

Since the first offshore installation in the Gulf of Mexico in 1947, the industry has designed, built and installed more than 6800 structures on the continental shelves of some 53 countries. By far the greatest number of these structures (about 87%) are steel jacket platforms in water depths of less than 75 metres. Steel jacket platforms in deeper water account for a further 9% of the total. The remainder consist of concrete and steel gravity base structures, floating production systems, tension leg platforms and compliant towers (see Table 1).

Very few structures are alike as each installation is site-specific, designed and built for a particular function, location, water depth and environmental envelope, and subject to the technology available at the time. The remoteness from land combined with deep water and harsh environmental conditions make some of the North Sea and North-East Atlantic offshore structures among the largest and heaviest in the world. Although the North Sea contains less than 10% of the world total of offshore installations, it has been estimated that decommissioning costs will be about 60% of the total because of the weight and complexity of the structures.

As well as site-specific factors, the geometrical diversity of steel jacket structures has also been significantly influenced by the original installation method (lift installed, barge launched or self-floater) and the installation date (dictating the maximum heavy lift). Basic topside configurations have been influenced by the size / weight of process and utility equipment or modules and by the craneage capacity available at the time of installation – the three basic topsides supports types being module support frame (MSF), cellar deck or integrated deck (ID).

A multitude of factors, and combinations of factors, therefore have an impact on the decommissioning method / approach, the associated risks and the costs for steel jacket platforms. Reaching a decision on the best decommissioning process for each installation and pipeline is a complex, rigorous process demanding care in order to balance protection of the environment and other users of the sea with health, safety, technological and economic considerations during decommissioning activities. The HSE is required to ensure that all risks to personnel associated with the decommissioning process are adequately considered, evaluated and controlled by the duty holder to as low a level as is reasonably practicable.
Until fairly recently, the offshore industry anticipated that many large steel structures would be disposed of in-situ by toppling using remotely controlled cutting techniques to produce a toppling mechanism at relatively low risk to personnel. However, OSPAR Decision 98/3

means that large structures within offshore areas covered by the Convention cannot be toppled and all but the base section of large structures with pile bottles (or "footings") must be removed. Inevitably, this will result in the closer proximity of personnel during the most critical decommissioning operations and a re-evaluation of safety priorities in the light of OSPAR has been imperative.

To date, only 30 or so installations have been removed from the North Sea (see Tables 2 and 3). As worldwide, the majority of installations decommissioned so far in the North Sea have been either small steel platforms, floating facilities or subsea systems. All (with the obvious exception of Piper Alpha) have been totally removed to shore for recycling or re-use. There is very limited experience to date of removal and disposal of large steel installations.

The first installation to be removed from the UK Sector was the BP West Sole WE Platform (in 1985), a small satellite installation (deck 180 tonnes; jacket 200 tonnes) situated 75km East of the Humber in 26 metres of water. Similarly, in 1988, the first platform removed from the Dutch Sector was a small satellite (Wintershall K13D: topsides 870 tonnes; jacket 700 tonnes) also in 26 metres water depth. The Odin Platform, removed from Norwegian waters in 1996-1997, is the only fixed platform of significant size and complexity so far removed from the North Sea – it was a gas facility located in a water depth of 103 metres, with a jacket weight of 6200 tonnes and a total topsides weight of 7600 tonnes.

There are upwards of 500 installations in the North Sea, of which over two-thirds are in shallow waters and must be removed under international guidelines for navigational safety. Under OSPAR Decision 98/3, the remaining one-third will also require complete removal with possible derogation applications for 41 "large" steel jackets (leaving "footings" in place) and 29 concrete structures (remaining in-situ).

Many of the safety issues arising during decommissioning programmes are associated with cutting and lifting operations. During decommissioning, lifting will take place either within the 500-metre exclusion zone or at the place of reception inshore / onshore. Offshore, if a large jacket is to be cut into sections, the upper section(s) must be supported during cutting operations to prevent collapse into an irretrievable position. This support is likely to involve a manned lift vessel and / or externally attached buoyancy. If explosive cutting were to be adopted, the crane vessel may have to stand away during the detonation and alternative support may need to be provided (and designed to be unaffected by explosion shock waves). Lifting onshore, within the sheltered confines of a harbour, generally poses fewer problems. However, if quayside access is limited and double handling of components is necessary, the potential risk increases.

The weight of the lifted item should always be carefully determined at the planning stage, but this may not always be possible accurately (eg. the pull needed to extract a pile). There are limits on the weight of sections that can be safely lifted. During an actual lift, the theoretical lift capability will be significantly affected by safety and operational constraints, the size and structural integrity of the pieces to be lifted, vertical and horizontal clearances, etc. It may be considered too risky to place the removed piece onto a heaving cargo barge, hence it will have to be placed onto the crane vessel deck. There are also limitations on the weight that can be lifted underwater when slings have to be attached below the sea surface. Anchoring requirements (e.g. proximity to other facilities, subsea wellheads, pipelines, etc.) will also affect vessel positioning.

Historically, there have been some notable failures during heavy lift operations at installations where purpose designed lifting arrangements, fully integrated into the design, have been provided. In many cases, during decommissioning, these lifting arrangements either will not be reuseable or will be expensive to reinstate and therefore lifting contractors may be tempted to use less well planned, engineered and constructed methods (ie. to take short cuts). Furthermore there may be considerable uncertainty associated with heavy lifting at the end of the field life due to lack of original design and construction information, changes made to the facility and structure during its operational life and lack of knowledge on the current condition and structural integrity.

Whilst many of the risks associated with deconstruction / removal of a steel jacket structure are either not applicable or less severe for Gravity Base Structures (GBSs), other areas of uncertainty are unique particularly if it is intended to refloat the GBS (as is the case with the Maureen steel gravity base substructure). It is anticipated that operators of most of the large concrete GBSs in the North Sea will seek derogation for removal of the substructures but the topsides must still be returned to shore. The size / weight and the method of installation (float-over and deballast) of some of these topsides will necessitate careful planning to ensure safety during decommissioning.

3   WORLDWIDE EXPERIENCE OF DECOMMISSIONING

Feedback on safety issues arising during platform decommissioning operations in US and Norwegian waters has proved valuable to complement the relatively limited UK experience so far.

These two offshore locations are considered to be of particular interest since:

- A great number of installations (approximately 1800) have been removed from the Gulf of Mexico over the last thirty years (albeit mostly single wellheads and small steel jacket platforms).

- A large / complex platform (Odin) has been removed from the Norwegian Sector of the North Sea. Apart from Odin, the offshore industry has very little first hand experience of complete removal of large offshore installations at the end of their field life.

One notable difference in the regulatory regimes between the two locations is the US "Rigs-to-Reefs" programme – offshore emplacement of jackets on the seabed (either at the original platform sites or in clusters at a designated offshore location, depending on State to State preference) is actively encouraged in US waters. This approach, directed at enhancing available breeding grounds for fish, is in sharp contrast to the North Sea requirement for removal and the concept of a waste hierarchy where re-use is preferred over recycling (and, in turn, over scrapping).

The lessons learnt have identified "soft" issues arising from working relationships and contractual drivers which can have as much influence on risk reduction (and safety of personnel) as the more commonly considered "hard" issues associated with such activities as cutting, lifting and transporting components.

Risk is greatly reduced if all parties associated with the decommissioning of an installation (oil company, contractor(s) and subcontractors) are fully aligned and aware of their responsibilities / accountabilities. The contractual arrangement should be such that, as far as possible, safety is not jeopardised by economic factors or compromised by conflicting interests.

Whilst many concerns on safety are common to both offshore locations, there are also some notable differences in approach. Opinions differ in particular over the use of explosives and divers.

In the Gulf of Mexico, divers are used extensively during decommissioning operations. Use of explosive leg / pile cutting is the norm, bulk charges are often used and charges are often installed by divers. However, it should be borne in mind that very little in-situ cutting of jacket structures into smaller units (for lifting and transportation) has been performed or required so far in US waters due to the size of the installations – small jackets are removed in one piece and larger structures have generally been reefed. [The US Minerals Management Service, which has regulatory responsibility for facilities and pipelines, does not allow explosives to be used at mid-water depths.] Whilst it is believed that there are "few" diver accidents considering the overall underwater manhours involved in decommissioning activities in the US, divers do sustain injuries from time to time and fatalities have occurred (possibly due to a culture of overconfidence which has developed over the years in handling explosives and cutting equipment).

In Norway, by contrast, it is difficult to obtain consent to dive from the regulatory authorities. Generally, all subsea operations during decommissioning are performed by ROV, including cold cutting of the jacket legs (using abrasive grit and diamond wire techniques). Explosive cutting of a jacket of any significant size into sections for retrieval to shore would require shaped charges which are difficult to position without divers and, hence, explosives are not considered suitable for such purposes.

Key safety issues identified by regulatory authorities, oil companies and contractors from their previous experience of decommissioning are generally universal (but, as seen above, there can also be some regional differences). The following represent some of these safety issues:

- Maintenance of "offshore" standards, practices and procedures. Although the structure is being removed, this must not engender second rate practices (a "scrap merchant" approach) with lowering of safety standards.

- Maintenance of safe operating procedures no matter how often a task has been performed previously. Overfamiliarisation can lead to overconfidence and a reduced perception of the associated hazards.

- Safety should take precedence over economic factors, for example:

- maximising the resale value of components should not add to the complexity of, or unnecessarily prolong, offshore operations
- unacceptable risks should not be taken to meet tight schedules and incentives should not drive operatives to take short cuts.

- Careful checking and rechecking of hydrocarbon inventories. Although comprehensive hydrocarbon cleaning may have been carried out during decommissioning, there is potential for subsequent build-up from hydrocarbons trapped in valves or impregnated into other materials. This will tend to be most noticeable when the time between platform shutdown and removal is significant.

- Comprehensive concept development and proving of purpose-built equipment to ensure that all application scenarios have been fully identified and appropriate safety factors / contingencies have been incorporated into the design. The decommissioning market is potentially lucrative for contractors offering leading technologies but, until fully proven, novel concepts present higher risks.

- Regular inspection and maintenance (or removal) of the secondary as well as the primary structures if the period between decommissioning and final abandonment is significant. Deterioration of a structure if left unattended can be rapid - access ways are particularly vulnerable. [The overall duration of some phased decommissioning / abandonment programmes for "supercomplexes" can be up to 10-15 years, with topsides removed many years before removal of the residual jacket structures.]

- Continuity of knowledge, including advice and input from the oil company's operating team into decommissioning and abandonment activities. Liaison and familiarisation needs to cascade down from one team to the next throughout the complete decommissioning / abandonment process. Unavailability of operating personnel familiar with the facility is of concern if a long period elapses between shutdown and removal.

- Integrity of lifting points and devices. The competency of a lifting device must be demonstrated for its specific intended application (not just by the provision of generic certification). New EU guidelines will regulate lifting padears / padeyes / lugs and specify NDT requirements.

- Accuracy of weight and C of G estimates for lifts to avoid overload or unexpected loading distributions (and potential failure) within the lifting system.

- Minimum handling of removed components and avoiding double handling. Each additional or unnecessary handling operation adds to the potential risk.

- Adequate provision of facilities / equipment for mechanical handling of equipment with hoses or umbilicals, and adequate seafastening of associated containers.

- Provision of safe transfer of personnel to / from the platform during abandonment.

- Minimum use of diving.

Many safety issues raised were associated with the efficiency and duration of offshore operations. The more complex the operations or the greater the number of offshore operations, the higher the exposure of offshore personnel to potential risk. Specific issues include the need for:

- Comprehensive onshore development and testing of equipment prior to deployment in the field. This is particularly relevant for ROV cold cutting equipment, where many problems were encountered during abandonment.

- Comprehensive contingency planning. During offshore installation, where components and connections are new and designed for easy assembly, it is not unusual to encounter unexpected conditions or scenarios. After 10-30 years in the field, with platform modifications and component deterioration, the potential uncertainties are far greater. Operations involving the cutting of conductor casings, piles and jacket legs seem to have encountered particular problems.

- Realistic assessment of weather sensitivity. If offshore operations (vessel movements, lifting, etc.) are extremely weather sensitive, activities may be deferred more frequently (waiting-on-weather) and the overall removal period extended. In addition, the weather lookahead and the prediction of weather windows becomes more important since the risk of weather deterioration once an operation has commenced is more significant. An accurate evaluation needs to be made of the duration of operations (e.g. the time from the decision to commence a lift until the lifted unit has been safely tied-down / seafastened on the transport vessel's deck).

4   CONCLUSION AND REVIEW OF RESEARCH

This paper is based on two project reports prepared by BOMEL Limited on behalf of the HSE and entitled "Decommissioning Topic Strategy" and "Worldwide Decommissioning Experience and Safety Cases". These reports will be published as Offshore Technology Reports and will be available via the HSE's website in due course. It is one of the objectives of the HSE that the results of research projects are disseminated to the industry. The Offshore Research Focus newsletter, which is published several times a year and is available on the internet (www.orf.co.uk), includes details of current / completed research and the availability of associated reports.

Ongoing projects on decommissioning issues being funded or jointly sponsored by the HSE include:

- Concrete Gravity Base Structures (GBSs): a joint industry funded project being carried out by WS Atkins examining the engineering safety and environmental risk issues of decommissioning and removing ageing "first generation" oil production GBSs from the North Sea.

- Hazards and Risk Control Mechanisms: identification and evaluation / re-evaluation of decommissioning hazards in the light of new techniques / equipment available to the industry and the removal / onshore disposal of large steel structures.

- Evacuation, Escape and Rescue (EER): the impact of decommissioning and associated platform changes to EER capability, particularly with larger and more complex UKCS installations where offshore decommissioning activities could extend over many months.

Table 1 Extent of offshore decommissioning

Offshore Structures Installed Worldwide	
Approximate numbers:	
Steel jacket structures in water depths ≤ 75 metres :	5920
Steel jacket structures in water depths > 75 metres :	610
Other structures (gravity base, floaters, etc) :	270
Total offshore structures worldwide :	6800
Offshore Structures Decommissioned	
Approximate numbers to date:	
Gulf of Mexico :	1800 *
North Sea :	30
* Note: largely single wellheads / small jackets	

Table 2 Facilities decommissioned: UK sector

Date	Installation
1985	West Sole WE (SP), Thistle SALM (FSU)
1985	Beryl (SPM)
1988	Piper Alpha (SP)
1990	Crawford (FPS+SS)
1992	Duncan (SS), Innes (SS)
1992	Blair (SS)
1993	Argyll (FPS+SS), Angus (FPS+SS), Forbes (SP)
1994	Fulmer SALM (FSU)
1995	Staffa (SS), Brent Spar (FSU)
1996	Esmond / Gordon (3 SP), Viking (4 SP), Emerald (FPS)
1997	Leman BK (SP)

Table 3 Facilities decommissioned: Norwegian & Dutch sectors

Date	Installation
1996	NE Frigg (articulated column + SS)
1997	Odin (steel platform)
1999	Mime (SS)

	1988	Wintershall K13D (steel platform)
	1989	Wintershall K13C (2 steel platforms)
	1998	Kotter (2 steel platforms), Logger (2 steel platforms)

FPS   =   floating production system
FSU   =   floating storage unit
SP   =   Southern North Sea steel platform
SPM   =   single point mooring
SS   =   subsea facility

© Crown Copyright 2001

C596/026/2001

Leveraging military technology for nuclear decommissioning

T YOUNG
QinetiQ, Chertsey, UK

ABSTRACT

This paper introduces military technology developed by QinetiQ for the UK MoD that can be levered into the nuclear decommissioning industry. This paper also outlines compliance legislation studies for the tele-operation of Remotely Operated Vehicles (ROVs) undertaken by QinetiQ on behalf of the UK Health & Safety Executive (HSE). Military ROV technology applied to decommissioning and potential innovative technologies and strategies for future decommissioning are discussed.

1 INTRODUCTION

QinetiQ's research continues in support of the UK Ministry of Defence's (MoD) Unmanned Ground Systems (UGVs) vehicle programme. UGVs have several applications within the UK MoD including Unexploded Ordnance Disposal (EOD), surveillance, reconnaissance, mine clearance and battlefield engineering tasks. Majority of technologies for EOD are applicable to nuclear decommissioning that requires tele-operated platforms.

Throughout these sections the key theme is the value of technology transfer from the military to industry. This "technology transfer" ranges from tele-operation to semi-automated remote control systems utilising real data to provide low latency virtual vision for the man-in-the-loop operation and the benefits of semi-autonomous operations using live data transmission. The benefits of using technologies such as Differential Global Positioning System (DGPS) and laser sensors to improve remote control equipment are discussed with the aim of developing semi-autonomous vehicles with greater intelligence, independence and versatility in order to greatly reduce operator input and stress for non-critical tasks.

The increasing use of GPS systems, teleoperation and computerised control systems to monitor and control mobile machinery movement and operation brings together both potential benefits and new hazards. The above trends will be driven by commercial considerations to reduce costs, through increased efficiency and manpower reductions. Hazards might emerge including inadequate command and control, lack of adequate safeguards, system latency, and poor man-machine interface. As technology advances there is the potential to produce systems for which there is insufficient legislation or safety design guidance. It is therefore necessary to review the state of the art of relevant technologies, explore their growth potential, use and future development and assess health and safety pros and cons in different operating environments. Specific environments envisaged are agriculture (including forestry), construction, extraction (i.e. mining) and utility supply.

2 REGULATION AND COMPLIANCE: HEALTH & SAFETY EXECUTIVE (HSE) LEGISLATION FOR TELE-OPERATION OF ROVS

QinetiQ is currently defining the UK's Health and Safety Executive (HSE) requirements - set out in "Mainstream Market Research 2000/2001". The Centre for Robotics and Machine Vision (CRMV), has amassed a unique body of knowledge from years of defence research in collaboration with the US and Europe from which the HSE study will benefit. The outputs from this study will increase the HSE's understanding of Global Positioning System (GPS) and machine automation and its impact on the health and safety of mobile machinery and allow the HSE to anticipate future developments. The study will highlight key safety issues arising from the introduction of new technologies and predict unforeseen uses to which they might be put. It will also forewarn industry by providing safety specifications and design guidelines and an assessment process/procedure against which new applications can be rigorously assessed by the HSE.

The principal aim throughout the study for the HSE will be to ensure that any new system can be assessed by the HSE for safety during operation or whilst in storage. The main output will be a safety assessment process or procedure to apply to the emerging applications considered during this study. The output will also provide specific guidance for system specification, design and development and to support legislation. The objectives and perceived benefits are to identify the state of the art of technology applicable to GPS control and machine automation both for reference purposes and to benchmark the technology. In addition the study will produce a forecast of technology growth and market trends to predict emerging safety issues, identify generic safety considerations and produce guidance for the specification, design and development of current and future mobile systems. The overarching aim is to improve safety and produce a safety assessment process in order to enable current and prospective systems to be evaluated.

3 IMPLEMENTATION: ROV TECHNIQUES FOR DECOMMISSIONING

Over the years there have been a number of concepts in developing de-commissioning tools for the remote control demolition of reactor plants. The original concept produced specialist, bespoke, machines. Each specialist system was built to undertake a single task within the active area. Such a concept is both extremely costly and very inflexible. Machines are unique, spares and support difficult to obtain and many of these bespoke machines failed in service, whereupon they were irreparable in the active environment so they too require disposal of as part of the decommissioning project.

This brought about the second-generation concept of using cheap Do It Yourself (DIY) tools grafted onto robotic limbs. Although these domestic or trade grade tools quickly failed in the active area through contamination or hard use they were receiving, their relatively low cost allowed them to be junked once they failed. However this concept produced relatively high amounts of low level radioactive waste, mostly composed of dead de-commissioning tools.

The third generation saw the introduction of mini diggers and other small industrial or builder's plant being fitted with remote control equipment. These tried and tested industrial grade units are relatively cheap, mainly due to the numbers being produced. As these units had been developed over the past 10 years their reliability had increased and so had the number of tool attachments. This increased their potential for use in the nuclear de-commissioning market.

4 OPERATOR CONTROL & NAVIGATION

The design of the RCV often negates the operators input and thus the HMI becomes an add-on rather than part of the holistic design process. In de-commissioning the largest instance of failure for the RCV is often the Human Operator. A number of RCVs have been trapped or lost in many reactor plants at some time. The prime failure modes are as follows:

- RCVs driven off steps or into pits
- RCVs picking up items too heavy for them, causing hydraulic cylinder or chassis failure
- RCVs driving into an unrecoverable position
- RCVs caught under falling equipment
- RCVs driving over and trapping/cutting their own umbilical power supply lead
- RCVs falling or toppling over
- RCVs caught by wire and pipes in track geometry
- RCVs catching fire or motor burn out

The other most common reason for many of the above is due to the operators lack of situational awareness. Operating an RCV and ancillary equipment(s) utilising only the onboard camera(s) views is very difficult. Third perspective cameras increase operator situational awareness and reduce the disorientation experienced by operators.

Preplanning via the use of synthetic environments is critical, as is route proving on non-active models prior to placement of RCV in an active area. Understanding of where the RCV is also critical to the safety of the RCV operation thus navigation and dead reckoning position fixing is a "must have", as is vehicle diagnostics. It is important to know the main drive motor is over heating before it burns out or that the centre of gravity is about to be exceeded prior to loosing communications with the RCV. Task planning using CAD and Synthetic environments is a great help and should prevent picking up items that are just too heavy or being trapped under the pipeline.

5 LEARNING FROM OTHER INDUSTRIES: MILITARY ROV TECHNOLOGY APPLIED

5.1 Robotics & remote control

The capabilities of CRMV include current robotics technology for applications within Unmanned Ground Vehicle (UGV) systems and have close links to QinetiQ centres of excellence within Unmanned Underwater and Air Vehicles. Our MoD customers have ensured and nurtured over 30 years of continued support, research and development within the field of Robotics and Remote Control. Figure 1 shows a future UGV concept.

Figure 1 Future Unmanned Ground Vehicle (UGV) concept

QinetiQ's continuing research in support of the UK MoD's Unmanned Ground Systems vehicle programme and future UGVs concepts with applications including reconnaissance, mine clearance and Battlefield Engineering tasks. This research and development is often collaborative with the UK MoD and US DoD.

CRMV has obvious links to the UK MoD and is increasingly active in the non-MoD, Surveillance, Police, Fire Service, Oil and Gas industry markets. The Groups activities within these markets have direct relevance to the Nuclear Decommissioning and Nuclear Services market and this area is already being successfully exploited. Current licensing agreements with both ABP Ltd and JCB demonstrate the successful commercial approach taken to leverage UK MoD funded technology into the UK commercial market. Figure 2 shows a concept RCV produced for the Fire Services.

Figure 2 RCV for the Fire Services

Practical expertise exists within QinetiQ in technologies applicable to nuclear decommissioning. These include: the development, installation, assessment and use of GPS on tele-operated vehicles; specifying, installing and testing communication and control systems for semi-autonomous vehicles; tele-operation for remote driving and vehicle intelligence; and the supply of appliqué kits for remote operation of existing vehicles. QinetiQ has developed a control methodology for operation over several kilometres for future Robotic Land Vehicles (RLV's). Further research into tele-operation for earth moving tasks including the use of mono/stereo camera vision; augmented/virtual reality; and vehicle tracking using real-time optical flow for obstacle avoidance. The growing requirement is to show that safety and operational benefits to the user and third parties outweigh the added costs.

5.2 Global positioning system (GPS)
Research into an "aid to digging" programme has been undertaken that uses GPS, can graphically represent a vehicle operating in a terrain in near real time. The system updates ground profiles, using bespoke software, by pre-scanning the ground with a laser system. Vehicle position attitude and bucket sensor information is relayed over a radio modem and converted into a graphical image. Data is referenced to a real time kinematic Differential GPS system having a resolution of a few centimetres offering data on both movement of the vehicle and vehicle operating systems relative to the ground profile. In digging trials, the ground profile, pitch, roll and bucket height is automatically updated as the vehicle operates.

5.3 Crew systems
This skill group investigates human-machine interface (HMI), crew workload to improve and enhance human operation, sustainability and effectiveness. Over the last 10 years the work has focused on the interface between operator and vehicle and covers Graphical User Interface (GUI) technologies, information presentation, multi-function crew stations, command and control, vehicle integration, platform information systems and human factors aspects. Facilities include a laboratory-based synthetic environment crew station; HMI prototyping and software development rig. Key staff have unique expertise in human factors, accident investigation, human reliability assessment, risk analysis and safety management with regular assistance provided to the DETR accident investigation branch. They have

recently developed improved methods for human factors risk assessment. Current work includes involvement on the Railtrack Paddington Inquiry

5.4 RCV Modularity
The CRMV has experience in defining and implementing the level of modularity required for a wide range of remote control systems. This has included modularity for operational, maintenance and repair purposes in harsh environments. Recent examples include: JCB 170, GROUNDHOG and BISON RCVs and JCB 410 RCVs.

5.5 Materials technology
The Mechanical Sciences Sector (MSS) in particular has extensive experience in specifying suitable materials for RCVs and similar structures. These include previous and recent studies into alternative materials for RCV design. The studies have concentrated on material types and manufacturing techniques and have also included possible design schemes using alternative materials. A number of ground vehicle projects in the recent past have utilised the extensive modelling capability to meet design criteria of new MoD equipment.

5.6 Command & Control (C2) architecture for RCVs
CRMV has experience in producing the type of C2 systems that may be required for the DFR ROV intervention. Conducted under research for MoD, CRMV has produced two new types of C2 system that could be suitable for the DFR ROV. One system is based on a distributed fibre optic data network, the other is a distributed CANbus system.

Drawing on earlier work and supported by the Centre for Human Sciences (CHS), CRMV has been involved in the design and development of a command and control consoles since early 1998. This has led to the development of two generic consoles, one fixed and the other portable.

In 1997, CRMV produced and integrated a common command and control system for the EOD RCVs GROUNDHOG and BISON, see figure 3. This system enabled both vehicles to be operated via the same hand controller-enabling savings to be achieved in training and in the logistic burden. This was the first time that this type of control system had been adopted for operational use and laid an early foundation on which CRMV could build and eventually specify the CIP.

Figure 3 BISON & GROUNDHOG

5.7 Sensors

There is a wealth of expertise across QinetiQ in a wide range of sensor and sensor-related technologies. These include acoustics, Infra-red and laser devices, radar, signal / imagery processing, simulation and pattern processing. This expertise, allied to a sound knowledge base in CRMV with respect to the selection and use of sensor technologies for remotely operated system applications means that CRMV are well placed to identify the correct level of sensing required for RCVs.

5.8 Manipulation

CRMV has a good level of knowledge and practical experience in manipulation for EOD RCVs, gained through practical experiment and design, theoretical study and assessment of third party systems. Making best use of a full range of design, modelling and simulation tools[1], CRMV are able to engineer fully integrated, yet cost-effective solutions that are closely matched to user requirements.

CRMV has previously produced concept demonstrator manipulator arm. The large robotic arm is Figure 4 was computer controlled and capable of resolved motion and pre-set configurations. A later, smaller version of this arm was successfully integrated onto BISON chassis and is currently undergoing software development.

Figure 4 RCV with Manipulator arm

5.9 Human machine interface (HMI)

HMI is one of the most important areas that directly influences the performance of equipment where man in the loop is required and assist to an efficient system-of-systems approach to the design. HMI's are generally given consideration towards the end of the design cycle. For decommissioning applications, optimised HMI's will help to ensure reduced mistakes due to

---

[1] Use of the Synthetic Environment at QinetiQ will also allow practical experiments and trials to be performed with fully integrated mobile concepts in representative operational tasks.

human error, help automate repetitive tasks and reduce operator workload and stress during complex tasks.

An example of experimental data collected during this year's research is contained in table 1 and figure 5. The experiment measured time and number of collisions of 20 operators as they completed a remotely operated task using a conventional hand controller and an optimised console work station. The results show that feedback displayed on a console can reduce the workload and allow the operator to concentrate on the driving or manipulation tasks more readily than hand controller and a bank of monitors.

Table 1 shows the average completion time of RCV driving and manipulation task for each controller. The average completion time for 20 operators for subsequent runs show the console workstation allowed the operator to complete the task faster than the hand controller.

	Run 1, s	Run 2, s	Run 3, s
Hand controller	526.16	454.06	397.97
Console workstation	440.42	356.61	344.26

Table 1. Table of transformed completion times for each controller type

Figure 6 shows the average number of collisions of RCV during the driving task. The results show statistically significant effect of run (F=5.32, df=2, 36, p<0.01), controller (F=5.90, df=2, 38, p<0.01) and a statistically significant interaction between run and condition (F=3.06, df=4, 55, p<0.05). Within Run 1 the hand controller had statistically significantly more collisions than the console workstation with a confidence level of 99.9%. Within the hand controller, Run 1 had statistically more collisions than Run 3 with a confidence level of 95%.

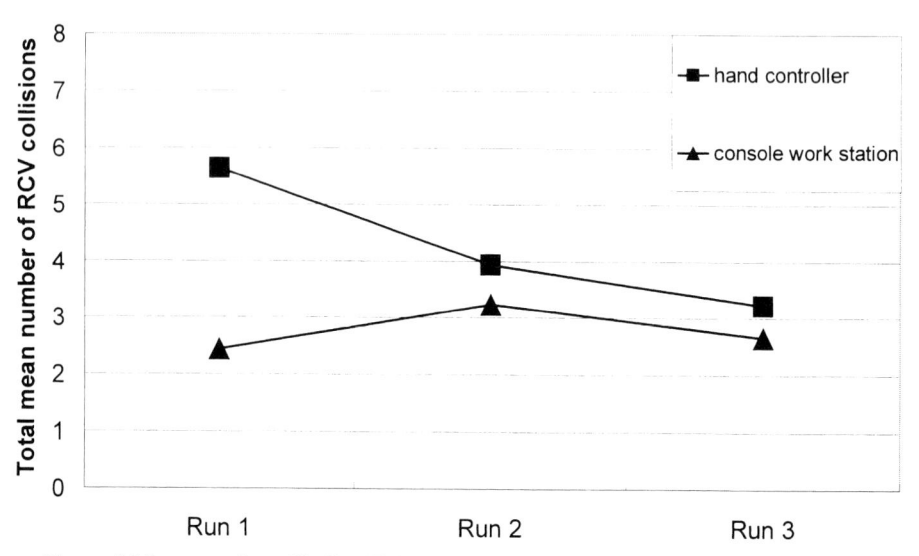

Figure 5 Mean number of hull collisions during RCV driving and manipulation tasks

5.10 Commercial-off-the-shelf (COTS) equipment

Optimising data fusion and data flow to the operator is of paramount importance. COTS Technologies that assist in reducing the operators work load are touch screens, Graphical User Interface (GUI) and hardware controls (Joysticks, space balls, teach pendants for control of manipulator arms). The HMI is kept to a minimum in order to present the operator with a simple, well-organised control station, ergonomically designed to the latest standards.

5.11 Computer vision

Current research within this area is the development of tele-operation kits for remote control of engineer vehicles. Technologies for remote control including both virtual and real vision systems, communications, DGPS, laser scanning, software and system integration. Engineer vehicles used for remote driving and digging where there is a necessity for vehicle safety and improved operator efficiency. QinetiQ has experience in the tele-operation of small wheeled and tracked vehicles, through to large combat engineer tanks and heavy earth moving vehicles including conversion of COTS JCB Robot plant vehicle to dual use (operator control/remote control) operation.

6 SUMMARY

This paper has demonstrated that QinetiQ has the capability and experience derived from past MoD projects that can be utilised to address most of the problems associated with nuclear decommissioning. Various technologies that range from sensors, materials and remote control vehicle design have been developed for military applications that have direct benefits and can be implemented to meet the challenges that face the nuclear industry today and in the future.

In particular we have shown that optimised Human machine interfaces for tele-operated control of remote platforms can reduce time and operator errors when compared to standard hand controller devices.

Contract Management

C596/043/2001

Contract strategy selection

R D NICOL
UKAEA, Dounreay, UK

ABSTRACT

UKAEA has developed a method for contract strategy selection for its forward programme of decommissioning, including major plant construction, based on many years of procurement experience. The method begins with the client's business model and objectives before moving to examine the project objectives. Each project is then analysed in terms of complexity and need for client involvement to give a 'first cut' guide on appropriate contract formation. The main contract models are described in relation to the analysis method. Other important factors are briefly mentioned – managing the risk, pricing and the market.

1. INTRODUCTION

The selection of a contract strategy for a project is a key decision which will have a major impact on the project's outcome. It is easy to get it wrong, and it is not uncommon for this to be identified, with the full benefit of hindsight, in post-project reviews. Getting it wrong can be very expensive, in time and money, and can even lead to failure to achieve the project objectives. Getting it right can help ensure success, and can bring wider benefits to client and contractor.

All clients need to develop their own approach, based on their own values, policies and organisational structure. The following is a methodology being used by UKAEA Dounreay to help choose appropriate contract strategies for its decommissioning and new construction projects, based on many years of procurement experience.

2. CLIENT BUSINESS MODEL

The starting point is with the client's business model, looking at the questions

- what is the core business?
- what must the business do for itself?
- what does it want others to do for it?

In other words, the classic 'make or buy' decision.

For UKAEA, the definition of core business has changed over the years, but since 1998 has been clearly focussed on 2 main areas – the programming and planning of the decommissioning and waste management associated with its nuclear liabilities; and the management of nuclear licensed sites. Whereas the former can be done by a relatively compact organisation operating at a strategic level, the latter requires an organisation with the skills, experience, knowledge and strength in depth to be in demonstrable day-to-day control of activities on its sites, and to be an intelligent customer for anything done for it by others. UKAEA has adopted a top-level policy statement which sets out the company's intent in using contractors. That is supported by a guidance note, in the form of a series of questions, to help decision-makers with the 'make or buy' decision. Some of these are safety-related (e.g. 'could safety be enhanced by a change in the current arrangement?'). Others are more commercial (e.g. 'is there likely to be a requirement to change the quantity or method of delivery of the service in the medium term?'). The aim is to make an informed decision about whether for any particular requirement it is best for the company to use its own staff, recruiting if necessary, or bring in external resources.

3. CLIENT BUSINESS OBJECTIVES

The next 'high level' factor to be considered, assuming there is a *prima facie* case for engaging contractors, is the business objectives of the company. This is addressing the question – 'is there something more we want from the client/contractor relationship than the delivery of a project?'. For some clients this may be a desire to involve their contractors and suppliers closely in their own marketing, sales and production activities. For UKAEA there has, from the early days of the decommissioning programme, been a stated intent to promote the development of a healthy market. UKAEA did this through

- strong emphasis on competition

- active communication links with contractors and suppliers, through published information on contract opportunities, supplier conferences, tender debriefing and key supplier management

- use of modern conditions of contract by adopting the New Engineering Contract (NEC, which encourages co-operative client/contractor relationships within the context of modern project management practice

- divestment of businesses and teams, thus transferring expertise to the private sector.

This approach led to a high level of contractorisation on UKAEA's programmes, and a lot of information on what is most successful commercially. It also helped define what is acceptable for a nuclear site licensee to devolve to others, and what needs to be kept under its own direct control.

4. THE PROJECT OBJECTIVES

These 'high level' factors help to define the context within which a contract strategy must be chosen for any individual project. At this stage it is necessary to start defining the sort of relationship with contractors which will best fit the project objectives.

UKAEA, and any other client, requires:

- A successful outcome

    To achieve a successful outcome there needs to be first of all a clear and approved statement of the project requirements and success criteria. Then the risk to that outcome needs to be identified and managed over the lifetime of the project, progressively reducing the risk of failure.

- Predictability of cost

    It is necessary to increase progressively the predictability of cost. Estimates will firm up at key stages of the project, depending on the contract strategy adopted. Fixed pricing is more likely to increase predictability, at least in theory.

- Value for money

    This partly depends on the competitive pressure on bidders, and the incentive on the contractor to innovate and reduce costs.

It is evident that attempts to increase the certainty in one of these areas can undermine another. For example (in admittedly simplistic terms):

- Requiring an early fixed price contract can appear to increase predictability of cost, but may undermine the others, due to risks not having been properly evaluated and allocated, and reducing the incentive to innovate and reduce costs.

- The client could take responsibility for detailed design, thus apparently increasing the certainty of outcome, but thereby maybe reducing value for money by stifling contractors' scope for innovation.

- Early contractor involvement may be seen to increase the likelihood of a successful outcome, but reduce predictability of cost and value for money because it is more difficult to subsequently transfer risk and achieve competitive pricing.

Each of these factors needs to be weighed up for each project. The objective should be to find the optimum balance between the 3 factors. This must be closely related to the generation and review of the risk management plan for the project.

This process can help determine the appropriate point at which to go to the market – for example, at an early stage when the project is still being developed in order to involve contractors before key decisions are made, and then keep them involved through to project completion; or, at the other extreme, 'salami-slicing' the project into small sub-projects with tightly specified contracts to match. A further degree of analysis is required though to properly understand the drivers affecting contract strategy.

5. PROJECT POSITIONING

For UKAEA there are two key drivers in establishing the best relationship with contractors in projects. These are the complexity of the project, and the need for client involvement. These will drive such decisions as:

- when to move from 'intelligence gathering' to implementation.

- when the client can feel confident about transferring risk, and contractors can understand it sufficiently to feel comfortable about accepting it.

- what sort of relationship to establish between client and contractors.

- and, crucially, how to 'package' the project – e.g. expand its scope into a 'super-project', for a large integrating contractor to manage, or divide down into smaller sub-projects with limited tasks for a range of contractors.

Characterisation for a project against the two factors requires consideration of the following issues:

- Project Complexity

    Performance requirements and associated constraints.
    Level of technical challenge (i.e. novelty).
    Development work required.
    Scope for innovation.
    Scope for cost reduction.
    Requirement for multiple specialisations.
    Opportunities for PFI/PPP

- Client involvement

    Strategic importance.
    Risk of failure.
    Stakeholder interest.
    Regulator interest.
    Proximity to existing nuclear facilities.
    Safety issues in implementation.

Plotting these on a matrix helps to give a first indication of the type of contract strategy required:

High	**B** Incentive for contractor innovation Maximise risk transfer Early contractor involvement	**A** Complex contract strategy Risk share
Project Complexity	**D** Simple contract strategy High risk transfer	**C** Late contractor involvement Fixed pricing possible, within strictly defined limits
Low		High

Need for client involvement

6. MAIN CONTRACT MODELS

It is not sensible to move mechanistically from this to a choice of contract type. It does though indicate what might be suitable. Four main models are now described, in relation to the matrix, with some of the advantages and disadvantages of each.

6.1 Alliance

This is likely to be most applicable to 'Box A' projects in that it is able to deal with complex risk sharing arrangements, and allows for a high degree of client involvement. The concept is relatively modern and has been used successfully by oil and gas industries as part of the strategy to reduce the cost of the exploitation of the North Sea.

The client selects a team of contractors to carry out the works, covering the key areas of the project. These may include an integrating contractor who will provide most of the management for the project. Alternatively, the client can adopt that role. The contractors may be appointed after competition, or chosen according to certain criteria from an existing set of partners. The client places a direct contract with each for the provision of their part of the works, but in addition client and contractors enter into an alliance agreement which sets out how they will work together to achieve the objectives of the project, with a target price and gainshare agreement. In negotiating this agreement there must be a building of trust between all the parties if the relationship is to have any chance of success. The working of the alliance is governed by an alliance board which may or may not be chaired by the client. If for any reason the alliance fails, the parties revert to the direct works contracts.

In a UKAEA application, competition is necessary in appointing the contractors, and chairing the alliance board helps to demonstrate UKAEA's overall control of the project.

For

- the client can demonstrate overall control through the direct works contracts and chairing the alliance board.

- competition can be used for appointment.

- successful application of alliancing is claimed to have achieved significant time and cost savings.

- there is a strong incentive on contractors to work together on all aspects of the project, including safety, and innovate to reduce costs.

- use of an integrating contractor removes much of the managerial load from the client, provided it is still able to demonstrate that it is in overall day-to-day control of activities, and has an adequate contingency plan in the event of failure of the contractor.

Against

- experience of this type of contracting is limited in the nuclear sector.

- it requires a high level of commercial capability to set up and manage.

- selection criteria have to include 'soft' issues which can be difficult to assess objectively.

- disintegration of the alliance is a risk to the project.

6.2 Traditional

This model was used for most of UKAEA's major projects in the past. Direct contracts are placed after competition for the major parts of the project, usually civil construction, major plant supply and installation, and some specialist services. The client takes design responsibility (and may do conceptual and even detail design in-house). This may suit projects in 'Box C'.

For

- maximises competition.

- permits mix of in-house/contract assets and resources.

- gives demonstrable client control over the project.

Against

- conflicting objectives of contractors and client, leading to claims.

- high degree of client involvement in managing the project.

- client may require large design team, in-house and under contract.

- client often carries most of the design risk.

- can lead to compartmentalisation of projects, to the detriment of overall value for money.

- poor record of delivery to time and cost.

6.3 Prime Contractor

The client places a single contract for project implementation with one contractor who manages the delivery of the objectives, placing sub-contracts with a range of other firms. The client only deals directly with the prime contractor. This may suit 'Box D' projects.

For

- transferred risk rests clearly with the prime contractor (providing the specification is sound).

- clear responsibility and accountability for all aspects of the project.

- relationships are comparatively straightforward.

Against

- tends to lead to long sub-contractor lines.

- weak overlap of client and contractor objectives (though incentivisation can help to produce common interest).

- tends to be weak overall commitment to the project from key sub-contractors.

- innovation and cost reduction are only likely if effectively incentivised.

- project management organisations in this role can filter information flows between client and the 'doers', thus reducing client control and influence.

6.4 Consortium

This is similar to prime contracting, except that contractors covering the main parts of the project come together in some form of consortium. This could be some form of contract, or could be a joint venture company (JVC). Even in the latter case it is most likely that the client's contract will be with one of the companies forming the consortium, rather than with the JVC which is unlikely to be sufficiently substantial to take significant liability. Formally the other consortium members are sub-contractors. This model is likely to be most applicable in 'Box B'.

For

- brings a mutual interest between the key contractors, with incentive for innovation and cost reduction.

- opportunity to transfer risk to the main contractor and on to the other members.

Against

- at least 2 layers of sub-contracting.

- weak overlap of client and consortium objectives unless strongly incentivised.

- a consortium formed for the purpose of successful bidding does not always work well for project implementation.

7. MANAGING THE RISK DOWN

The fact that an initial analysis of a project puts it in 'Box A' does not automatically mean that a complex contract strategy is the best option. The other approach is to reduce the risks before entering into the major contracts. This links back to the discussion in Section 4 about the appropriate time to go to the market.

8. DEFINING THE PRICING BASIS

There are two primary ways of pricing contracts – fixed pricing, and cost re-imbursement. There are many permutations within and between these broad categories. UKAEA has traditionally had a strong preference for fixed price contracts, in common with other public sector bodies. This is appropriate where the requirement and scope are fully understood, the risk is well characterised, the desired project outcome is clear and there is a healthy competitive market. The absence of any of these will tend to undermine fixed pricing as the best basis for achieving value for money. Moving towards some form of cost re-imbursement requires a more complex client-contractor relationship, which could involve target price negotiation, incentivisation and open-book accounting. The contract strategy proposal needs to establish the pricing mechanism for the contract, demonstrating how best value for money can be achieved.

9. THE MARKET

Development of a contract strategy cannot be complete without an understanding of the market; a client cannot assume the ability to dictate terms. The strategy needs therefore to be underpinned by an assessment of the extent of competition, the range of skills and experience required, the availability of these resources, the managerial capabilities of relevant firms, and the wider aspirations of the players.

10. CONCLUSION

UKAEA is adopting a flexible approach to contract strategy selection, believing that no single model is appropriate for all projects. The methodology described in the paper is helping managers to think through the relevant issues and make a rational decision in the interests of the company and the project.

C596/008/2001

Partnering experience at BNFL

T J CARR
Histiric Waste Management Group, British Nuclear Fuels plc, Sellafield, UK

ABSTRACT

The Project delivery programme in BNFL covers a wide range of schemes both in value and type, from Greenfield construction, through asset maintenance to decommissioning and clean-up. The large number of concurrent projects and variety of formats presents a constant challenge to the Engineering community. With a forward workload forecast of several £BN it is important to gain maximum leverage from a finite resource pool. A range of supply chain strategies has been developed to address this challenge. The approach allows BNFL to concentrate on developing and strengthening its resource base through development of constructive relationships with suppliers that complement the core skills.

The objectives of the approach is to have a flexible, skilled and experienced resource available to form project teams in the most efficient way. One way this can be achieved is by building long term, stable partnering and alliancing agreements. This paper describes the methodology and experience of the approach. The areas covered include:

- Environmental influences
- Historic Waste Management
- Operating Models
- Next steps

The paper shows that the potential benefits from Partnering can repay the effort of introducing it into an organisation, however, it is essential to clearly understand the business modcl, be clear about the desired value and to approach the exercise with an holistic strategy.

INTRODUCTION

BNFL have expended over £5BN over the last 20 years on capital projects, the vast majority of which has been in support of the Sellafield Site. Much of this expenditure was historical Nuclear Liability. The Thorp, SMP and Magnox Asset Stream projects also represent major commercial investments by BNFL and customers. In 1998 BNFL made the decision to implement a Partnering methodology for delivery of its capital projects and initiated a selection process that was completed in July 1999. Three Engineer, Procure, Construct (EPC) Partners were chosen as well as five specialist suppliers of equipment and services. Over the past two years BNFL has deployed Partnering in a wide variety of situations from greenfield construction, through asset maintenance to decommissioning. In this period there has also been significant changes in BNFL's business, with the publication of the HSE's Team Inspection audit of Sellafield, a new Chairman and Chief Executive and a redesigned project approval and delivery process. Many of the original drivers have changed and new ones have emerged, however the basic need to have access to high quality resources to support BNFL's business remains intact.

We have recently completed a review of the first two years of the Partnering and Alliancing initiative which includes feedback from our business organisations, our Partners, the wider supply chain, the industry and other stakeholders. This process has proved invaluable in setting out our plans and expectations for moving forward.
During the last 18 months significant progress has been made with some good learning and some tangible early examples of benefits accruing. Most of this has been around the creation of alliances for specific projects. There has been particular benefit for the drive to fulfil the recommendations in HSE's Team Inspection report through the capabilities for rapid deployment of quality resources. There has, however, been little application of true Partnering and the potential gains originally expected are yet to be fully realised.

THE ENVIRONMENT

Over the last 18 months BNFL has undergone a dramatic change in both business focus and organisational structure. The Business analysis shows a continuing major capital and revenue projects programme at Sellafield for the next ten years and beyond, however, the shape of the portfolio of project work is changing dramatically from that of the past 20 years.

The future profile is likely to be much more focussed on operating and maintaining existing plants, improving the performance and life expectancy of the ageing plants and decommissioning the plants which have reached the end of their productive life.

Engineering in BNFL has also undergone major restructuring with firm emphasis on primarily supporting the Sellafield Business drivers and those of the Nuclear Decommissioning and Clean Up Business. They will focus on building and developing core competencies and nurturing centres of expertise, particularly in front end and nuclear technology capability. Project Management and Engineering Design will lead the regeneration of Best in Class processes and capability.

HISTORIC WASTE MANAGEMENT

A new organisational group has been formed to provide the facilities and organisation to deliver the Legacy Waste Strategy and Programme for the Sellafield Site. This work is of utmost importance to BNFL, and the new HWM Group will provide a single focus to maximise the efficiency of the Historic Waste delivery programme within the Sellafield Site Organisational structure. The scope of historic waste management covers the development of strategy to deal with the historic waste currently within the Integrated HWM Project and associated orphan streams as well as liquid wastes either associated with current solids, or arising from their treatment. These wastes are typically in redundant spent fuel storage ponds or waste silos, stored as sludge from ion exchange processes, or from other treatment facilities. The Historic Waste Management organisation has been formed to integrate all of the interdependent projects and facilities and will support the current facilities as well as providing new assets to manage this significant project.

CONTRACTORS, DESIGNERS, SUPPLIERS

The last 10 years have also seen a major shift in the Process and Energy Engineering Industry. The early 90's saw a sharp downturn in economic activity in the Engineering Industry sector leading to significant over capacity in the industry and 'cut-throat' margins. Many of the household names in major construction have disappeared and those that remain have radically changed their business operating model to a much more selective customer focus with emphasis on collaborative working, value creation and enhanced margins. The specialist Nuclear Industry capability has reached very low levels in many areas and in some cases is on the verge of extinction.

BNFL has a heavy dependency on its supply chain and many of our suppliers have a heavy dependency on BNFL. Several years of undervaluing these relationships in favour of a strictly competitive, cost driven and often adversarial contracting regime have accelerated that demise. Developing a strategic approach to this mutual dependency will be crucial over the next five years and beyond in enabling BNFL to deliver its overall work.

These industry trends and the demographic shift in career profiles has led to a chronic shortage of skilled and experienced people from graduate level to blue collar. BNFL has a major technological and delivery challenge to successfully discharge the legacy waste programme. We must learn from the past, but live in the present and plan for the future.

NEXT STEPS

Over the past two years we have built a core team with the skills and experience to enable the business to implement Partnering. This has been essential to understanding the grass roots practicalities which make every business unique. Factors such as the size and diversity of BNFL's business activities, the range of supplier capabilities and the possible commercial frameworks have all got to be tested against the best practice model. This process is now almost complete, and we are preparing to move to phase 2. The following headings are brief summaries of the detailed planning exercise that we believe will move BNFL nearer to the potential breakthrough possibilities of Partnering.

Integrate the delivery model
Every interaction between a Partner and BNFL should be part of the overall development of a win-win relationship. This covers the entire spectrum from Directors to accountants, engineers, etc. Each level will have specific objectives, but they should be aligned to the overall goal

Review the current Partners' fit within the new business environment
The selection process, which was completed in 1999, should be revisited in light of our changing needs to ensure that we have the right numbers and types of Partnering organisations.

Every interface should have a relationship manager
A recent survey of over 100 alliances in the US found that 52% of failures are due to broken or damaged relationships. The clear message was that all interactions should be consistently managed, and this means that Relationship Management will have to become a standard skill in the new environment.

Limit the deployment of Partnering to where it adds value
Ultimately, Partnering could become the predominant model for relationships with our supply base, however, there is significant internal and supplier development required to make this happen. In our business, we must have a clear vision of what we are trying to achieve before we can adequately measure where value has been added. The diversity of potential applications will be managed through the further development of pilot alliances in each class.

Take a change management approach
Partnering will be a progressive change, and as such it should be part of a structured process to move the business model to the desired End State. Executive and senior management commitment is required in the form of integral, aligned programmes to put the relevant capabilities in place. A specific accountability has been established to deliver the programme.

Partner Selection
The selection of the right organisations that have the required number of SQEP skills and competencies to satisfy the characteristics and needs of the particular workload required is vital to success. In tandem with the review we are developing a competency profile for each of the key business areas.

Measurement of Performance
There has to be a clear value proposition which underpins the use of partnering as a strategy which must be collaboratively developed between the business and chosen suppliers/ contractors. These must be capable of being easily and transparently measured. The diagnosis of the underlying trend of these measurements will influence the direction in which the relationship will go.

OPERATING MODELS

The review recommended two primary operating models to be adopted in BNFL, these are:
1. **An alliance which is defined as;**
    "A commercial arrangement between two or more parties for the delivery of a single, or group of, specific objectives through an integrated team".

Which has the following characteristics:

Specific, hard objectives supported by a commercial arrangement with agreed terms and conditions, identified obligations, responsibilities, deliverables and rewards. A defined end point based on satisfactory completion of specified goals (e.g. a specific business solution). Made up of selected organisations with complementary skills needed for delivery (not necessarily Partners). Managed at the most appropriate level in each organisation.

2. **Strategic partnering relationship**
"A long term, beneficial, strategic relationship between two organisations, which is founded on a business results based commercial framework"

Which is characterised by:
Long term, open-ended relationship based on defined obligations, responsibilities, goals and rewards linked to delivery of the BNFL business case. Typically not less than 5 years and driven at the highest level with aligned corporate strategic objectives. Culture of mutual interdependence and continuous improvement with the potential to change either organisation's operating model.
It is vital that Partnering is not merely a collection of alliances, but a structured, focussed, development of each organisation's capability to deliver to their potential.

CONCLUSION

There are many challenges facing BNFL but we are confident that strengthening our organisation and processes in key areas and building long-term strategic relationships with our supply chain, will enable us to maintain and develop our capacity as a leader in commercial nuclear capabilities and to fulfil our obligations in legacy waste management and clean up.

© BNFL 2001

C596/017/2001

WOMAD – the use of NEC–EEC (New Engineering Contract – Engineering Construction Contract) on a major decommissioning contract

S J PARKINSON and **K P GREGSON**
NUKEM Nuclear Limited, Winfrith, UK
A STAPLES
UKAEA, Winfrith, UK

SYNOPSIS

WOMAD (Winfrith Operations Maintenance and Decommissioning) is a major restoration project at the UKAEA's Winfrith site. NUKEM Nuclear Ltd (NUKEM) was awarded the implementation contract in March 2000. The contract value is approximately £30M with a duration of 8.5 years. Briefly the WOMAD contract comprises:-

- Further operation, decommissioning and finally demolition of the A59 Active Handling and Decontamination Facility.
- Taking over the care and maintenance and decommissioning of various other facilities on site including the SGHWR (Steam Generating Heavy Water Reactor), ZEBRA (Zero Energy Breeder Reactor) and Dragon.
- Retrieval and processing of active sludge from the SGHWR's sludge tanks

The contract awarded to NUKEM is based on a functional specification and NEC – ECC (Option A Fixed Price with activity schedule) Contract. This approach allowed both UKAEA and NUKEM to have a clear understanding of the project deliverables and a proper apportionment of risk whilst using a recognised standard set of contract conditions. These conditions have clear remedies and obligations, which should both assist with the management of the difficulties, which often beset decommissioning contracts.

This paper sets out to describe the process from tender inception to the completion of the first year of the Project.

1 INTRODUCTION

What is WOMAD? The acronym 'WOMAD' is derived from 'Winfrith Operations, Maintenance and Decommissioning'. UKAEA has employed NUKEM to undertake the WOMAD contract at Winfrith and the scope of work includes:

- Operation, decommissioning and demolition of Building A59
- Maintenance and decommissioning work at other facilities including SGHWR, ZEBRA and Dragon
- Retrieval and processing of the active sludge from the SGHWR sludge tanks

WOMAD as a UKAEA project is sanctioned at £66m over 10 years. This covers all aspects of the project, a major part is the implementation contract which NUKEM won in a competitive tender valued at over £30m. A further significant contract yet to be let but still part of the WOMAD project is refurbishment of the Treated Radwaste Store (TRS), the medium term repository for the treated EAST sludges.

The A59 Active Handling Building is a UKAEA Category 1 facility that was constructed in the early 1960's for the post irradiation examination (PIE) of reactor fuel. The facility contains two large, heavily shielded cave lines constructed from reinforced concrete, a decontamination centre and a pressurised suit area together with a range of supporting of workshops and laboratories. Building services include an active ventilation system, an active drainage system and overhead cranes. Following the cessation of PIE work in the early 1990's, NUKEM's A59 staff have carried out a number of waste management tasks including re-packaging Dragon fuel and processing of sea dump packages. After completion of some further waste management operations in the early stages of the WOMAD contract, the building will be completely decontaminated, the building fabric removed, the cave line carcases demolished and finally, the floor slab removed, leaving the site as a de-watered hole.

Under the WOMAD contract, NUKEM is also responsible for the care and maintenance of A59, together with care, maintenance and some decommissioning work on other Winfrith facilities including Dragon, ZEBRA and SGHWR reactors.

Throughout the operational lifetime of SGHWR, radioactive sludge was accumulated in the external active storage tanks - EAST. The tanks are constructed from reinforced concrete and are located away from the main reactor building, adjacent to the site boundary. The sludge is considered to be LLW not acceptable for disposal at Drigg, and consists mainly of powdered ion exchange resin from primary circuit water clean-up and primary circuit decontamination wastes. The sludge contains significant quantities of cobalt and caesium isotopes and thus shielded facilities are required for processing. The sludge will be mobilised and hydraulically transferred to a new waste processing plant, which NUKEM will design and build. Here, the sludge will be conditioned, immobilised in cement in 500 litre drums and delivered to a storage building. A total of ~ 1000 drums will be produced. On completion of operations, the EAST tanks will be decontaminated and demolished and the waste treatment plant decommissioned.

2 WHY THE NEC

Within the UK Nuclear Decommissioning market it is recognized that often projects do not complete to time and budget, this can occur for a variety of reasons such as:-

- Ill defined deliverables
- Ill defined risk apportionment
- Lack of knowledge about the items to be decommissioned
- Regulatory interactions
- Poor project management/performance by either Client and/or Contractor

Quite often, disputes occur when there is no proper contractual framework for their resolution, frequently because the form of contract used is inappropriate for the works being undertaken. UKAEA aims to be an expert procurement organisation and recognising that this was a real problem for both client and contractor alike, UKAEA undertook to trial the New Engineering Contract (NEC) form of contract in the mid 1990's on a number of contracts with the objective of developing their expertise and reviewing its suitability for widespread use on UKAEA's projects. NUKEM was an enthusiastic participant in these trials and, for example, the DIDO decommissioning contract was successfully completed under the NEC. Following this trial period the NEC was introduced as the contract of choice within UKAEA in 1997.

The NEC family of contracts includes the Engineering and Construction Contract (ECC), used for major works projects, and the Professional Services Contract (PSC) for consultancy contracts. There is also an Adjudicators Contract, a Subcontract, and a Short form of Contract. (Ref 1- NEC publication details).

2.1 Benefits of the NEC

The NEC aims to stimulate good Project Management, which gives certainty to all parties, and sets out generic processes for dealing with unforeseen issues that inevitably arise during implementation.

The NEC was developed to accommodate trends that were taking place within the construction industry in the late 1980's, such as:-

- A move to milestone payments
- Partnering and target cost
- Design, build and operate contracts
- Multi disciplinary construction projects
- Moving away from the confrontational approach
- Avoidance of the adversarial disputes that occurred on many large construction projects.

The benefits to both employer and contractor are numerous, the most significant ones being:-

- Payment by Activity Schedule, which is linked to Programme. This ensures that the contractor has thoroughly researched his programme and it allows both parties to understand their cashflow requirements. It also ensures that the project is programme driven – payment only being made against completed activities.

- The agreement of the value of compensation events in a contemporaneous manner, which avoids issues being allowed to languish unresolved.
- The assessment process, being based on activities that are complete within the assessment period, is far swifter and does not involve an army of quantity surveyors and their concomitant costs.
- As it does not use specific engineering terms it is applicable to a wide range of activities.

3   ITT PREPARATION

In June 1998 UKAEA proposed an Activity Schedule covering the award of the WOMAD Project with the key dates being:-

March 1999   -   Issue ITT
March 2000   -   Appoint successful bidder
July 2000   -   Handover.

This Activity Schedule consisted of a very detailed list of activities and the timescale upon which they would need to be completed to achieve a July 2000 handover date. A multi-disciplinary UKAEA team carried out the work.

The main elements with this schedule were

3.1 Programme Management
This was under control of the UKAEA Project Manager and included the significant issues of interfaces with regulatory bodies (NII, EA), with existing contractors (NUKEM, Vectra) and other internal UKAEA interfaces.

3.2 Development of a cost model
In order to obtain a realistic cost model for the project UKAEA employed the services of their corporate planning department to produce a cost estimate.

This was prepared using UKAEA's PaRametrIc Cost Estimating System or PRICE, which is a knowledge based cost estimating tool that has been specifically designed and developed by UKAEA for use on nuclear decommissioning projects.

Briefly this system is based on the aggregate relationships between cost and physical and performance parameters, in which standard man-hours rates "norms" and labour "unit rate" are applied to the quantities of physical components undergoing decommissioning.

This took a period of 4 months and allowed the UKAEA Project Manager to develop his cost model required to justify the proposed project expenditure both with UKAEA and at the DTi, the ultimate funding body.

3.3 Integrated technical and commercial specification
In order to tie in the principles of the NEC – ECC contract is was recognized that a very clear specification was required that would reflect the terminology of the NEC contract.

Early agreement was reached that as the risks were quantifiable and should be placed with the party who was best able to manage the risk that option A – priced contract with activity schedule was the best form.

The main requirements that this form requires are:-

Works Information	-	which specifies and describes the Works.
Site Information	-	describes the Site and its surroundings.

This was then developed into a document comprising General, Site and Works Information, A59 specific site and works information, EAST site and works information, care and maintenance site and works information, pre-tender health and safety plan and personnel transfer works information.

The personnel information was required as secondary TUPE obligations would be placed on the winning contractor.

All these sections were given in standard layout and form and were checked against each other for conflicts.

This integrated approach took some time to develop, but the investment was considered worthwhile as a benefit to the avoidance of inaccuracies at the tender stage, greater confidence in the tender price returns and avoidance of costly disputes during the contract.

This was a significant departure from decommissioning specifications received by NUKEM over the last few years, which have tended to be a collection of standard forms bolted together.

Key features of the specification were:

EAST Peformance specification
   Well defined interfaces
   UKAEA responsible for product recipe

Organisation Requirement to identify 'key people'
   Clear interfaces

Risk A developed risk register that placed risk where best managed

A59 Well defined end point
   'extra over' rates for decontamination risks, i.e. extra depth of contamination removal
   Incentives to decommission early to reduce maintenance costs & to reduce waste volumes.

3.4 OJEC Prequalification

As the cost estimate for this works project was estimated to be in excess of 5.0M Euro UKAEA were required under European legislation to publish details and invite bidders for the project via the European Journal. From this pre-qualification exercise many pre-qualifications were received from which 4 potential bidders were selected. The selection criteria used required demonstrable experience in all elements of the works required and a knowledge and understanding of TUPE transfers.

3.5 Red Team Review

During production the contract specification was subject to internal review by UKAEA personnel external to the project. On completion of these reviews an external team including external independent members headed by UKAEA's Head of Planning and Performance reviewed the specification.

3.6 Enquiry Film

As the project covered a variety of facilities on the Winfrith Site a professional video film was made showing the facilities and describing the key features of the contract. This allowed the tenderers to review areas of work from their own offices without having to make repeat site visits.

3.7 Tender Period

As this was a complex project, which involved design, a six month tender period was allowed. During this period questions were raised by bidders with all question and relevant answers being passed back to each company. Other activities that were carried out in this period were an QA/Commercial/Safety audit of each bidders operations and visits by UKAEA staff to similar projects proposed by the bidders. The results of these activities were fed into the later tender assessment phase.

4 TENDER PREPARATION

As users of the NEC began to accrue experience of this contract form, one of the aspects that became apparent was that if the tender had been under-priced, the NEC makes it more difficult to recover the situation. A number of reasons contribute to this:-

- The regular updated programme means that original cost and timescales are more transparent.
- The structured method for the compensation event procedure, including what is allowed for in the Schedule of Cost Components means that justification of the additional costs is more rigorous and transparent.
- The timely settlement of compensation events prevents a contractor from waiting until the end of the contract to build up his claims.

Taking these factors into account and the knowledge that this was a significant tender (initial estimates £25-30 million), NUKEM undertook to carry out a very comprehensive tendering exercise.

Prior to receipt of the ITT the marketing department produced a strategy paper putting forward the need for a "Project" team to manage and produce the tender along with the identification of the key resources needed to achieve success. This also estimated the tendering budget to be in the range £150K – £200K.

A bid manager was appointed, the Winfrith Manager who had project managed Winfrith's previous A59 Project, who in turn produced a Tender Plan and a Tender Preparation Programme.

The Tender Plan was similar to a Project Quality Programme and identified outputs, roles, interfaces, reviews and organisation.

The Tender Programme identified all the major activities to be undertaken, the most significant activities being:-

- Bid Brainstorm
- Site visit
- Commercial
  - Bonds
  - Insurances
  - Pricing methodology
  - Cash flow
- Monthly review meetings
- Red team review
- Client requirements
  - Visits to reference projects
  - Client audit of management systems
- A59 Decommissioning
  - Outline decommissioning plan
  - POCO methodology
  - Deplanting methodology
  - Demolition methodology
  - Subcontract packages
- EAST
  - Outline design
  - Design review
  - Detail design
  - Subcontract prices
- Risk Workshop
- Documentation
- Tender documentation production
  - Technical text
  - Designs and calculations
  - Tender programme
  - Activity schedule
  - CV's and organisation.

In terms of setting off the Project on a sound footing, if won, significant attention was paid to the Programme, Activity Schedule, Risk and Design.

As the whole ethos behind the ECC is to be programme driven with payment only being made against <u>completed</u> activities the data contained within the programme had to be accurate enough to allow a detailed resource breakdown to be applied. This both gave confidence in the pricing and allowed an Activity Schedule to be produced which would allow a near enough neutral cash flow.

Risks were identified by holding a multidisciplinary risk workshop that applied keywords to the main programme phases of the project. The output from this was a risk register comprising 49 items, the treatment of these risks is discussed in the next section.

On completion the tender submission comprised 3 volumes, incorporating detailed programmes, detailed decommissioning methodologies, EAST plant designs, option proposals for alternative processes, CV's of all key personnel and a fully priced activity schedule.

As discussed in the above a detailed risk register was produced for the tender, this was submitted with the tender and identified risks that lay with UKAEA or NUKEM. The risk register formed the basis of the later negotiation process.

5 ASSESSMENT/PRESENTATION/NEGOTIATION/CLARIFICATION

Following the tender submission UKAEA then commenced a 6 month period which covered Tender Assessment, Clarification, Presentation and Negotiation. On receipt the tenders were assessed by the UKAEA's Tender Preparation Team against the Tender Specification. From these assessments numerous questions were raised which were passed to the bidders. The next stage of the process was Tender Presentation. For this UKAEA required the Project Team, not the Marketing and Sales Team, to present all aspects of the tender submission to the UKAEA Project Team. From this process NUKEM were chosen as preferred bidder in early January 2000 and invited to attend a series of pre-Contract Negotiation Meetings.

The intention of this negotiation was to develop a single contract document, following the format of the Tender specification that both parties were content with. This single concise document could then be used as a ready reference by both parties management team.

This also avoided the common practise of having a large contract document which comprises Specification, Tender Offer, Clarification Letters and Contract Award Letter.

The Negotiation Phase also allowed any remaining risks or clarifications to be discussed and accommodated within the contract document.

As the Project was of a significant size the Chief Executive of UKAEA, Dr John McKeown performed a formal signing ceremony with the NUKEM Marketing Director. Local media attended the event which signified a major step in UKAEA's medium term plan to decommission the Winfrith Site.

6 CONTRACT IMPLEMENTATION

Following the signing ceremony the contract formally commenced on the 13th March 2000 with a possession date of 27th July 2000. During this Mobilisation Phase, Safety and QA documentation was written and approved to allow the Works to progress as well as the necessary recruitment required. After one year the Project is generally running to the Project Programme.

In particular the Decommissioning Works within the A59 facility have been very successful with many areas being completed ahead of Programme.

Significant A59 milestones achieved to date:-

- Removal of workshop facilities
- Creation of waste processing facilities
- Transfer of all remaining ILW from A59
- POCO of both North and South Cavelines
- Removal of Caveline Roof Laboratories.

The first year of the EAST Project has been dominated by the tasks of Design and Safety Case documentation with the key milestone of the planning application being submitted to Programme in May 2001.

7 LEARNING CURVE

As this is a new form of contract it is important that both teams buy into the ethos of the contract and understand that it is essentially enabling rather than prescriptive that forms the framework for how to deal with a situation.

Issues which need to be considered when embarking upon along the NEC trail are:-

7.1 Programme
A soundly developed programme containing all necessary activities is an absolute necessity. This is required at tender stage and the resource required to produce a realistic document should not be underestimated. For instance the WOMAD Tender needed a full time planner to construct a tender programme of 2000 activities which has now grown to an accepted programme of 3500 activities. The programme must carry enough detail to allow an activity schedule to be produced which will allow a balanced cashflow. Each activity item must have a defined completion point.

7.2 Activity Schedule
This must be clearly and unambiguously constructed to allow a balanced cashflow. Completion statements for each activity accepted by the client are a clear advantage to the assessment process.

7.3 Period of Reply
The standard period of reply is 10 days, this places a strict management practice on both parties but does require a slick document control and administration system. The resource requirements should again not be underestimated that will be required to comply with this requirement.

8 CONCLUSION

One year into the contract has seen. The issue of 20 Early Warning and Compensation Events have been recorded, but notably in full harmony, and the objectives of both client and contractor are being met. A recent UKAEA audit commenting that the integration of the two teams was highly visible and was paying dividends in terms of productivity.

The overall programme is being met along with the intended payment profile. Given this current state it is the opinion of the authors that the NEC was the most suitable contract form under which to carry out the WOMAD works.

It must however be borne in mind that this is one particular project with a clear strategy and objective. With out this the NEC would fare no better than other contract forms where more time is spent on contract clauses rather than achieving the objective.

© With Authors 2001

Implementation

C596/033/2001

The Greifswald decommissioning project – strategy, status, and lessons learned

H STERNER and **D RITTSCHER**
Energiewerke Nord GmbH, Lubmin, Germany

Abstract

On the Greifswald site, 8 units of the Russian-designed reactor WWER 440 are located, including several facilities to handle and store fuel and radwaste. Shortly after the reunification of Germany in 1989, the operating units 1 - 5 were switched off and the construction work at the nearly completely installed units 6 - 8 was stopped. After serious investigations to restart some of the units, a decision was finally taken to decommission all units. Due to this decision, massive personnel reductions were unavoidable.

Technically the work was primarily focused on the removal of fuel and treatment of operational waste to provide the preconditions for decommissioning and dismantling. To solve this task of management of the spent fuel, operational waste and the large amounts of dismantled material, a large interim storage facility was built at the Greifswald site. This facility was built in order to guarantee that continuous dismantling could be achieved throughout the lifetime of the project.

Following extensive but necessary restructuring the owners of the Greifswald site, Energiewerke Nord, have planned and are implementing a challenging nuclear decommissioning programme which is proceeding effectively. This paper describes the strategic approach taken, provides detail of the current status of the work and offers some key lessons which have been learnt to-date.

1. GREIFSWALD SITE AND INITIAL SITUATION

At the Greifswald site (KGR), there are in total 8 reactor units of the Russian pressurised water reactor type WWER 440. The units 1 - 4 are of the model 230 and the units 5 - 8 of the more recent model 213. There are also a wet storage for spent fuel, a warm workshop and additional buildings for the treatment and storage of radioactive wastes, see site layout in figure 1.

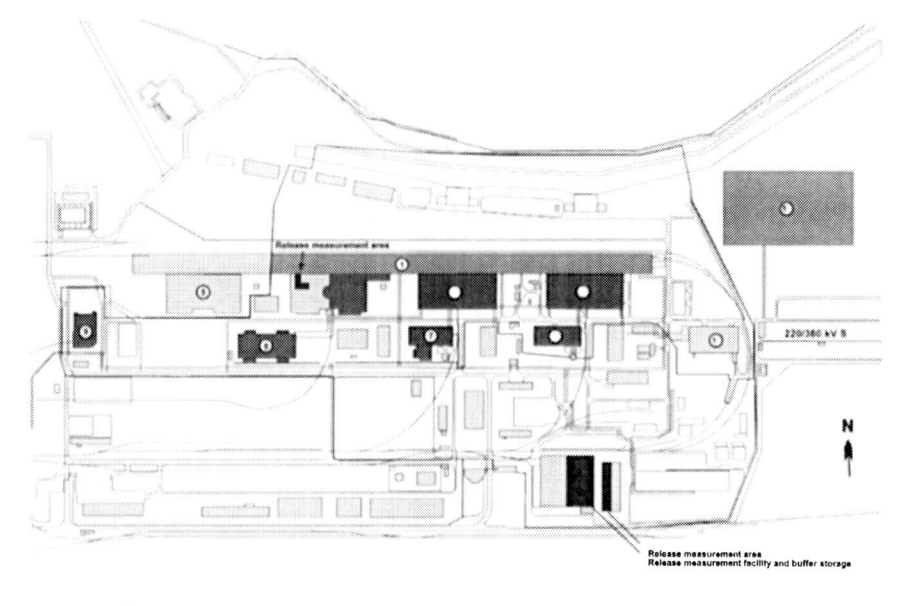

1 Turbine hall
2 - 5 Reactor units 1 - 8
6 Storage of liquid and solid waste
7 Liquid waste treatment and storage
8 Liquid waste treatment and storag
9 Central storage for fuel elements
10 Central warm workshop
11 Interim Storage North

Figure 1: Site layout NPP Greifswald

After the reunification of the German States, the 4 operating units of the Greifswald Nuclear Power Plant were shut down, the trial operation of unit 5 and all construction work for the units 6 - 8 were stopped, see table 1. Investigations in view of the reconstruction of some units showed no acceptable economical solution [1; 2]. Finally, in 1990 the decision was taken to decommission the units 1 - 4, followed by the same decision for unit 5 in 1991. The Energiewerke Nord GmbH (EWN) was created as new owner to replace the combinate "Bruno Leuschner". The sole share holder is the German state (today the Ministry of Finance).

TABLE 1: Basic data on the reactor units in Greifswald (KGR)

unit	type	power [MW(e)]	operation start	shut down	produced energy [GWh]
1	WWER 230	440	1973	18.12.90	41321
2	"	"	1974	15.02.90	40040
3	"	"	1978	28.02.90	36028
4	"	"	1979	02.06.90	32077
5	WWER 213	"	1989	29.11.89	240
6	"	"	ready for commissioning		
7	"	"	building erected, major components installed/delivered		
8	"	"	"	"	"	"	"

At this time (1991) there were on site ca. 5000 employees and a further ca. 1000 in research groups in Berlin, Rossendorf and Leipzig, the ca. 8000 construction workers had already left. This amount of employees can only be understood within the context of the previous socio-economical system. The site is also located in a basically agricultural region without any major industries, which made job relocation very difficult.

After the reunification the society had to be transformed from a commando driven planned economy to a free market economy. In this connection also the complete legal system had to be changed, i. e. also the nuclear licensing authority and authorized expert system had to be renewed. Obviously also the people and employees had to get adjusted to this new environment. Furthermore, it was necessary to introduce "Western" planning and management methods. Thus, it can be understood it was and is a major task to perform this major project within such boundary conditions.

Thus, it was and is a major task to perform this major project within such boundary conditions [3 –5].

2. STRATEGY

As can be understood from the initial conditions mentioned, this decommissioning project is multi-facetted and it was necessary to develop a strategy covering the following key areas:

- personnel,
- decommissioning/dismantling,
- licensing,
- waste/material management.

All these issues are interrelated and had to be solved in an integrated and iterative manner. In one respect EWN is in a good position, namely that the German State has taken over the plant and thus a certain financial basis is secured.

Since no preparatory decommissioning planning had been performed before the decision to shut down, it was absolutely necessary to prepare a basis for the project and also the company.

First of all, a strategic analysis of the company was performed with the aim to:

- establish and evaluate all possible alternative company developments, considering all prevailing boundary conditions (technical, political, legal, economical and social),
- evaluate personnel needs and qualifications,
- transfer the company from an operational structure to a decommissioning project structure.

As a result of this analysis the following corner stones were fixed:

- complete direct dismantling (i. e. no safe enclosure period),
- construction of an interim storage for waste and fuel onsite (to achieve independence),
- transfer of operation license into decommissioning license,

- removal of nuclear fuel from the reactor units into the wet fuel storage on site and later dry storage in the interim storage,
- conditioning and removal of all operational waste,
- establish on overall technical concept,
- perform as much as possible of all activities with existing personnel,
- reuse of the site.

After agreement on this company analysis, it was possible to introduce a project structure and begin the planning of the decommissioning in a well defined manner.

2.1 Personnel

First of all measures had to be taken to reduce the number of employees, since this under all circumstances was much too high. To solve this problem, the following measures and principles were introduced:

- No major contractors,
- Retirement scheme,
- Privatisation/outsourcing,
- Education,
    - decommissioning,
    - better position on labour market,
- Dismissal with economical support,
- Re-industrialization of the site.

In this way it was possible to reduce the personnel from ca. 5000 to only 1100, which is still high, but justifiable. This value will be slowly reduced due natural fluctuations over the project life time. In this way the remaining personnel has become clear perspectives and thus the basis for a motivated workforce has been laid.

Social impact

At the time of the reunification ca. 4900 people employed on the Greifswald site. In addition to this personnel was also employed on other sites and institutes:

-	NPP Rheinsberg	675
-	Morsleben final storage	280
-	Stendal NPP construction	980
-	Research Inst. Berlin	ca. 200
-	Research Inst. Leipzig	ca. 800.

These were all employees of the combinate "Bruno Leuschner", i.e. in total 7800 employees. The site in Stendal and the Research Institute in Leipzig were closed/privatised before we started, and the Morsleben site was completely taken over by a Federal Institute. During the years 1990 - 1994 our strategy for the remaining staff resulted in the following:

- education for new job   ca. 400
- retirement scheme   ca. 275

- new employer[1]   ca. 150
- privatisation and departure with
  economic support   ca. 1600
- dismissal jobless   ca. 1600
- remaining personnel NPP
  Greifswald and Rheinsberg   ca. 1700.

The privatisation efforts concentrated on the activities:

- building maintenance,
- catering,
- cleaning service,
- site supervision,
- hotel,
- educational centre,
- operation simulator,
- research institutes.

In connection with these privatisation efforts, very interesting conditions were given for taking over property and goods.

Of the employees dismissed; there is no detailed statistics as to what happened afterwards. It is estimated that ca. 25% managed to obtain a new job or privatise themselves, ca. 25% went into retirement and roughly the half remained unemployed.

To sum up, it can be concluded that approximately

- 1/3 of the employees remained in the company,
- 1/3 got new jobs,
- 1/3 got jobless or retired.

Once these major efforts had been initiated the psychological work commenced, i.e. to explain to the remaining personnel the new perspectives. It is clear that it is difficult for the personnel to change from operation to dismantling. Especially when the reason is not so clear or well understood. In addition to these internal issues also the society is in transformation. This issue needs time and patience and can only be resolved when living conditions improve.

---

[1] mainly previous West-Germany

2.2 Decommissioning

The second major decision to take was to decide on the decommissioning strategy, i.e. direct or deferred dismantling after a safe enclosure phase. Taken the overall boundary conditions on site, this was clearly a main issue with major implications. In order to resolve this issue on a technical-economical basis, it was necessary to perform a complete project planning and calculation for both alternatives. Hereby, it could be shown that the direct dismantling is ca. 20 % cheaper, produces less radioactive waste and results in less total dose commitment. In order to understand this it must remember that the earlier Russian plants have a limited design life time (especially the buildings), have no containment (i.e. are not tight) and have inadequate stores for operational waste. Obviously the direct dismantling option also had a positive influence on the job situation on site.

The timely planning on the basis of a thorough technical and radiological registration of the plant and the organisation of the overall waste management are absolutely necessary preconditions for a successful project. Due to the lack of disposal facilities in Germany in the near future, the Interim Storage North (ISN) was erected on site as an independent, integrated treatment and storage centre of radioactive waste and dismantled material, as well as a storage of spent fuel in CASTOR casks.

2.3 Licensing

The licensing strategy is an intricate issue, since on one hand, it is easier to execute the project with one licence, on the other hand for a large plant, this represents an enormous effort from the applicant and also from the licensing authority and its authorised experts. If an unplanned plant shut down takes place it is necessary within a short time to prepare the licensing documents, i. e. the initial amount of documents must be limited in order not to seriously delay the project start. Furthermore, it is normally necessary to proceed in an iterative manner with the licensing authority in order to agree on the amount of licensing documents necessary and the needed degree of detail and finally on a licensing time schedule. Since our provisory license ended 30 June 1995 - as a result of the transition agreement on laws between both German States in 1989 - it was tried to obtain an as large as possible license and then to complement this with part dismantling license applications. In this way, the consistent use of personnel capacities, a continuous planning work and the continuity in the licensing procedures and in-process control could be guaranteed.

It can be distinguished between the following 7 main license applications for the dismantling project, see also figure 2.

1. Application for possession, decommissioning, waste management and dismantling of plant parts (turbine hall, controlled area unit 5); granted 30 June 1995.

2. Application for model dismantling of inactive components in unit 5; granted 23 May, 1997.

3. Application for dismantling of plant parts in the controlled area units 1/2; granted 18 July, 1997.

4. Application for the dismantling of plant parts in the controlled area units 3/4; granted 16 July 1998.

5. Application for remote dismantling of the reactors 1 – 4.

6. Application for the dismantling of waste storage and treatment facilities and the post operational systems.
7. Decommissioning and dismantling of the wet interim storage for spent fuel.

After lengthy discussions with the authority and authorised experts, it was furthermore agreed that no public hearing is required, since there is no real public concern. However, the importance of informing the public of progress and developments on the project is well recognised and this is achieved through liaison committee with representatives from politics, NGO's and the public who meet regularly.

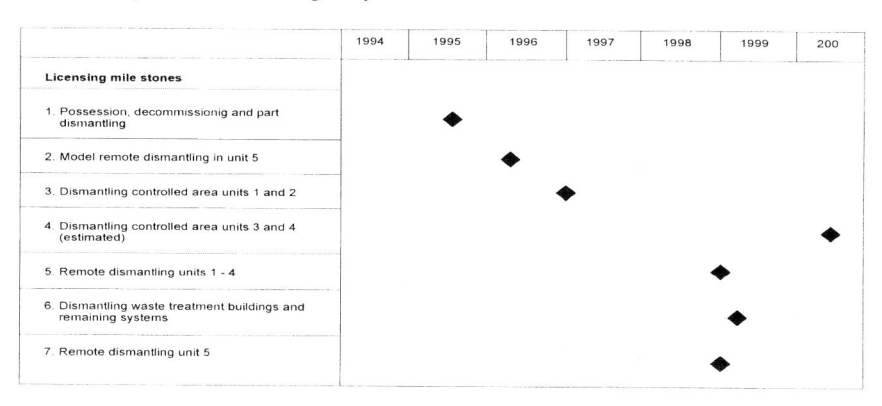

Licensing mile stones	1994	1995	1996	1997	1998	1999	200
1. Possession, decommissionig and part dismantling		◆					
2. Model remote dismantling in unit 5			◆				
3. Dismantling controlled area units 1 and 2				◆			
4. Dismantling controlled area units 3 and 4 (estimated)							◆
5. Remote dismantling units 1 - 4					◆◆		
6. Dismantling waste treatment buildings and remaining systems						◆	
7. Remote dismantling unit 5						◆	

Figure 2: **Licensing mile stones Greifswald**

2.4 Waste/material management

The waste management concept is mainly based on the following boundary conditions and principles:

- Provision of sufficient buffer and intermediate storage capacities to achieve a high flexibility in the logistics and waste management, including the construction of the Interim Storage North (ISN).

- Removal of the spent fuel from the reactors and cooling ponds to the wet interim storage (to obtain easier boundary conditions for dismantling activities) and later transport in dry CASTOR casks to the ISN.

- Installation of equipment for the treatment of dismantled material using modern technologies for the reduction of dose exposure and increase of efficiency.

- Further use of the existing waste facilities, upgrading or extension, as far as it is economically justified.

- Use of the final storage capacity of the ERAM disposal facility as far as possible (closed 1998).

2.4.1 Fuel elements

After the shut down of the plant in 1990, there were 5037 spent fuel elements - 3 of them are defect - and 860 fresh fuel elements on the Greifswald site. At that time, 1011 fuel elements were in the reactors, 1628 in the cooling ponds of units 1-5 (incl. 3 defect ones), and 2398 were stored in the wet interim storage.

The fuel elements from the reactors and the cooling ponds have been transferred as far as possible to the wet fuel storage in order to create easier conditions for dismantling activities. Subsequently the fuel will be transferred to CASTOR casks for dry interim storage. A part of the low burn-up fuel (elements) has been sold to Hungary (Paks NPP). The 860 fresh fuel elements have also been sold.

2.4.2 Operational waste

The already produced radioactive operational waste and the waste which will be generated during post and decommissioning operation was planned to be disposed of in the Morsleben disposal facility (ERAM). Up to the closure of this disposal facility in 1998, due to license formalities, approx. 2584 m^3 of conditioned waste could be disposed of.

The operational waste consists of activated and high-contaminated solid waste, low-active and medium-active resins as well as liquid evaporator concentrates, sludge and sludge-resin mixtures, sludge from outside the controlled area, low-activated resins in the turbine hall and solid radioactive mixed waste in storage bunkers as well as in temporary buffer stores.

The major part of this operational waste can be treated by common conditioning techniques. For this purpose, the following facilities and devices are available on the Greifswald site: rotary thin film evaporator plant (max. 400 l/h), drum drying facilities, equipment for drying of ion exchanger resins, compaction facility (20 Mg), sorting facilities, and high efficiency suction devices for granular material. Treatment can also be performed in the new interim store with high pressure compaction, cutting and shearing.

3. DISMANTLING

3.1 Strategy

The basic principles for the dismantling can be summarised as follows:

- it will be performed from system/areas with lower contamination/radiation to higher and finally activated plant parts,
- it will start in unit 5 and the turbine hall and then continue in unit 1 - 4 in order to use the experience from work in a low dose rate/contamination unit,
- as far as possible market equipment will be used,
- as large components or parts as possible will be dismantled and transported to the ISN for decay storage and/or further treatment,
- it will take place on a room basis, i. e. not on a system basis.

Preparatory to execution of dismantling works, measures are taken to reduce the dose rate. First of all parts, of the primary loops were decontaminated electrolytically and secondly hot spots were removed with high pressure water jet or mechanically. Before dismantling activities start asbestos containing isolation material is removed in a controlled manner [6].

The dismantling itself in the monitored and the controlled area takes place with conventional, preferably mechanical tools.

The dismantling rates achieved are:

- monitored area   46 kg/manhour
- controlled area unit 5 39 kg/manhour
- controlled area unit 1 - 4 25 kg/manhour

These figures give the pure dismantling rate without any infrastructure resources.

3.2 Remote dismantling

The reactor pressure vessels (RPVs) and internals of the units 1 - 4 will be remotely dismantled. The activated components of unit 5 will be dismounted and transported to the new interim store. The dismantling activities will take place in the steam generator room which is situated around the RPV. Here cutting caissons (dry and wet), package and transfer areas will be installed, see figure 3. The complete system will be dismountable and will first be installed in unit 5 for inactive testing and afterwards in unit 2 and finally in unit 4. The inactive testing started mid 1999 and will be terminated end of 2001. The selected techniques to be applied are summarised in table 2.

TABLE 2: Cutting techniques for the remote dismantling

Cutting area	Components	Techniques
dry	• reactor pressure vessel • upper part of protection tube unit • upper part of reactor cavity	band saw disc cutter plasma arc oxyacetylene burner
wet (pool)	• core basket • lower part of protection tube unit • lower part of reactor cavity • cavity bottom	band saw CAMC[1] plasma arc fret-saw

1) CAMC = Contact Arc Metal Cutting

After the dismantling of the operational equipment, the preparations for the installation of the cutting equipment in steam generator room unit 5 were executed from June 1997 until March 1998. Approximately 780 t have been removed from the wall and floor areas to have space for the installation of the cutting equipment. From March 1998 up to September 1999, the main equipment was installed, and since December 1998 the systems have been under commissioning. In October 1999, the reactor pressure vessel was put into the dry cutting place and thus, model dismantling could start.

Arrangement of cutting places
For the cutting of the activated reactor components, one cutting place is foreseen for each of two units (see figure 3). For the units 1 and 2, this cutting place will be arranged in the steam generator room of unit 2 and for the units 3 and 4 in the steam generator room of unit 4. For each dismantling area, one dry cutting place is foreseen for the cutting of the reactor pressure vessels and the lower activated parts of the reactor cavities and protecting tube systems. For the cutting of the core basket, cavity bottoms and the higher activated areas of the reactor cavities and protecting tube systems, a wet cutting place will be installed.

The two cutting caissons as well as the packing station are implanted below openings to the reactor hall, so that the loading and unloading of these areas can be performed without problems with the overhead crane in the reactor hall.

Dry cutting caisson
The dry cutting station is divided in a pre- and post cutting area divided by a shielding wall. Through a vertical gate in the shielding wall, it is possible to move a transport vehicle between the cutting areas. The dry cutting caisson is equipped with an under-pressure control system.

The component to be cut is placed in the pre-cutting area on a turntable. The turntable is mounted on the transport. The component is fixed with a wire hoist. In the pre-cutting area, the horizontal cuts are performed. With the transport vehicle, the piece is subsequently transported to the post-cutting area. There, the vertical cutting takes place into pieces suitable for packing and later final storage. The normal cutting tools are band saws. In the framework of the testing programme also autogenous and plasma cutting will be tested.

In the packing station which is directly connected to the caisson, the containers are loaded and covered with a manipulator. During filling, the container is equipped with a protective sleeve in order to avoid exterior contamination.

Wet cutting caisson
This caisson consists basically of a cutting pool with cutting devices, different transport and handling devices, water purification system, emptying and filling connections and a ventilation system.

The caisson is separately ventilated and kept under negative pressure. As a consequence, neighbouring rooms are not radiologically affected.

The component to be cut is placed in the caisson on a turntable. The cutting is normally performed with band saw, horizontal as well as vertical. Optionally, also plasma-arc and CAMC-cutting devices are foreseen. For the handling, two manipulators are foreseen. One is used for the handling of the cutting tool and the other one for the handling and transport of the pieces, which are placed in wire mesh baskets. The loaded baskets are transported to the packing station with an overhead crane and placed in a container. Subsequently, the container is automatically closed.

Experience
Despite the use of mainly conventional techniques, initial difficulties were encountered with some equipment parts and e. g. the video system. It is to be noted that all equipment has been designed for the subsequent active operation. Up till now the dry cutting installation has been tested with the reactor pressure vessel and the wet cutting pool with the core basket. The results are so far very positive: the ventilation system and the water cleaning system are very effective, the under water plasma cutting technique, the dry and wet band saw are operating very well. The CAMC technique is also working very well, however, the mechanical stability of the electrode had to be improved. Due to residual tension in the material, in some cases the band saw was stuck. For this case careful planning of the cutting procedure and a mechanical devices to prohibit the closing of the cut have solved the problem.

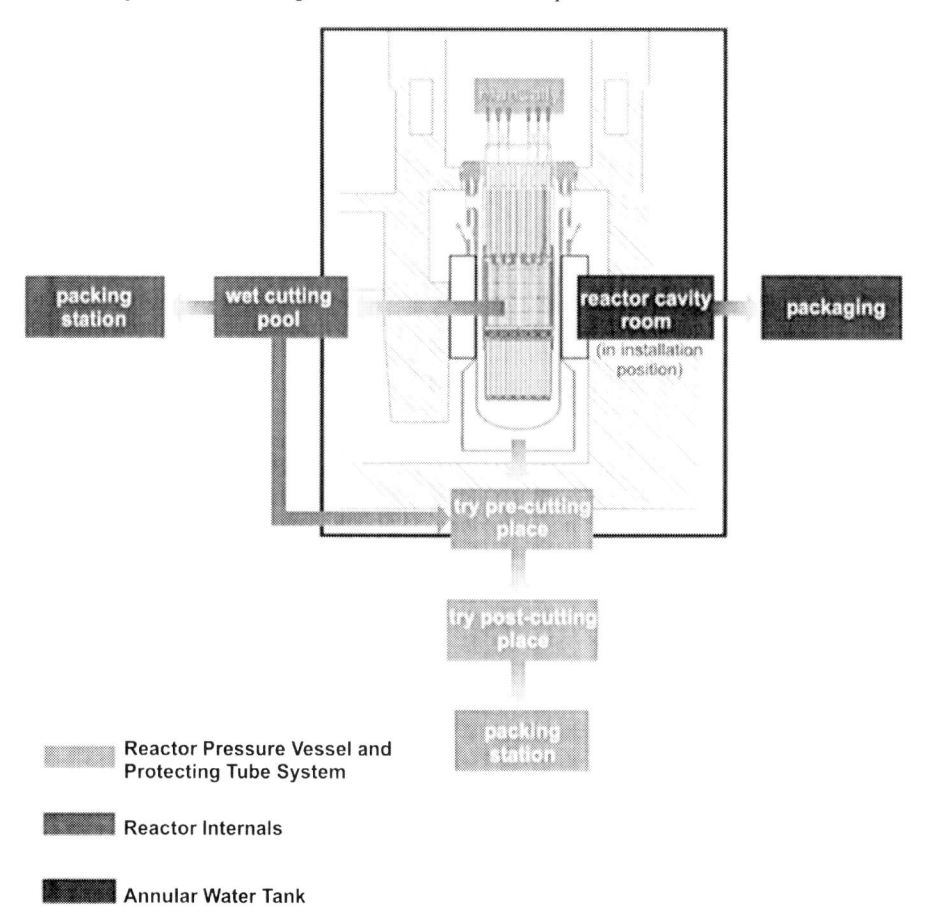

Figure 3: Dismounting, cutting and packaging of reactor components

4. DECOMMISSIONING WASTE

4.1. Inventory

For the planning, it was mandatory to record the actual plant state. This particularly included the registration of masses and materials, dose exposure rates in the plant rooms (ambient) and at the main components, as well as the preparation of a contamination catalogue, i. e. the interior contamination of the systems [2].

In total, **1 800 000 Mg** of material is expected to be produced during the dismantling of the Greifswald plant (complete site). The definitively non-radioactive masses can be treated as conventional residues and reutilised or removed according to the valid waste regulations. The remaining **570 000 Mg** have to be classified as radioactive residuals or possibly contaminated material and must be treated accordingly. The resulting material flow in this dimension is extraordinary and difficult to manage logistically. Figure 4 shows an overview of the mass flow during the decommissioning and the prospective classification to different disposal routes, based on the mass and radiological inventorisation.

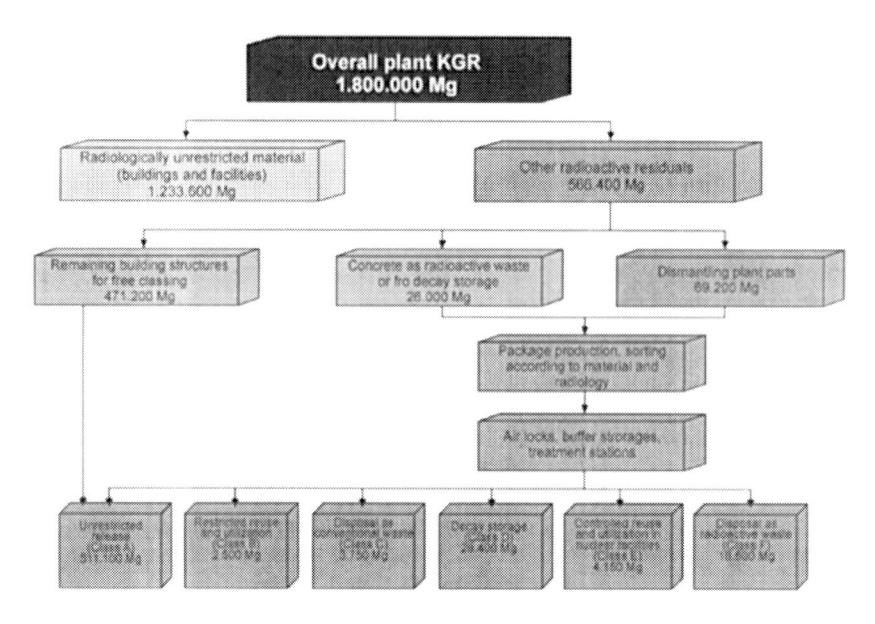

Figure 4: Mass flow at the Greifswald site

The activity inventory of the Greifswald site in 1990 excluding spent fuel was approx. 3.5 E+17 Bq. The share of the activated components is 99%. The remaining percentage includes the operational waste, contaminated plant parts, and the building structures of the controlled areas.

---

[2] Eventual contamination detected in rooms, is directly removed.

4.2. Disposal routes

The timely definition of the disposal routes under inclusion of the collection and sorting of the produced masses and their classification makes it possible to proceed systematically and to reduce the amount of radioactive waste. It is distinguished between the following classes (compare with figure 4):

Class **A** Free release
Class **B** Reuse
Class **C** Disposal as conventional waste
Class **D** Decay storage
Class **E** Reuse in nuclear facilities
Class **F** Disposal as radioactive waste

The limit values for the classes A, B and C are an integral part of the license for decommissioning from 30 June 1995, see table 3.

TABLE 3:   Licensed limit values for different disposal paths

Class	Disposal path		Corresponding limits		
A	Unrestricted release of metals	specific activity: surface contamination:	< 0.1 Bq/g < 0.5 Bq/cm^2	all nuclides e.g. Co 60	10 000 Mg/y
	Unrestricted release of other residuals	specific activity: surface contamination:	< 0.2 Bq/g < 0.5 Bq/cm^2	(Co 60-equiv.) (e.g. Co 60)	100 Mg/w
B	Restricted release and utilization (melting) of metal scrap	specific activity: surface contamination:	< 1.0 Bq/g < 0.5 Bq/cm^2	(all nuclides) (e.g. Co 60)	10 000 Mg/y
	Release of debris for further utilization	specific activity:	< 0.2 Bq/g	(Co 60-equiv.)	100 Mg/w
C	Disposal as conventional waste	specific activity:	< 2.0 Bq/g	(e.g. Co 60)	(100 Mg/y)
D	Decay storage	Materials which cannot be classified as Class A, B or C due to a too high contamination, but which will surely reach these classes within the following 10 - 15 years due to radioactive decay.			
E	Controlled utilization	Materials which can be used in other nuclear facilities.			
F	Radioactive waste	All materials which cannot be classified as class A to E and which will be orderly removed as radioactive waste.			

5.   TREATMENT AND STORAGE

After dismantling, the plant parts are sorted and packed in appropriate packages for transport, further treatment and storage. Geometrically homogeneous parts or individual large components can also be declared as a package. When a package is ready, it is immediately provided with a routing card containing all important data for registration and tracking of the package over the whole treatment process until the corresponding disposal goal has been reached. This system is supported by developed software routines. The main stations of the material flow will be shortly described below.

The free classing [7] can be executed with surface contamination monitors and a free classing facility. The free classing facility is used for all smaller plant parts and pieces which can be filled into drums, pallets etc. These can be measured as a whole in an equipment consisting of a shielded chamber with several large γ-detectors. Materials suitable for free classing with surface contamination monitors, i.e. thick walled components with a clear geometry, and others with additional sampling (wipe tests, material samples) will be measured in free classing areas.

The warm active workshop accommodates machine tools, baths for chemical decontamination, cranes etc. These facilities are used and now extended with additional facilities for decontamination; abrasive blasting and high pressure water cleaning facilities as well as electrochemical procedures. The aim of this decontamination is the unrestricted or restricted release of the treated material (classes A and B).

The interim storage facility (ISN) [8] on the Greifswald site started operation in March 1998. This store serves as a treatment station, a buffer storage and an interim storage. In this way, the logistic security needed for a continuous dismantling operation is guaranteed. The building comprises eight halls, a loading corridor and a treatment/conditioning area. The store has a theoretically usable storage volume of 200 000 m^3. Storage hall 8 will house spent fuel elements in CASTOR casks. Storage hall 6 and 7 will be used for big components from the primary circuits, awaiting further treatment. Halls 1 – 5 is utilised as interim and buffer store for all kinds of packages. The treatment/conditioning area consists of five caissons, one of which is intended to be used for maintaining fuel element casks. The other caissons will be used for treating and conditioning activities, e.g. cutting, shearing, high pressure compaction, concentration of liquid waste, drying and packaging.

6. PROJECT MANAGEMENT

On the basis of an analysis of the company development and personnel strategy, a technical concept was worked out and the project was broken down to work package level.

The EWN Management Information System (MIS) is a software package consisting of the following main parts:

- decommissioning information system,
- documentation management and service event tracking system, and
- environmental information system.

The most important part of the MIS is the decommissioning information system covering the complete process from project planning to project progress supervision. The other parts have been developed as a support to increase the functionality of the system.

The objectives of a successfully working MIS are:

- to make the decommissioning process transparent,
- to get real and convincing information about all important subjects concerning the decommissioning process at any time,

- to support decision-making processes in the course of decommissioning to increase safety and efficiency,
- to collect, systematize and evaluate experience in the course of decommissioning.

To meet all requirements, the MIS built up in several modules which fulfil the following criteria:

- use each module separately,
- provide the complex application with a maximum integration of each module,
- implement each module step-by-step,
- implement other programmes or particular functions defining appropriate interfaces,
- integrate already existing software applications or various measurement facilities defining appropriate interfaces,
- rapidly react on changing conditions and to integrate new functions.

Based on these general considerations and basic requirements on a software system for a nuclear decommissioning project, a market investigation was performed. Because there was no available commercial software, it was necessary to develop a new software in house with some assistance by contractors.

Thus, with this developed software it is possible, in addition to the normal project control tasks to support the technical planning, work preparation planning, tracking and control of dismantled material and radioactive waste etc. Actual data from the dismantling operations are registered, evaluated and fed back into the system, see figure 5

The project was optimised from the cost and personnel point of view in order to achieve a smooth personnel curve over the project lifetime. The basic time schedule is shown in figure 6. As can be seen the critical path goes over the mock-up remote dismantling in unit 5 to the remote dismantling in the units 1 - 4 and the dismantling of auxiliary systems and finally the building demolition.

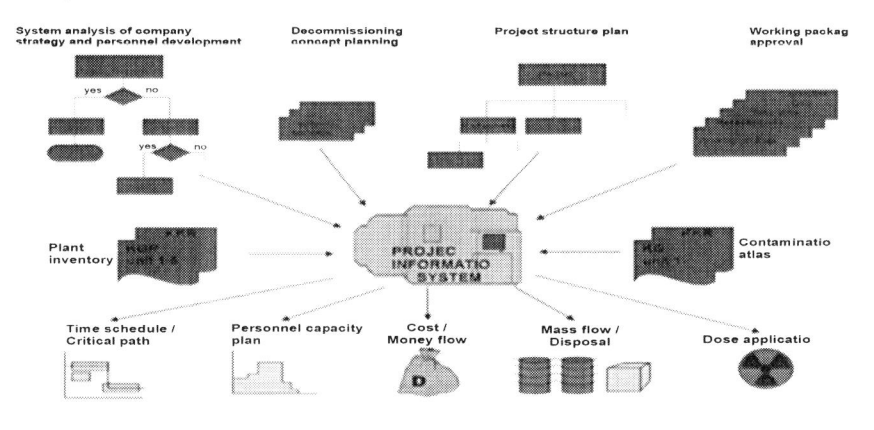

Figure 5: **Project management system**

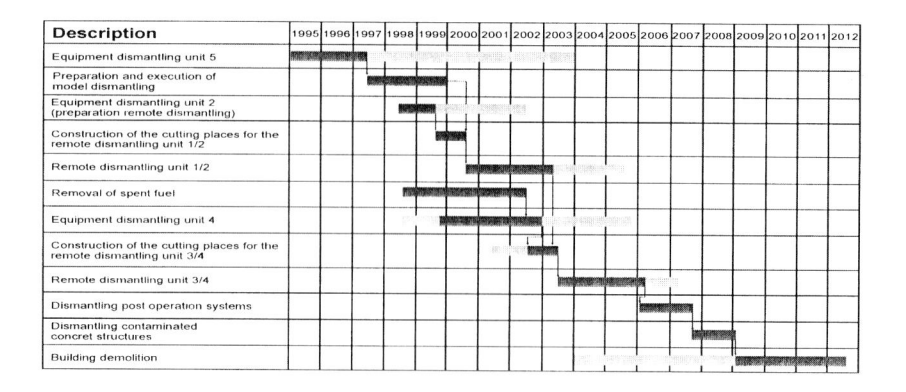

Description	1995	1996	1997	1998	1999	2000	2001	2002	2003	2004	2005	2006	2007	2008	2009	2010	2011	2012
Equipment dismantling unit 5																		
Preparation and execution of model dismantling																		
Equipment dismantling unit 2 (preparation remote dismantling)																		
Construction of the cutting places for the remote dismantling unit 1/2																		
Remote dismantling unit 1/2																		
Removal of spent fuel																		
Equipment dismantling unit 4																		
Construction of the cutting places for the remote dismantling unit 3/4																		
Remote dismantling unit 3/4																		
Dismantling post operation systems																		
Dismantling contaminated concret structures																		
Building demolition																		

Figure 6: Basic time schedule and critical path

7. STATUS OF THE PROJECT (MAY 2001)

7.1 Dismantling

The dismantling activities started in October 1995 in the controlled area of unit 5 and in the turbine hall. Today dismantling work is going on in the controlled areas of units 1 – 4 and in the turbine hall. Totally, 46,962 Mg (30.04.01) plant parts have already been dismantled, packed and stored on the site or in the ISN, or if possible free classified. The operation with the free classing facility started at the beginning of 1996 and is, due to the produced masses operated in two shifts. Up to the end of April 26,067 Mg have been measured and released by the authorities. Up to end of April in the hall 7 of the ISN 11,480 Mg radioactive waste and residuals have already been stored, including steam generators (each 166 Mg) In hall 8 of the ISN 14 CASTOR casks with spent fuel have been stored including 4 casks with all spent fuel elements of KKR.

7.2 Costs

The costs only for dismantling are calculated with 400 Mio DM per unit, excluding the conventional demolition. Due to the serial effect and the availability of the ISN, the expenditures are clearly below the amount of 600 - 700 Mio DM which is estimated by the energy utilities for the dismantling of one NPP unit in the FRG. For the conventional dismantling of the overall plant in Greifswald, additional costs of ca. 400 - 500 Mio DM have been planned.

7.3 Site reuse

During the initial personnel reduction phase it was possible to establish a number of small and medium size enterprises on the site. In total slightly less than 1000 work places were created. Recently 2 contracts were closed for the construction of gas fired power plants (total capacity 2400 MW(e)). The creation of 400 new jobs in new firms has been guaranteed by these 2 companies. Thus, despite the rather isolated location of the site, it has been possible to keep the site as industrial and energy producing site. The efforts in this area are continuing.

When closing down the plant a political decision was taken to keep the site as industrial and energy producing site. And the work started then to try to find interested power utilities. Despite the existence of infrastructure, e.g. switch yard and grid connection and cooling water channel, all efforts reached only a conceptual stage.

During the initial personnel reduction phase it was possible to establish a number of small and medium size enterprises on the site. In total slightly less than 1000 work places were created.

However, as a consequence of the deregulation of the German electricity market in 1998, the interest of foreign investors on the German market has increased dramatically. This has lead to the purchase of two pieces of land on site of two Scandinavian utilities for the construction of two 1200 MW (3x400) G&S power plants, see figure 5. Both investors have also to guarantee the creation of 200 new jobs on site by allocation of other companies. The aim is to find energy consuming enterprises which can benefit from the power stations and coastal location. Presently negotiations are going as with wood processing industries which would get their raw material by ship from the Baltic States.

In order to prepare the site, the infrastructure is being improved. A major part here is the creation of a harbour area at the cooling water outlet channel.

7.4 International activities

EWN is trying to convert the know how from operation and maintenance of WWER reactors and the presently ongoing decommissioning activities into new long term jobs, especially in the engineering area. Due to the previous socialist structure, it was necessary for the personnel to perform all activities on their own, e. g. planning and execution of modifications, maintenance etc. This means that there is a very well-founded engineering know how which now has been mixed with Western experience, rules etc.

Thus, EWN now has the ideal background to perform engineering work in the previous East European countries, especially in the decommissioning and waste management field, notably the following specific areas:

- Decommissioning planning,
- Preparation of licensing documents,
- Environment assessment analysis,
- Project implementation and management,
- Waste management and material flow,
- Remote dismantling operation,
- Development and implementation of software routines.

In this framework EWN is involved in decommissioning projects in Chernobyl (Ukraine), Kozloduy (Bulgaria), Russian submarines in Bolshoi Kamen. Other projects include sampling of WWER pressure vessels, delivery, education and commissioning of photogrammetry equipment for Russian power stations and R/D-projects for the European Commission.

8. CONCLUDING REMARKS

After initial difficulties caused by massive personnel reductions combined with the introduction of a market economy and West German laws and procedures, EWN has succeeded in restructuring the company to arrive at a size suited to the task of decommissioning. A positive atmosphere has now been created to enable work to proceed effectively and to prepare part of the personnel and the site for the new tasks.

The decommissioning and dismantling of the Russian WWER type reactors do not pose specific problem when compared with the Western PWRs. However, the size of the project and the resulting mass flow is extraordinary. It can be concluded that dismantling of nuclear facilities is basically not a technical problem but a challenge to project management and logistics, once the legal and economical boundary conditions have been clarified. In order to achieve a safe and cost effective project, it is necessary that all stakeholders, i.e. EWN, authority and authorised experts, and public achieve a positive co-operation.

The project has proceeded very well: major licenses have been obtained, agreement on licensing strategy with the authority has been achieved, fuel elements have been transferred, interim storage of radioactive waste, dismantled material and spent fuel is running. For the future, the 2 gas fired power stations will keep the site as an energy producing site and thus, we will have a nucleus for further industry settlements.

To sum up the lessons learned are:

- The development of a comprehensive inventory (radiological, material) is a necessary prerequisite for all planning especially waste management logistic.
- Social aspects and psychological effects must taken into account.
- Clear licensing structure - one license better if not too large project.
- Clear and realistic requirements from the licensing authority (related to real safety risks) - avoid self-service shop mentality.
- The overall project must be planned, i. e. from shut down to disposal.
- Establish a project structure and integrate all site activities.
- The dissemination of open public information is a key activity.
- Decommissioning is basically not a technical problem, but rather a management and waste management logistic issue.
- Simple and industrial sturdy tools and equipment; mock-up tests if new or complicated technology.
- ALARA-principle must be strictly applied already in the planning phase.

REFER(ENCES

(1) Sicherheitsbeurteilung des Kernkraftwerkes Greifswald, Block 1 – 4
    Gesellschaft für Reaktorsicherheit; June 1990; GRS-77

(2) Sicherheitsbeurteilung des Kernkraftwerkes Greifswald, Block 5
    Gesellschaft für Reaktorsicherheit; August 1991; GRS-83

(3) Sterner, H. etal. Stilllegung und Abbau des Kernkraftwerkes Greifswald
    Atomwirtschaft (40) 1995, pp 247 – 252

(4) Sterner, H. etal. Greifswald and Rheinsberg, East European WWER's with a new mission
    Nuclear Engineering International (40) 1995, pp 19-21

(5) Sterner, H. etal Decommissioning and waste management at the Greifswald NPP site.
    7th International conference on radioactive waste management and environmental remediation, Nagoya 26-30.09.1999

(6) Cornelius, M. etal. Experience by the decommissioning of WWER reactors in Greifwald
    1st European ALARA network workshop on ALARA and Decommissioning
    Saclay 1-3.12.1997. Proc EC radiation protection 108., 1999,
    ISBN 92-828-7107-X.

(7) Clearance of materials, buildings and sites with negligible radioactivity from practices subject to reporting or authorisation
    Recommendation of the Commission on Radiological Protection
    SSK Heft 16; 1998; ISSN 0948-308X

(8) Hartmann, B. and Leushacke, D.
    The role of the world's largest interim storage for waste and fuel within the Decom project of 6 WWER reactors in Greiswald and Rheinsberg (Germany)
    Proc. waste Management conf. Tuscon 1999

© With Authors 2001

C596/016/2001

Retrieval of intermediate level waste at Trawsfynydd nuclear power station

S WALL and **I SHAW**
AEA Technology Nuclear Engineering, Harwell, UK

1. INTRODUCTION

Trawsfynydd nuclear power station has two Magnox type reactors which were shut down in 1995 at the end of their working life. During normal operations two types of intermediate level waste (ILW) have accumulated on site; namely Miscellaneous Activated Components (MAC) and Fuel Element Debris (FED). MAC is predominantly components which have been activated by the reactor core and then discharged. FED mainly consists of fuel cladding produced when fuel elements were prepared for dispatch to the reprocessing facility. As part of the decommissioning programme for the site, these waste streams are to be retrieved from storage vaults, monitored, packed and immobilised in a form suitable for on site storage in the medium term and for eventual disposal in the long term waste repository to be commissioned by NIREX at a later date.

In 1996, BNFL Magnox Generation, defined the waste storage strategy and awarded two contracts to AEA Technology Nuclear Engineering for the design, supply, commissioning and operation of equipment to retrieve, pack and immobilise the two waste streams. This included waste recovery from vaults in both reactor and pond locations and the final decommissioning and removal of plant from the site after successful completion of the waste recovery programme.

2. PROJECT REQUIREMENTS

2.1 The Waste
Operation of the Trawsfynydd Nuclear power station has resulted in the production of radioactive waste material of both low and intermediate level activity. Low level wastes have been packaged and transferred to the Drigg disposal site in Cumbria, but solid intermediate level waste is stored in below ground concrete vaults on site.

2.1.1 MAC

MAC waste mainly comprises of irradiated core components such as flux flattening bars and pantagraph rods, but also includes a wide range of items discharged from the charge face, such as thermocouples, cables, grabs and steel components from stand pipe closure assemblies. The majority of the waste comprises of flux flattening bars, which are cylindrical, 75mm in diameter and 760mm long.

MAC is stored in two vaults located in the basement areas of the bioshields of reactors 1 and 2. The waste was discharged through two chutes that start at the reactor pile cap and control rod service bay and run approximately 35 metres down into the vaults. Some of the smaller items were packed into canisters and it is thought likely that some of them would have split open as a result of the drop.

2.1.2 The MAC Vaults

The two MAC vaults are approximately 7.5m by 6.5m by 3m deep. They are constructed from concrete and have a wall thickness of 1m and roof thickness of 1.5m. The walls of the vault were protected from the falling waste by mild steel crash plates. These are part of the vault structure and as such will not be retrieved as waste but will be decontaminated, as necessary, during final decommissioning. The vaults are located beneath the carbon dioxide inlet duct chamber 3 for reactor 1 and 10 for reactor 2. (see Figure 3).

2.1.3 FED

FED waste comprises the Magnox cladding and end fixtures stripped from the fuel elements prior to their dispatch for reprocessing. This involved separating the fuel rods from ancillary items including magnox splitter cages, nimonic springs, housing cups and thermocouple attachments. This material was discharged into two sets of 16 storage positions called the north and south fuel element debris vaults situated in the ponds buildings.

2.1.4 The FED Vaults

The FED vaults are each approximately 2m by 2m by 4m deep and most are between three quarters and completely full. The waste was discharged using a vibratory conveyor system housed in the chambers directly above the vaults.

MAC waste in storage vaults

FED waste in storage vaults

2.2 Scope of the Projects

The projects includes all activities necessary for the safe recovery and packing of the defined ILW waste streams in a form suitable for on site storage and eventual disposal in the NIREX waste repository.

Primary activities required to be completed included;
- Detailed site survey of the vaults and adjacent areas including the layout of plant and equipment and available services.
- Scheme and detail design of the plant and equipment and the production of calculations and manufacturing drawings
- Preparation of the safety case (BNFL Magnox obtain necessary approvals)
- Manufacture, assembly and off site works testing of the plant and equipment
- Site preparation, delivery, installation and commissioning
- Training of operators and operation of the facility.

Requirements for waste storage included the use of 'NIREX' $3m^3$ stainless steel boxes together with separate concrete overpack boxes which provide the necessary shielding. The $3m^3$ boxes and overpacks are to be stored in the reactor sub-basements and special handling equipment is needed accordingly.

The MAC and FED waste streams are recovered from concrete waste vaults at four separate locations on the site. These locations are within existing buildings and provide very limited access for the deployment of recovery equipment. Appropriate access routes are cut through existing structures and the waste recovered to a contained facility where it is safely catalogued and packed into $3m^3$ stainless steel boxes. The waste is immobilised in grout and placed inside a reinforced concrete overpack. The completed package is transferred to one of four temporary ILW stores constructed in the reactor sub-basements.

3 TECHNICAL SOLUTION

3.1 Outline

Because of the difficulties of accessing the waste a conventional solution to waste retrieval would be to provide local recovery and transportation of wastes across site (Flasking) for input to a large multi-waste processing facility. This approach was considered unattractive for the following reasons.

- **Cost** - A large facility would be expensive to build and decommission.
- **Programme** – the design/construction programme would be comparatively long compared to the operating life.
- **Production rate** – the required production rate would be difficult to achieve due to the number of potential waste movements across site.
- **Secondary waste** – potential for large quantities of secondary waste during final decommissioning.
- **Public relations** – difficult to obtain approval to construct a major new facility on a decommissioning site

An alternative and innovative approach was adopted, whereby the plant size was minimised and incorporated into the existing spaces above and adjacent to the vaults. This had advantages in the control of contamination and secondary waste and also reduces the subsequent decommissioning required by making use of areas already scheduled for decommissioning. This strategy has resulted in two facilities (one each for MAC and FED) designed to be operated at the point of waste recovery. At the completion of recovery operations at the first location the plant design enables them to be dismantled and transferred to the second location to complete operations.

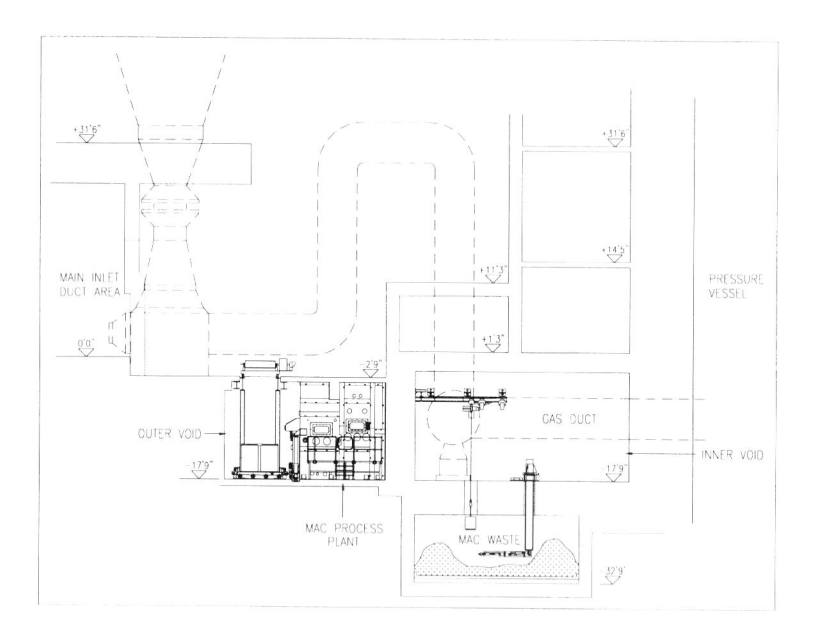

Figure 1. MAC plant : Elevation

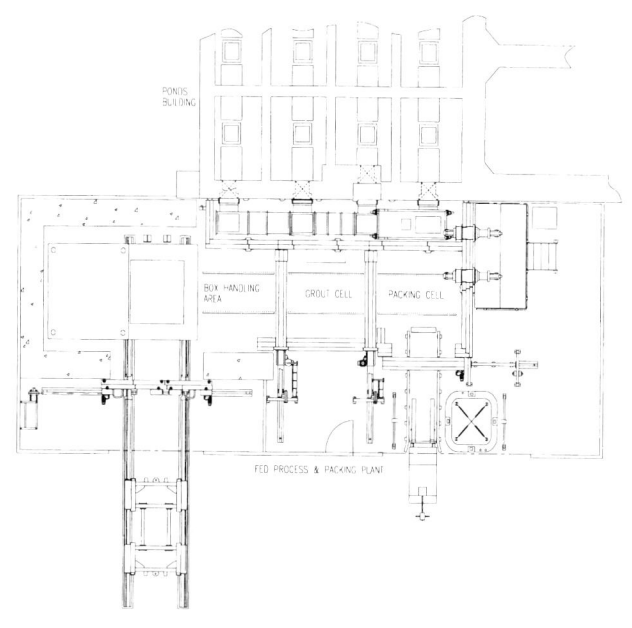

Figure 2. FED Plant : Plan view

3.2 Waste Retrieval

The treatment of MAC and FED waste after recovery is similar but the method of waste retrieval is quite different, reflecting the different problems involved. The most significant differences in the retrieval solutions are;

- MAC recovery is achieved with a hydraulic Artisan manipulator arm deployed through a series of new access points in the vault roof.
- FED recovery utilises two retrieval units which deploy petal grabs through tubes originally connected to the waste discharge conveyors.

3.2.1 MAC Retrieval

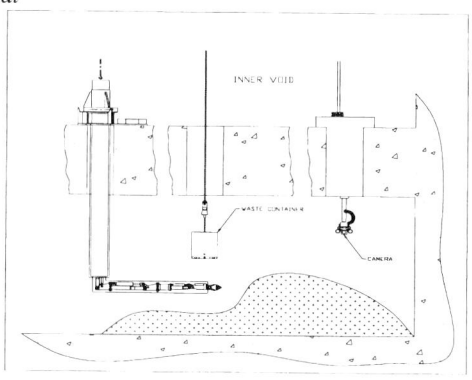

Figure 3. Artisan deployed in the Vault

Individual items of MAC are picked up in the jaws of a hydraulic manipulator, located in the inner void, operating through a access holes in the roof of the vault. The waste is loaded into a basket suspended in the vault from a travelling hoist. The manipulator is a proprietary unit developed for use in a radioactive environment. It can operate from any one of six newly cut holes in the vault ceiling to give full coverage of the vaults. The manipulator is operated from a remote control room by means of two 3-axis joysticks, linked to a PC controller, and two CCTV monitors. The radiation tolerant cameras and associated light units, mounted from plug units in the roof of the vaults, provide pan tilt and zoom functions.

The travelling hoist also operates through a hole in the roof of the vault from the inner void. The hoist is mounted on an overhead monorail, forming the waste transfer system, so that when the basket is full, it can be raised through the access hole and transferred to the packing cell via the waste transfer tunnel. Three gamma monitors in the transfer tunnel measure the exterior dose rate of the container as it passes through.

3.2.2 FED Retrieval

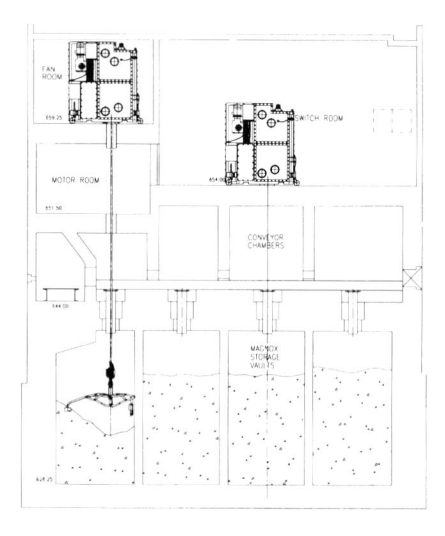

Figure 4. FED Retrieval Machines

FED is picked up by a petal grab(s) deployed from a retrieval unit(s) situated in the rooms above the vaults. A new waste transfer trolley system has been installed in the chambers below where the vibratory conveyor systems were housed The old conveyors had to be cleared of waste and removed before the new trolley systems could be installed.

The grab is withdrawn to a point where it releases the waste into the trolley. The trolley is then driven to a the waste conveyor that, in turn, transports the waste to the packing cell, via an assay station.

The Retrieval Units
The mobile retrieval units comprise a high integrity glove box which houses an electric winch, pneumatic hose reeling system and grab storage station. Each unit is equipped with 3 different retrieval grabs allowing waste to be recovered from any position in the vault. The retrieval units can be moved over any of the 16 positions in the conveyor chamber floor which give access to the vaults. They locate on 10 inch diameter stand pipes which cover each of the holes in the floor and are fitted with blanking shield plugs to maintain containment.

The Transfer trolleys
The trolleys run on rails in ducts installed in the old waste conveyor spaces. Each trolley serves four of the vault access shafts and there are four trolleys at each of the two pond locations. Stand pipes, ten inches in diameter, line the access holes and form a containment between the vaults and the trolley ducts and also between the trolley ducts and the retrieval units. A ventilation system maintains the ducts at a depression to produce an inward airflow to minimise the potential for spreading contamination.

3.3 Waste Processing
The two waste streams undergo similar processing consisting of the following activities:
- Monitoring the retrieved FED waste so that fuel fragments can be identified and separated.
- Monitoring dose rates and waste sorting so that efficient packing and self shielding in the boxes can be achieved
- Packing waste into $3m^3$ boxes for safe storage
- Fitting a lid and grouting the waste to immobilise it in the boxes
- Applying a capping grout to completely fill each box, after the initial fill has cured.
- Swabbing the boxes to ensure they are free from external contamination
- Placing the boxes into overpacks and fitting an outer lid
- Moving the overpacks to interim storage areas in suitable existing buildings on site.

3.3.1 The Processing Cells
The MAC and FED processing cells are similar in design and incorporate the following features.

Shielding
The cell shielding panels are constructed from laminated mild steel plates which interlock, to eliminate shine paths at the joints. Access to the cells is via gamma gates in the Shield walls. Each cell is equipped with a lead glass window and a pair of tongs to enable manual operation of incell equipment.

Waste packing and grouting cell
The packing cell is part of the primary containment and incorporates a 'double lidded' port which enables the lid of the $3m^3$ box to be removed whilst preventing the spread of contamination to the sides of the box outside the containment area.

The cells are equipped with XY hoists, tongs and manipulators for waste packing and operation of in-cell tools and equipment.

Monitoring equipment is provided to enable the activity of the waste being packaged to be controlled.

The grout cells are provided with equipment to bolt the $3m^3$ box lid and encapsulate the waste in grout whilst maintaining full containment. The grouting cells are designed to remain free from contamination.

A robotic swabbing system checks the exterior surfaces of the box to confirm they not contaminated.

Box Transfer System
Trolley systems transport the 3m^3 box between cells and to the overpack handling area. The trolleys incorporate jacking and rotation devices to facilitate bolting and swabbing. An integral weigh station is used to ascertain the weight of packed waste.

Overpack Handling Area
The overpack handling area is a shielded area where the finished waste package is placed in a concrete overpack. An overhead gantry crane has been designed to handle both overpack lids and 3m^3 boxes. The overpack lid is first removed and placed on an adjacent stand. The grouted waste package is then lifted and placed inside the overpack and the lid replaced ready for transport out of the facility

FED Assay System
The FED process cells incorporate a fuel monitor which assays the waste on the tray. By monitoring the data produced the operator is able to maximise the packing density and control the surface dose rates of the box.

3.3.2   The Grout Plant
The grout plant produces a high shear, high fluidity grout in a commercially available colloidal mixer. The grout plant output is 2m^3 per hour. The plant has systems for:
- cement powder delivery
- chilled water delivery
- grout mixing
- mobile grout storage with agitation
- grout discharge
- wash down

The grout is transported to the packing plants in mobile grout hoppers which discharge the grout through cell delivery pipes into the 3m^3 box.

3.3.3   The Ventilation System
The active ventilation system comprises a mix of existing plant and ductwork and some additional plant and ductwork for the new facilities.
The ventilation system provides the following design features:
- Air flows from areas of lower risk to areas of higher risk of contamination.
- Primary containment facilities are held at a depression relative to other areas.
- Air velocities across openings are controlled to prevent back diffusion.
- HEPA filters prevent back flow from primary containment facilities.
- All extracted air is drawn through HEPA filters and is sampled for airborne active particles before discharge to atmosphere.
- The extract plant is failsafe and has appropriate monitoring and control systems.
- The ventilation system is monitored and ventilated for hydrogen where appropriate.

Hydrogen
The installed instruments will monitor the flow rate and will check for the presence of hydrogen. The design flow rate will keep hydrogen levels within specified limits during grouting and curing.

3.3.4 Fire Precautions
The main potential fire hazard is on the FED project. There is a potential to ignite the Magnox through the disturbance of hydrided fuel.

Two sophisticated fire detection systems have been provided. The first samples the waste storage vaults and triggers an automatic Argon based fire suppression system. The second system monitors the sort cells, conveyors and retrieval tunnels and generates an alarm to alert the operators to the problem. In cell fires are extinguished by the manual discharge of Graphex powder.

3.4 Waste Storage
The Trawsfynydd ILW waste will be packed and immobilised in $3m^3$ boxes which in turn are placed in overpacks manufactured to designs approved by Nirex.
The following systems have been provided by BNFL Magnox generation and interface with the retrieval and processing plants.

3.4.1 $3m^3$ Boxes
The boxes are fabricated from stainless steel for corrosion resistance. Lifting features are provided in the top of the box walls for accepting standard twist locks. A separate lid bolts to the box. The maximum allowable weight of a filled box is 12 tonnes.

3.4.2 Overpacks
The waste packages will be placed inside concrete overpacks to facilitate safe handling and storage. They are thick walled reinforced concrete boxes that provide shielding sufficient to reduce exterior surface dose rates to levels in accordance with the transport of radioactive materials regulations. The weight of the overpack and lid is approximately 30.5 tonnes.

3.4.3 Straddle Carrier
The overpacks will be moved across the site from the overpack handling areas adjacent to the process cells to the interim storage area by a straddle carrier. This has been modified from cab to pendant operation and has a capacity of 45 tonnes.

3.4.4 Air Transporter
Within the Overpack handling areas and the storage areas, the overpacks will be moved on an air transporter which is a steel frame fitted with air pads. It has very good manoeuvrability and is very compact.

3.4.5 Interim Storage
The current strategy assumes that the waste will be stored securely on site until NIREX provide a suitable repository. To this end, four of the circulator hall sub-basements cleared of all plant, will act as interim storage.

4   REVIEW

At the time of writing the MAC facility is successfully recovering and encapsulating waste from its first reactor location and the FED facility is nearing completion of its commissioning phase.

By developing an innovative approach to the project it is believed the following has been achieved.

- Access and waste handling problems solved
- Construction work on site reduced
- Costs of final removal and disposal minimised
- Environmental impact negligible
- Significant reduction in investment required
- Solutions can be adapted for other stations

C596/015/2001

Windscale advanced gas-cooled reactor decommissioning – hot gas manifold dismantling strategy and tooling

G J WALTERS, S J BATCHELOR, and M J STEELE
BNFL, Windscale, UK

ABSTRACT

The Windscale Advanced Gas-Cooled Reactor (WAGR) is the UK's demonstration project for power reactor decommissioning. The reactor is being dismantled on a 'top-down' basis and a major task recently completed has been to dismantle the hot gas manifold (hot box). The UKAEA hold the site licence for Windscale and exercise direction through their Safety Management and Control Team.

BNFL has been contracted by UKAEA to project manage, engineer, supply and operate equipment to size reduce and encapsulate the WAGR reactor, which is one of a number of reactor decommissioning projects currently being undertaken by BNFL. The strategy for dismantling the hot box makes use of manual, semi-remote and remote techniques. Semi-remote tools were built and deployed and then set to work with the assistance of personnel entering the radiation area. The equipment was then operated remotely from a control room located in an adjacent building.

This paper describes the strategy for size reducing and removing the WAGR hotbox structure and discusses some of the challenges that have been successfully managed and the actual plant dismantling operations.

1. DESCRIPTION OF WAGR

The WAGR facility is located close to the northwestern boundary of the UKAEA Windscale Nuclear Licensed Site in Cumbria, United Kingdom. Since the shutdown of WAGR in 1981, many of the systems used for the operation of the reactor have been removed or disabled with some systems being adapted to meet the requirements of the dismantling operations. Many of the services, in particular the ventilation systems, which served the operating reactor, have now been changed to systems more suited to the dismantling operations. The current programme for decommissioning of the WAGR facility involves the removal of the reactor and associated plant and equipment. A decision about the removal of the concrete bioshield, the reactor containment building and associated equipment and buildings will be made at a later stage.

The reactor containment building consists of an all welded steel hemispherical dome 41 metres in diameter surmounting an inverted truncated conical section at the base. This containment is a controlled area, which totally envelops the reactor and the heat exchangers' biological shielding. It was originally designed as a pressure vessel capable of withstanding the significant internal pressurization, which would have been caused by the total release of the reactor coolant gas during operation. A schematic outline of the internal structures of the WAGR containment building is given in Fig. 1. Significant decommissioning modifications and structural changes have been made since shutdown of the reactor, which include the installation of the Remote Dismantling Machine (RDM) and the removal and disposal of the heat exchanges as LLW to BNFL Drigg.

The reactor pressure vessel originally consisted of a steel vessel 16.3 metres high and 6.5 metres diameter with domed ends. The condition of the reactor vessel at the start of the hotbox dismantling campaign is shown in Fig. 1. Internally, there is a graphite core surrounded by thermal shield plates and surmounted by a neutron shield and hot gas collection manifold (hot box). Two hundred and fifty-three fuel channels ran from the core upwards to refueling tubes, which penetrated the neutron shield, hotbox and ultimately the top dome, running up to the refueling floor level. The approximate weight of the pressure vessel was 820 tonnes. Around the reactor, and where the heat exchangers used to be, are concrete biological shields of 2.7 m thickness and 600 mm thickness respectively. The internal section of the reactor biological shield concrete structure forms the reactor vault.

WAGR and associated buildings can currently be defined as a defuelled, partially decommissioned power reactor and is a 'Category 2 Plant' under the UKAEA system of classification of buildings on the Nuclear Licensed Site.

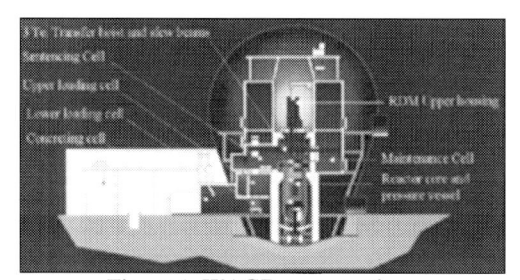

Figure 1. WAGR cross section

2. WAGR DISMANTLING SYSTEMS

The RDM has been constructed over the reactor to facilitate the proposed overall dismantling strategy. Some of the main components of the RDM are a containment structure mounted on a rotating shield floor, which houses a deployment mast for the dismantling manipulator, which could not be deployed due to height constraints for dismantling the hot box). Working in conjunction with the manipulator is a 3te hoist mounted onto a slewing beam. The slewing beam aligns with beams mounted in the sentencing cell and maintenance cell allowing the 3te hoist to travel into those areas.

The 3te hoist is capable of deploying a range of cutting tools and grabs, which can be controlled via an integral, grab control system, operated from the main control desk (CS1). A test facility has been utilised which has a simulation of the WAGR grab control system, which has been used for hot box tooling development trials and training.

The reactor vault area also has additional services available, fed through auxiliary penetrations constructed through the bioshield wall adjacent to the hoist well.

A number of camera systems have also been mounted in the reactor vault area, which provides good viewing for the CS1 operators to control the dismantling equipment.

3. HOT GAS MANIFOLD (HOT BOX)

The 'hot box' is a large flat-ended cylindrical vessel, approximately 5 m in diameter and 0.9 m high being constructed from 30 mm thick mild steel plate (see Fig. 2 and Fig. 3). It is penetrated by the 247 refuelling channels and six loop tube channels and is a complex structure, the internals of which are insulated with stainless steel foil known as 'refrasil'. Its purpose was to distribute the hot coolant gas emerging from the reactor fuel channels to the four heat exchangers and as a result of this it has become internally contaminated with fission products.

The hotbox structure has a number of components, which make up the overall assembly. The section of refuelling channel left projecting from the upper surface of the hotbox top plate has been designated the Upper Refuelling Tube (URT) flange. The section of refuelling channel contained within the hotbox structure is known as the Fuel Element Guide Tube (FEGT).

Each FEGT was fitted with Burst Cartridge Detection (BCD) pipework, which was linked to detection instrumentation. This pipework was spirally wound around the FEGT and linked to a network of BCD pipes running underneath the hotbox.

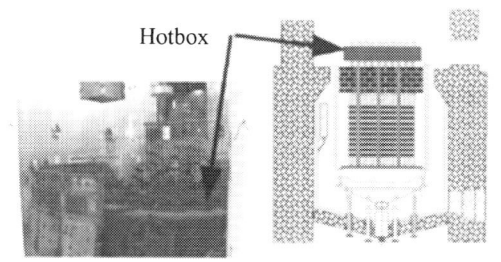

Figure 2. Disposition of hotbox

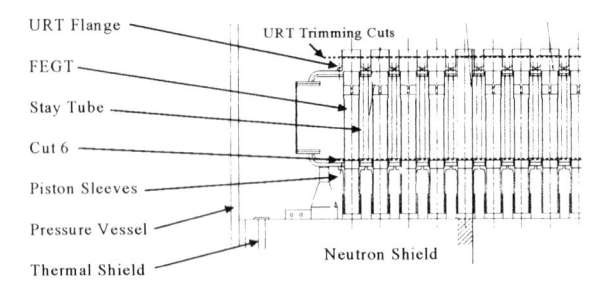

Figure 3. Cross-section of hot box

4. ENVIRONMENT/RADIATION FIELDS

In considering size reduction strategies, one of the key factors affecting the choice was control of operator dose uptake. It was agreed that semi-remote working with the operators assisting directly with the operations was the best practical option, but before this could be finalised a better understanding of dose rates in the reactor vault, once the hot box top plate is removed, would be required to demonstrate ALARP. It was established that a clear understanding of dose rates and component waste categories could only be achieved by sampling and direct radiation measurements on the hotbox.

Sampling of hot box components was carried out and the results of the analysis of the samples were used to predict the dose fields at various stages during the hotbox dismantling campaign. These values, in turn, were used to estimate the dose budget for the proposed hot box dismantling programme. The results of the sampling analysis predicted that a considerable amount of radiation from the hotbox was associated with the FEGT's. It was therefore decided to remove the FEGT's at an early stage in the hotbox dismantling campaign.

5. PROPOSED STRATEGY

As a result of a number of trials, it was concluded that a proposed dismantling strategy for size reducing the hotbox by semi-remote means was feasible and ALARP. Also, the proposed dismantling strategy for size reducing and removing the hot box would meet the two main objectives in that it:

- Would minimise the release of active particulate into the reactor, which may result in widespread contamination of the RDM, dismantling manipulator and reactor ventilation system.

- Would minimise any impact on subsequent removal of the neutron shield. To implement the strategy it was decided to size reduce the hot box, in sections, by a number of 'mini' campaigns by semi-remote operations.

6. DESIGN, TESTING, DEVELOPMENT AND TRAINING

To help manage the potential risks and safety hazards associated with the proposed strategy, various brainstorming sessions were set up to help identify any potential risks and to propose methods of eliminating, mitigating or managing the risks. The results of the risk assessments and HAZOPS (Hazard and Operability Study) investigations were fed into the design phase of the tooling development and into the production of the Safety Case.

A major factor to consider when designing the equipment was the limited headroom from the maintenance cell where the equipment was deployed and the vault. As the maximum headroom was only 1600 mm from the hoist pin to the floor, the tooling and lifting equipment needed to be compact. Where the equipment could not fit through the available opening then the it needed to be modular with simple methods of attachment so that the man entry time in the vault remained as short as possible.

To test the various items of equipment that had been developed for each of the 'mini-campaigns', a number of representative mock-ups were manufactured. The main function of the mock-ups was to provide a section that was truly representative of the actual plant and was sufficiently large enough to allow a number of trials to be performed to gain confidence in the reliability of the equipment. At the same time, the risk of performing adequate development trials had to be balanced against cost.

Another factor that needed to be taken into consideration when performing development trials was the operational environment. Issues such as viewing, lighting, access, deployment and control systems had to be represented. The level, to which these issues were represented, again was a compromise between gaining confidence in the adequacy of the equipment and the time and cost of set up.

The main cutting technique selected for size reducing the hotbox structure was plasma arc (see Fig. 4). To help minimise costs, as much as possible, "off the shelf" equipment was used or modified to perform the cutting operations. This principle also applied to the motorised deployment equipment using standard automated welding drives where possible. The use of standard equipment modified for remote working, also gave the added advantage of proven reliability and minimal cost of consumables and spares.

Figure 4. Plasma cutting trials

A non-active test facility was used to set up and test the various items of equipment. During the trials, various cuts were performed to optimise the cutting speeds and plasma current settings for size reducing the various hot box components. Different techniques were tested for piercing a plate and 'edge starts' with the remotely controlled plasma cutting equipment.

Just before each of the 'mini-campaigns' was implemented on the reactor, the plant operations team underwent a detailed training session. Training was broken down into five modules covering general training, details of the equipment, equipment set up, operation of the equipment and maintenance. As part of the general training module, issues such as strategy, campaign objectives and safety hazards were addressed. The training in the operation of equipment involved cutting and handling representative mock-ups under non-active operational conditions.

7. IMPLEMENTATION/PLANT OPERATIONS

Fuel element guide tube (FEGT) removal

The technique for removing FEGTs using an internal ball grab was well proven during earlier preparation work in the reactor vault. A number of damaged guide tubes were removed and replaced to provide continuity of passage for removal of waste stored in the final channels.

That work, however, revealed a problem, which had to be resolved for the mass removal of the FEGTs. Each guide tube was wrapped around by two BCD small bore stainless steel pipes each of which was attached at the top to the guide tube and at the bottom to the bottom plate of the Hotbox. The configuration was such that when the hoist deployed ball grab was engaged in the FEGT and raised, the pipework formed a very strong retention spring.

When removing odd ones the technique was to manually cut the pipes during the lift using long reach hydraulic shears. This was not considered to be ALARP for the production line removal of 247 tubes and a more remote method of overcoming the problem was deemed necessary.

The method selected was to remotely deploy a plasma torch down each of the channels in turn. The torch, mounted on a rotating frame, was then used to cut through the wall of the tube close to the bottom. In the process, the plasma arc cut through the two BCD pipes thereby releasing the FEGT for a clean lift. The application proved successful and only a handful of the 247 tubes required additional manual intervention to release them. Completion of the mini-campaign, however, resulted in three findings, which were to have significant effect on subsequent methodologies for Hotbox removal.

Calculations had predicted that the FEGTs formed a significant proportion of the Hotbox's radiological inventory and that their removal would allow more semi-remote work to proceed. Investigations following removal of the tubes revealed this not to be so and that infact the rates inside the box were upto 30 times higher than predicted. The source was now believed to be the stainless steel refrasil insulation which lined the entire inner surface of the steel shell of the Hotbox. The effect on subsequent methodologies was to reduce planned manual intervention, in some cases, to zero. This necessitated a rapid review of the planned methodologies and the development of more remote techniques.

The second finding was with regard to the plasma cutting torch assembly. It was believed that the cutter assembly had been put through adequate works proving trials, its deployment in the harsh reactor environment quickly proved otherwise. The configuration and close confines of the reactor did not dissipate the local heat as quickly as the trials mock-up had done. Whereas the trials involved a number of cuts in as many hours, the actual work rate of six cuts per hour caused heat to build up in the cutter, damaging drives and consumables. The net effect was a large increase in operator dose uptake in maintaining and repairing the equipment.

Thirdly, removal of the FEGT's now gave access for the deployment of a remote camera unit, which for the first time gave views of the internal of the hot box and led to the identification of a number of internal features which had not been anticipated or accounted for in the methodology to date.

Turnbuckle & Side Wall bracketry removal
Previous decommissioning operations had left a number of brackets still attached to the sidewalls of the hot box, which had to be removed to accommodate the sidewall semi-remote cutting equipment. The majority of the bracketry was in the form of small sectioned pipe supports but four turnbuckle brackets posed some difficulties due to their size and configuration.

Cutting trials using hand held oxy propane and plasma torches concluded that the turnbuckle brackets were best removed by taking away the section of sidewall to which the brackets were attached. As the sidewall was backed by the stainless steel refrasil the hand held plasma torch was the preferred cutter. The technique proved very successful but in the event plasma was supplemented by the use of oxy propane for some of the work. The bulk of the plasma torch nozzle restricts its use where access was limited.

Side Wall Vertical Cuts
In preparation for the later removal of the sidewall, 13 vertical cuts were to be made using a semi-remotely deployed plasma torch. A deployment system was developed which could be located on the thermal shield wall and outer neutron shield adjacent to the hot box. Two remotely operated drivers could then move the torch to set its stand off distance and then move vertical to complete the cut. Due to the profile of the sidewall, the joining cuts around the top and bottom radii were to be completed by hand.

Once again circumstances in the vault exposed limitations in the workshop trials of the equipment. Whilst the actual cutting caused few problems, and helped reduce the exposure time of the operators, the setting up of the equipment at each location took three times longer that the cutting and was in a relatively high dose area. In terms of dose uptake the maths speaks for itself and the result was another review of the forthcoming semi-remote mini-campaign and the effectiveness of workshop trials.

Removal of Upper Refuelling Tube Flanges (URTs)
In order to prepare the top plate for cutting, 189 of the 253 URTs were to be removed using a 200 amp plasma torch. A purpose built deployment frame was developed which could be deployed remotely to position the torch inside the bore of the URT. The frame positioned the torch at the interface between the URT flange and the top plate and a rotational drive then traversed the torch around the bore. The trials demonstrated that by cutting through the interfaces, in effect a path of least resistance, the plasma arc could cut through each of the six 16mm diameter bolts attaching the URT to the top plate.

The released URT could then be picked up individually by a sequencing grab and transferred to a waste basket and ultimately into a waste box for encapsulation in grout. In implementation, however, the reactor configuration raised problems which had not been replicated in trials. The major flaw in the design of the torch deployment system was the need to position the torch so precisely in terms of its vertical position and how level it needed to be to maintain alignment with the flange/plate interface, and concentricity, to effect an acceptable stand off distance. How earlier decommissioning operations had left deposits of cutting dross and debris around the URTs, which could not be removed. The consequence was that even small amounts of dross displaced the cutter from its optimum position. The effects included torch burn out, due to lack of concentricity (i.e variable stand off), and angular cuts through the flange/plate interface which failed to release the URTs.

The increased maintenance on the cutting equipment added considerably to the rising dose budget and two thirds of the way through the operation a review of the strategy was held. The conclusion was that manual intervention to remove the bolts attaching the remaining URTs was more ALARP than to proceed with the originally intended remote strategy.

The balance of the URTs were then successfully removed using a pneumatic impact wrench and, where more appropriate, oxy-propane cutting by teams of skilled operators. It would be wrong, however, to assume that this technique should have been used throughout the mini-campaign. The difficulties which arose, using both techniques, could not have been predicted.

Figure 5. Hotbox top plate

Top plate removal
A combined grab and 200 amp plasma cutter, remotely deployed, had been developed and tested for cutting and removing the carbon steel top plate which was backed on the underside by stainless steel insulation foils. The grab was to locate on each of the remaining 64 URTs in turn and the integral plasma torch then rotated to release a nominally circular section of plate and insulation. The combined grab and cutter then transferred the section to a buffer position within the vault. A second grab was then deployed to later remove the section to its waste box, along with a number of other sections.

By this stage, however, a number of hard lessons had been learned, leading to the following concerns to be taken into account. Whilst plasma cutting was effective in good operational conditions, reactor configuration caused inconsistency and failed cuts. If an equipment failure occurred during a cut the grab/cutter could not be released without the risk of the work piece becoming irretrievable. Plasma cutting through the top plate and the insulation pack would be susceptible to failure due to the likelihood that the internal insulation had dropped and created an air gap between the top plate, which the plasma arc would not jump.

Once removal of the top plate commenced, the very high dose rates from the exposed internals of the hotbox would preclude man access to assist in recovery of a failed cut or stranded grab/cutter.

To overcome these significant challenges, a unique device was conceived, designed and a limited number deployed in a very short timescale. The device was effectively a remotely deployed prop, which could be inserted down through the bore of a URT into the inside of the hotbox. Once through the constriction of the URT bore the prop's internal mechanism automatically deployed legs to stand on the bottom plate and latches to sit under and support the top plate and its refrasil. As each section of the top plate was removed the prop was relocated to another position. Eight such props were deployed at any given time. Use of the device proved invaluable and improved confidence and techniques in using the plasma cutter resulted in the efficient removal of the 64 sections of top plate.

Removal of hotbox internals

Prior to decommissioning, the top and bottom plates were joined by a number of tubes, clad in stainless steel insulation and open at each end to the environment outside the hotbox. Commonly referred to as stay tubes, the tubes provided passage for reactor instrumentation and stability for the hotbox structure. Previous plasma cutting operations had been intended to sever all the tubes from the top plate and all but twenty from the bottom plate. However some of these cuts were incomplete and had to be repeated.

The stainless steel insulation on the outside and the relatively small bore of the stay tubes provided considerable difficulties for cutting techniques but previous operations had provided some experience. A remotely deployed 40 amp plasma torch and drive system was developed and tested.

The system had two drives, one of which lowered the torch down the bore of the tube to a selected level and the second rotated the torch around the bore. The close confines of the tube resulted in a high attrition rate of the torch nozzles but, with prior trial knowledge, sufficient spares were available to complete the task. The fallen tubes, and significant amount of collateral debris from earlier operations, were recovered by simple grab techniques. However from the earlier internal views of the hot box a number of features and some debris had been identified which could not be removed so simply. The decision was made to design a remote deployment system for a proprietary manipulator (Slingsby TA40) which would pick up and dispose of this debris. This proved to be a very successful tool and was utilised further to accelerate progress of the campaign.

Removal of bottom plate insulation

The original strategy had proposed that at this stage of operations the sidewall of the hotbox was to be removed. It was, however, demonstrated that although the sidewall was internally clad in the stainless steel insulation it was providing significant shielding for essential man entries to the surrounding areas of the vault. Hence the sequence was changed to that of bottom plate insulation removal.

Trials had proved that plasma was the most effective cutting method for the 38mm thick covering of foils and a remotely deployed cutter assembly, using 40 amp plasma, was deployed. A rigid beam, which could locate in any two of the 253 channels in the bottom plate, formed a base for the torch and its x, y & z axis drives.

Good overview visibility and the three axis drive for the torch made for an easy task in cutting the insulation into manageable areas. Removal of the cut areas proved more difficult. In some instances the lower layer foils had become fused to the bottom plate by the plasma arc. Additionally, insulation retention studs had been damaged by oxidation during operational life, thereby unexpectedly releasing or partially releasing sections of insulation. Such problems had been anticipated and the "Slingsby" robotic arm proved extremely useful in supplementing the planned sequencing grab for recovery of the cut insulation.

Sidewall removal
Cutting of the sidewall into manageable sections had been planned to be a semi-remote operation using a 200 amp plasma torch. The deployment assembly was a development of that used for the vertical sidewall cuts and had been successfully tested on a reactor mock-up.

Experience thus far gained, however, suggested that the semi-remote approach would carry too high a dose penalty. A dose assessment revealed that it would be more ALARP to manually cut through the mild steel wall plate and stainless insulation.

Using a 200 amp hand held plasma torch in a carefully planned sequence the sidewall was cut into fourteen sections each approximately 1.0m high and 1.5m long. The sequence of cutting and the use of a remote grab greatly assisted by the robotic arm, prevented exposure of the operators to the high dose rates emanating from the insulation.

Figure 6. Sidewall removal

Bottom plate removal
The bottom plate, 32mm thick mild steel plate, was penetrated by 253 holes 150mm dia forming the fuel paths. Bolted to the underside of the plate at each hole position was a flanged cylinder (a piston sleeve) which engaged in the upper Neutron Shield, furthering the fuel path. Connected to each flange were two BCD pipes.

A number of methods for cutting the bottom plate had been conceived, trialed and discarded in favour of plasma cutting in a pattern based on the centre of each of the fuel channel holes. A 200 amp plasma torch assembly was remotely deployed to a given channel and the torch driven by a rotational drive. The plasma arc cut through forming a roundel of bottom plate with the piston sleeve attached.

It had been planned that the plasma arc would also cut through the two BCD pipes thereby allowing the whole cut section to be lifted free using an internal ball grab. Unexpected reactor configuration, including secondary attachments of the BCD pipes, necessitated manual intervention with long reach hydraulic shears to facilitate final release of most of the sections.

This relatively intensive manual intervention was achievable only because of a very significant reduction in the local dose rates resulting from the removal of all highly active insulation. The reduction also allowed hand cutting, using oxy-propane, for removal of the periphery of the bottom plate omitted from the rounded cut pattern.

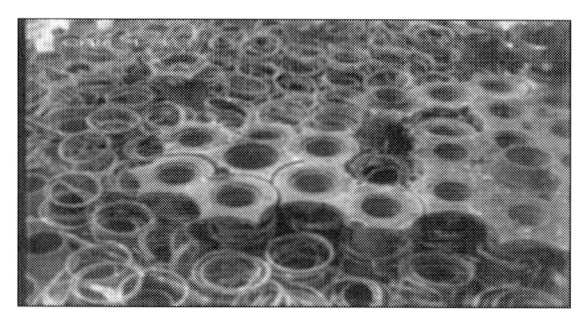

Figure 7. Bottom plate removal

BCD pipework removal
Once the bottom plate had been removed the interwoven mass of the BCD pipework, some 1500M in all, was exposed. Little was known about the details of the pipe runs and means of clamping and supporting the pipes until this stage.

The complex, and now partially collapsed, pipe array quickly demonstrated that the planned removal strategy of hand held and remotely operated hydraulic shears was impractical. Similarly, a hand held plasma torch, ideal for the stainless steel, was impractical to use on the small diameter pipe.

It was found that a hand held oxy-propane torch, while not the ideal cutting method, could effectively release large bundles of the tangled pipework for removal by remote grab.

Waste Packaging
The dismantling methodologies had been regularly challenged to optimise the techniques against the number of waste packages which would be produced. From the initial methodology which predicted nineteen WAGR waste boxes would be generated during the hotbox dismantling campaign, each containing two waste baskets, this was ultimately reduced to fourteen, hence making a significant saving in final disposal costs for the client (UKAEA). The total weight of waste produced was approximately 31 tonnes. The specific activity of the hotbox components meant that all this material was designated Low Level Waste and was dispatched to the Drigg storage facility for final disposal.

8. DOSE BUDGET

The total radiation dose uptake for the Hotbox campaign was 63 mSv. This covered all operations including equipment setting up, operations, maintenance and support operations such as filter changing of the building ventilation system and erecting temporary shielding to allow torch consumables to be changed in the reactor vault area. An EPD system is used to monitor and manage/operate dose uptake on a real time basis, including daily and monthly reviews. Also included in the total dose uptake was the waste box grouting and concrete lidding operations. The total number of personnel involved in the hotbox dismantling campaign included the operations team working on three shifts, Health Physics support and operational support from AEA Technology and other organisations who provided camera systems and personnel to perform specialist hand burning operations. The number of people involved in the dismantling operations at any one time varied over the period, but in total amounted to about 40 people.

9. PROJECT REVIEW

The dismantling of the WAGR hot box has been successfully completed. From the review of the development trials and the actual dismantling operations it has been highlighted that the strategy and equipment developed for remote and semi remote dismantling tasks should be as versatile and robust as possible. This allows changes to be made quickly in response to differences in the assumed condition of the structure and environment as the work progresses.

ACKNOWLEDGEMENTS

The work reported in this document has been funded by UK Government DTI and UKAEA as part of the SAFER programme.

Colin Dixon, Roger C Bloor, Steven Chester
AEA Technology Nuclear Engineering, Windscale, Seascale, Cumbria

Terence Benest
UKAEA, Windscale, Seascale, Cumbria

C596/018/2001

Decommissioning the Berkeley vaults – remote operations

P K J SMITH and **R A PECKITT**
NUKEM Nuclear Limited, Warrington, UK
N P SALMON
Magnox Electric, Berkeley, UK

SYNOPSIS

The paper describes key technical elements of the project to retrieve, process and store approximately 1900m^3 of Intermediate Level Wastes (ILW) stored underground at the Berkeley Power Station, Gloucestershire. The facilities are now undergoing commissioning and approaching remote active operations. The emphasis is on describing the remote handling systems provided for dealing with the waste material, and the experience obtained in breaking into the heavily shielded waste bunkers during construction. Details of the purpose designed waste recovery machinery and the waste processing remote tooling are given to illustrate how the project has progressed from concept to practical realisation.

1. INTRODUCTION

Operational wastes from the Magnox Electric plc owned Berkeley Power Station and Berkeley Technology Centre have accumulated in a suite of underground bunkers known as the Active Waste Vaults from the time the power station was commissioned in 1962, up to the present date. The waste is radioactive and is largely Intermediate Level Waste (ILW) with some Low Level Waste (LLW), comprised of solids and mobile wastes. The waste is to be retrieved from the vaults, processed in a new process facility and placed into safe storage. To this end the mobile ILW and LLW is to be transferred to the Caesium Removal Plant Storage Facility (CRP) for storage. The solid ILW is to be packed into boxes, grouted and transferred to a new purpose built storage facility constructed on the foundations of an existing reactor building circulator hall. The LLW is to be processed and packaged in a manner suitable for storage at the Drigg facility in Cumbria.

2. PROJECT BACKGROUND

The contract to recover the wastes from the vaults was initiated in 1996 and since that time a team managed by NUKEM has designed and constructed facilities that include a Retrieval building, a Process building, an ILW store for some 800 Nirex 3m³ boxes and 500l drums, and a facility for storage of approximately 100m³ of ILW mobile wastes (sludges and ion exchange resins).

The wastes consist of power station Fuel Element Debris (FED), removed at the power station to reduce the transport volumes sent to Sellafield for reprocessing, which was discharged loose and consists mainly of graphite and Magnox (approx 1000m³) and a miscellany of canned items including sludges and resins (approx 150m³), plus various Miscellaneous Contaminated Items (MCI) which arose from power station operations and from the adjacent Berkeley Laboratories. The weight of the items to be removed varies from a few kilograms up to 200kg in the case of a number of 45gallon drums. This waste resides in three underground bunkers each approximately 18.3m long by 4.7m wide by 6.4m deep. A layer of gravel chippings some 305mm thick covers the floor of these vaults. The shield roof of the bunkers is approximately 1.8m thick.

A fourth bunker known as the Chute Silo contains various heavy items including 78 control rods, 4 charge chutes and one trailing lead thermocouple chute, plus a set of "bomb doors" which were cut away remotely as part of the preparations for retrieving the Chute Silo wastes. The charge chutes are 12m long and 300mm in diameter with a mass of 1780kg.

The recovered wastes are to be assayed, processed, packaged and conditioned in a Process facility. This facility contains equipment for dealing with ILW solid wastes by grout encapsulation and mobile ILW wastes by hydraulic transfer into a set of storage tanks. LLW wastes can also be segregated and disposed of. The facility handles the 3m³ Nirex box package and the 500l drum Nirex package in a four-array stillage.

At the time of writing (May 2001) all facilities have been constructed, equipment installed and non-active commissioning is well advanced.

3. FIRE DETECTION AND SUPPRESSION SYSTEM

The Fuel Element Debris stored in the vaults contains materials (Magnox, graphite) that under certain conditions can be induced to burn. The possibility of a uranium hydride ignition source has been postulated which in turn could create a vault fire. The circumstances in which this event might occur are remote, nevertheless the safety case has recognised this possibility and measures are in place to guard against a vault fire. There is also the possibility of a hydrogen build up due to Magnox corrosion: this is also catered for.

The vaults have been covered with a watertight building and a continuously operating ventilation system has been installed and commissioned. This system prevents any hydrogen build up as well as providing protection against the release of contamination. The ventilation extract ducting for each vault is fitted with a sensitive smoke detector, known as a VESDA alarm. A completely separate detection system is also provided, based on Cyrus cloud chamber devices. These instruments have probes mounted directly into the vault via vault roof shield plugs. Both instrument types have several levels of alarm, and should both give a confirmed high-level signal (safeguards are built in to allow for an instrument fault condition and for loss of power) an alarm is sounded and an automatic sequence is set in train to initiate fire fighting.

A reservoir of bottled argon gas is stored adjacent to the vaults and on completion of the fire detection sequence motorised valving is automatically energised to blanket the alarmed vault with argon. Override controls are provided for manual argon injection also should this ever be necessary. The fire detection system annunciator panel is located in the Retrieval building control room, and a duplicate system is provided in the Berkeley centre shift control office so that 24 hours site supervision is immediately informed of any situation.

4. REMOTE SYSTEMS FOR WASTE RECOVERY

The waste is recovered remotely and filled into a waste skip with a volume capacity of approximately $2m^3$ (Figs 1 & 2). Apertures for remote machinery access routes have been cut through the vault roof by core drilling. The plugs remaining were remotely lifted and shield doors deployed to cover the holes. Two remote retrieval machines are deployed through holes cut through the vault roof. Four holes per vault are provided (Two for the Chute Silo). The size of these apertures is approximately 920mm diameter. The waste skips are deployed through rectangular apertures 1.7m x 3.2m, by the skip mast, one per vault, the entry route for the Chute Silo differs and has a vertical entry 1.7m x 0.9m through a side wall utilising the chute silo hoist.

4.1 Retrieval Machines

The retrieval machine design is required to cater for the need to be deployed through a minimally sized opening, to reach down into the vault and to grab a diverse range of items. It has to be capable of reaching out radially from the entry point and yet its overall height must be kept to a minimum to keep the building roof height at a sensible limit. A telescoping mast system provided the solution and a modified proprietary industrial arm commonly used for vehicular mechanical handling forms the basis of the design of the vault end of the manipulator. Full coverage requires four access points per vault due to the reach limitations.

The vertical telescope of the machine is achieved in two stages. The outer mast assembly is raised/lowered by means of a hydraulic cylinder. The outer mast in turn carries an inner mast assembly that is raised/lowered by a hydraulic winch. A carriage supports the retrieval manipulator deployed vertically down. This enables the machine to be moved into position on rails over one of the three waste vaults, or at a higher level the Chute Silo. The carriage has a 120mm thick steel "bell housing", which is suspended from the underside of the carriage, to maintain continuity of shielding over the vault. It also provides an access point for tool changing and maintenance of the manipulator arm section. The entire manipulator mast and arm can rotate +/-190° supported on a slewing ring mounted on the carriage. The manipulator maximum payload is 250kg at a reach of up to 5.6 metres.

The usual end effector is a special petal grab (Fig 2) designed for picking up the waste, which fits within the deployment envelope, and handles the different types of waste present in the vaults. Alternative tooling can be fitted for ad-hoc operations.

The waste retrieval manipulator is hydraulically powered and is remotely operated. The operator controls the machine via joysticks actuating proportional directional electro-hydraulic valves. An on-board camera is provided. Recovery features are provided to aid extraction of the machine from the vault in the event of a malfunction.

The retrieval machines have been tested and in-actively commissioned in the manufacture's works in a mocked-up vault environment. Simulated wastes were remotely grabbed, viewing via CCTV. During the commissioning trials operator training was carried out using the vault mock-up that included a mock–up of the control room. Initial training was completed retrieving wastes whilst allowing the operators to view the simulated vault and retrieval machines by eye. Once they had achieved this task further training was completed simulating the actual remote condition using only cameras and screens to deploy the retrieval machines to retrieve waste and fill skips. These tests proved the operability, versatility and dexterity of the retrieval machines.

4.2 Skip Mast Machine

The skip mast machine is designed to lower a $2m^3$ skip (smaller capacity skips are available for the initial operations into an over filled vault) into the vaults to collect the waste deposited from the retrieval machines. It has the capacity to lift a laden skip weighing up to 3te. Three sets of forks are located in the tunnel roof, held by a set of closed shield doors, positioned above each vault penetration. The skip mast is towed into position over an individual set of forks and located and locked into the fork set. The shield doors are opened releasing the forks, allowing the skip mast to be lowered to pick up a skip from the transfer bogie and then to be lowered into the vault. The extension of the inner and outer masts is hydraulically actuated. Once in the vault the skip mast can be slewed to present the skip to the retrieval machine for filling. Control systems prevent overloading of the skip mast and withdrawal of the skip mast from the vault unless the skip is in the correct orientation.

When filled the skip is retracted into the tunnel and withdrawn from the forks by the transfer bogie. The skip mast vault shield doors and transfer bogie are remotely operated from the control room by trained operators using joysticks and mimic panels with buttons, aided by views from cameras positioned within the tunnel and through the vault roof. The operator is able to change the view on the screens in the control room by moving cameras and choice of camera.

4.3 Chute Silo Hoist

The chute silo hoist is designed to introduce a $1/3^{rd}$ height skip into the Chute Silo to retrieve reduced lengths of activated components. During the early stages of the works a penetration was made into the wall of the Chute Silo and covered by a pair of hydraulically actuated shield doors, at the same time a beam was introduced and fixed to the Chute Silo roof soffit.

Due to the limited available headroom a compact design was developed. The skip hoist is suspended from a beam attached to a horizontal travelling frame which itself is suspended from a vertically travelling frame. Through a number of horizontal and vertical movements the skip hoist grab mates with the skip delivered by the transfer bogie, the whole assembly is aligned with the pre-installed beam and the skip hoist mechanism powered into the Chute Silo from where the skip can be deployed. Remotely size-reduced waste is placed into the skip via the retrieval machines; the skip is removed from the Chute Silo and returned to the transfer bogie for export.

The Chute Silo waste retrieval mechanism and shield doors are remotely operated from the control room aided by cameras positioned through the Chute Silo roof. The operator controls the chute silo hoist via an individual control mimic panel and TV screen.

4.4 CCTV Camera Systems
Three separate camera systems have been installed for viewing the remote retrieval operations. All camera systems employ telemetry for remote operation of pan, tilt, zoom, focus and halogen lighting. All telemetry signals are superimposed on the video signal, reducing field-wiring density. Switching matrices provide the operator with full control and choice of camera views.

The first system is for general viewing, giving views on four monitors of all parts of the building and also views from within the skip transfer tunnel to assist with the skip movement operations. The second and third systems are identical to one another, each being used for the retrieval operations on both the East and West sides of the Retrieval facility shielded transfer tunnel. Monitors and camera controls are mounted on the respective retrieval machine operator stations. Each retrieval machine has a full remote function camera with lighting installed. A further three remote function cameras are pole mounted for insertion into the vaults through penetrations in the concrete shielding to give a full view of the remote retrieval operations.

Video recorders record all aspects of the operations using time-lapse recording techniques. Further monitors are installed in a viewing gallery for the benefit of visitors to the site.

5. WASTE TRANSPORT

5.1 Bogie and Rails
The waste skip bogie is a vehicle which runs over the vaults within a shielded tunnel on rails between the Retrieval building waste skip import/export door and the Chute Silo (Fig 3). It accepts waste skips and locates them on any of the skip mast forks and also under the chute silo hoist. Operation of the bogie is remote and semi-automatic. It is equipped with cameras and its location is indicated on the control room panel. Waste skips can be positioned at the export/import door and the bogie's inbuilt roller system used to either export full waste skips from the shielded tunnel into the transport overpack or to import empty waste skips from the overpack.

5.2 Buffer Store Crane
Full and empty skips are lifted off and on the bogie with the remotely operated buffer store 4.5te crane. It is equipped with an electrically operated grab. In normal operation the crane can only travel when the grab is fully raised and can only be raised or lowered when positioned over the bogie or when over one of the four buffer store bays. The position of the crane is monitored electronically and also by cameras and fixed targets. The crane has remote digital readouts: -

- The location of the crane East/West
- The distance the grab has been lowered
- The weight being lifted
- Slack rope indication
- The status of the grab, open or closed

This information is displayed on the control room panel. The grab cannot be opened whilst there is a load indication and the crane will automatically stop should the system detect a slack rope.

5.3 Overpack and Tow Vehicle

There are two overpacks. One is used to transport filled waste skips between the Retrieval building and the Process building and to return empty wastes skips. The other transports grouted Nirex boxes from the Process building to the ILW Store. Fully shielded and each located on its own trailer, the overpacks are positioned at the import/export doors of the Process facility, the Retrieval facility or at the ILW Store. The overpack door is then locked on to the facility shield door. The doors are raised together and the overpacks built-in roller system is used to either transport or accept waste skips and Nirex boxes into or from the facility.

Transport of the overpacks is by a diesel powered towing vehicle. There is an on board camera to aid the driver in locating the overpack accurately. Heavy-duty guide rails assist. Lights indicate when the correct alignment is achieved. Once the transport system is docked the driver vacates the facility whilst the waste is remotely transferred.

6. WASTE PROCESSING

6.1 Process Cell

The shielded waste-processing facility, 8m wide by 30m long by 6.5m high, has been constructed within the existing Decontamination building, which provides a secondary containment. It has shield walls ranging from 0.6m to 0.9m thick. This facility contains arrangements for handling both solid wastes, principally FED, and mobile wastes extracted from containers. The shielded overpack docks with the import shield door and the skip is transferred into the cell by a roller conveyor, emptied and returned to the overpack.

6.2 Skip Input Conveyor

The waste skip is imported into the Process cell from the overpack. The overpack and skip input roller conveyors have their speeds synchronised to ensure a smooth transfer. The skip input conveyor transfers the waste skip to the correct part of the cell for emptying either by grabs on the cell crane in the case of canned waste, or for FED to a skip tipping machine, which lifts and tips the waste onto a vibro feeder table.

6.3 Vibro Feeder and Waste Sorting

The waste passes along the vibro table and under the assay instrument. The vibro-feeder is driven by two motors rotating in opposite directions and fitted with eccentric weights. The speed can be controlled outside the cell using an invertor drive. The unit can process a skip of waste in approximately 3 batches, each batch taking 2 minutes. The operation of the vibro feeder is controlled from a control panel at a cell window adjacent to the feeder. The operator has in cell manipulators at this station, which can be deployed to spread the waste and to sort out any items that are not to be placed in the Nirex box/drum.

6.4 In cell Cranes, Grabs and the Condor.

A 5 tonne overhead travelling crane and a 1 tonne wall crane are used in the waste receipt room of the Process cell. These cranes are fitted with several features to aid remote operations. Duel long cross travel motors are fitted which can be simply lifted off, to permit changing through a roof plug. All cables are terminated with plugs and sockets. These can be connected using the master slave manipulators. The main crane is fitted with main and auxiliary hoist drums, which allows a load to be handled in the event of failure of a hoist.

Two magnets are provided for use with the cranes. These are used to handle canned waste and baskets containing loose waste. They are operated either by hydraulic or pneumatic systems controlled by the operators at the cell windows. The magnet on the overhead crane can be replaced using an in-cell "Condor" manipulator. This is a remotely operated arm, deployed from the crane hook, which is used chiefly for maintenance purposes e.g. changing light bulbs.

6.5 Mobile Waste Processing

The mobile waste cans containing ion exchange resin and sludge are opened (Fig 5) and the contents placed in a re-suspension vessel. A computer-controlled system adds a predetermined amount of water and controls the air sparging ensuring that the waste particles remain suspended. The pH is measured and adjusted to minimise corrosion and the waste is transferred to buffer tanks prior to despatch to the Mobile Waste Store.

7. WASTE ASSAY

The waste assay system consists of a network of computers, which collect data from both manual input and sensors. Its purpose is to identify waste types, i.e. LLW and ILW recovered from the vaults and Chute Silo. The assay system also produces the records required to identify the contents of the Nirex box and 500l drum packages. Every item of waste that enters the Process building is assayed and tracked until it eventually leaves as one of the following:

- FED waste encapsulated in a stainless $3m^2$ Nirex box
- Mainly alpha contaminated materials encapsulated in a 500l stainless drum
- LLW items removed for size reducing and dispatched to Drigg
- Mobile LLW and ILW waste to storage awaiting further treatment

The in-cell equipment consists of: -

- Radiation monitors – To determine the radiation levels of the waste.
- Weigh sensors – To determine the weight of the waste.
- Gamma camera – To visually determine the High-level emissions from the waste.
- Germanium detector – To determine the isotopic contents of the waste.

8. WASTE PACKAGING

8.1 Box/Drum Grouting
The grout is mixed in a PLC controlled mixer unit situated in a separate grout room outside the Process cell. It can mix between 125 and 500 litres of grout per batch. A 500l drum can be filled from one batch, a process that takes about 30 minutes. A $3m^3$ Nirex box requires between 4 and 6 batches each batch taking approximately 20 minutes to mix and inject. The grout is transferred from the mixer to the box or drum by a mono-pump located within the grout room.

The $3m^3$ Nirex box and 500l Nirex drum are both bottom filled via an inbuilt 25mm diameter dip tube. The injection lance (a length of rubber hose fitted with a special nozzle) is connected and held in place by the manipulator inside the cell. The grout level is measured using an ultrasonic sensor and the control system provides automatic shut off at the correct level. Overfill protection is provided by a conductivity probe lowered into the box or drum. These instruments are positioned by a hydraulically operated arm, which also carries the lid bolting equipment. (Fig 6).

The cap grout, applied 24 hours after the main grout, is pumped via the lance directly on top of the main grout. Pigging and wash out facilities clean the grout delivery pipework. These are operated manually from outside the cell. The pig used to clean the in cell lance is expelled into the box at the end of grouting. Very little splashing was observed during commissioning trials.

The operator controls the grout operations via a touch screen and a control panel positioned by a cell window viewing the in-cell grouting station. Cameras provide additional viewing.

8.2 Lid Handling
The Nirex box and drum lids are secured in place with stainless steel bolts. These have modified heads, incorporating a circumferential groove. This is used to locate a detent pin, which retains the bolt in a bolt holder. The lids and bolt holder are imported into the cell grouting area using a series of cranes and roller conveyors and positioned on the box or drum. The operator guides this into position from the control panel using a camera target mounted on the ceiling. The bolts are tightened using a pantograph-type arm incorporating a socket drive. The operator, using a master slave manipulator, manoeuvres the arm into position. The bolt is tightened using an air motor. A torque limiter is installed to ensure the correct bolt torque.

8.3 Box/Drum Swabbing
The Nirex box is transported to the swabbing area using a roller conveyor and is picked up using an in-cell crane. This crane incorporates a rotating grab, which allows each side of the box to be offered for swabbing. The swabs are attached to manipulator friendly carriers that are posted in and out of the cell via a pair of interlocked gamma gates. The operator picks these up using the in-cell manipulators. Sample swabbing is performed on all sides of the box and the swab is posted out-cell for monitoring. The process for swabbing a drum is generally the same as for a box, except the drum is held in a drum-rotating device during swabbing. After swabbing the waste packages are exported to the ILW Store.

Fig 1 Skip Mast and Retrieval Machines

Fig 2 Retrieval Machine under Test

Fig 3 Transfer tunnel showing Skip Mast Forks

Fig 4 Skip Overpack and Tow Tractor

Fig 5 Process Cell Canned Waste Opening Area

Fig 6 Process Cell Grouting Room

C596/011/2001

Big rock point successfully employs innovative process for fuel pool cleanout

K FORRESTER
NUKEM Nuclear Technologies Corp, South Carolina, USA

ABSTRACT

Consumers Energy's Big Rock Point Plant is currently undergoing decommissioning returning the area to a natural state by 2007. This project has many steps to completion of which the spent fuel storage pool was a critical task to support the pending dry fuel storage and large component removal work.

Due to the uniqueness of the Big Rock Point Plant, the scope of the work required specialty equipment to be designed, engineered, and fabricated in addition to the on-site processing and shipping services. Onsite work commenced in February 1999 with completion in March 2000.

1. INTRODUCTION

Consumers Energy's Big Rock Point Plant is a General Electric designed 67 megawatt BWR. It went on-line in August 1962 and operated successfully until being shutdown in August 1997. The plant is currently undergoing decommissioning with a planned completion date of 2005, returning the site to a natural state.

In preparing for this large undertaking, Consumers Energy determined they would oversee and manage the project rather than following other decommissioning projects that selected the DOC model. Selecting this owner managed model meant that they needed to develop partnerships with key service vendors that could perform specialized tasks or a multitude of tasks under one contract.

Consumers Energy selected NUKEM Nuclear Technologies based in Columbia, SC as their teaming partner due to their proven performance and commitment to volume reduction technologies. NUKEM had performed previous planning work at BRP and used that knowledge to build a project plan that produced a very successful fuel pool cleanout project.

2. PROJECT PLANNING

One of the tasks determined to be key to the success of the overall project was the cleanup of the spent fuel pool necessary to facilitate spent fuel removal and large component removal work. Early on, NUKEM was called in to evaluate Big Rock Point's (BRP) current situation since they had successfully performed the previous fuel pool cleanup work 1996 and were very familiar with the current status of the fuel pool and its contents. In performance of the 1996 fuel pool work, NUKEM discovered that BRP had made commitments to the Public Utility Commission (PUC) that placed significant restraints on any work being performed in the fuel pool when fuel was stored in certain racks. Since the pool was storing all of the spent fuel generated since startup, the racks in question contained spent fuel causing significant concern whether or not any work could proceed in the pool. Additionally, the fuel pool is very small with the area not already used for spent fuel filled with irradiated components. No room other than the limited weight that could be suspended from the equipment rail was available for processing or shipping equipment.

In order for BRP to support the removal of the spent fuel for placement on a remote storage pad, the fuel pool needed cleaning to make room for the dry fuel storage cask loading operations. Without any available floor space in the fuel pool for staging and operating volume reduction processing equipment or waste liners, an alternative processing area was deemed necessary given the extremely high radiation fields associated with these components. Working together, NUKEM and BRP settled on using the empty reactor vessel as the processing area because it was empty, it offered similar safety features, and its proximity to the fuel pool.

Although this solved the primary problem of how the work could be done, it created several smaller logistical ones some of which were i) the size and shape of the vessel was not compatible with the industry standard processing equipment, ii) the weight restrictions imposed for equipment placed within it was significantly less than standard equipment and casks, iii) reactor vessel was isolated with no water filtration, iv) the water depth and limited maneuverability determined a rather slow process, v) general area and work area dose rates were expected to be significantly higher than the industry norm.

The waste material inventory indicated the components that required processing for volume reduction were mainly the 91 control rod blades (CRBs), 16 incores, and the 241 fuel channel support tube assemblies (CST). These components along with an unknown amount of miscellaneous contaminated and irradiated material were all stored in the fuel pool. Approximately 50 of the 91 CRBs in the fuel pool were stored in the spent fuel racks designed for CRB storage while the remaining 41 CRBs along with other miscellaneous equipment and waste was suspended in the pool by cables attached to the equipment rail (a permanently mounted rail on the upper inside area of the fuel pool). The combined weight of the equipment and waste hanging on the equipment rail at the onset of the project was within 500 pounds of the maximum allowable for the rail. The CST was a significantly longer component with no planned storage arrangement as part of the plant design. As such they were stored side by side in a vertical fashion in the area of the fuel pool not containing spent fuel racks. This proved to be the most efficient storage configuration during plant operation however, after shutdown, the balance of the fuel channel support tube components were moved to the fuel pool completely filling any remaining area in the pool.

To limit the costs associated with performing a fuel pool project in the reactor vessel, any work that could be performed in the fuel pool rather than the reactor vessel was pursued. This limited the amount of design changes and fabrication of equipment necessary to the extent practical. In the end, the removal and packaging of the stellite rollers and spud ends of the CRBs and the processing of the incores were the only tasks that were performed in the fuel pool. The Universal Velocity Limiter Shear (UVLS) was a machine developed to remove the spud ends. Other similar machines to remove the velocity limiter on standard BWR CRBs existed but Big Rock Point's CRBs were designed differently and did not fit in the standard machines. The UVLS and the NUKEM Control Rod Punch (CRP – used to remove the stellite rollers) designs were altered such that they could be suspended from the equipment rail. A matrix was developed and components moved accordingly to distribute the weight on the equipment rail sufficiently such that the processing equipment could be distributed within the requirements. Additionally, a redesign of the suspension mechanisms was necessary to distribute the machines weight load across a greater distance.

All of the balance of the industry proven processing equipment normally used for irradiated component volume reduction or packaging was originally designed for use in fuel pools rather than reactor vessels where the walls of the pool were flat. The radius of the reactor vessel walls combined with the relatively small working diameter (roughly 68 inches) prevented hanging processing equipment off of the walls.

The work area in the vessel was available after the upper guide bars were upended exposing the lower core support plate. This area was small enough to require some design changes to existing equipment and the manufacture of specialty equipment in-order to fit and still have room for a waste liner. The depth of the reactor vessel made the work area even less accessible with the remote tooling and maneuverability restrictions caused by the upended guide bars.

The overall layout of the BRP refueling deck is unique in that the fuel pool is completely independent of the reactor vessel although it shares the same floor elevation. Components are transferred between the two via a specialized transfer cask (original plant equipment) designed to raise the components into the cask as it is suspended by the crane at the surface of the water. This cask had to be used since it was designed specifically for the components and it provided a redundant cabling system that was unique in design. With NUREG-0612 (Control of Heavy Loads in Nuclear Power Plants) in force and the BRP gantry crane that serviced the fuel pool/reactor vessel area having a single failure crane hook, the transfer cask was required to meet the requirements. No alternative design could be used without extensive upgrades to the crane system. Using the existing BRP transfer cask as the mechanism to move the components between the fuel pool and the reactor vessel insured that this project would truly be a team effort.

3. PROJECT EXECUTION

The project began in February 1999 and was completed in March 2000. When completed, approximately 130,000 curies of activity, in the form of channel assemblies, control blades, stellite rollers, in-core detectors, and other miscellaneous components, were removed from the pool and shipped for disposal.

3.1 Control Rod Blade and Incore Processing and Shipping

It was determined that BRP wanted to remove the bulk of the curies in the waste inventory as early on in the process as possible. This worked well with NUKEM's plan to remove the CRBs first since they contained the majority of the curies and they were more readily accessible in the fuel pool than the channels. Additionally it provided more time to complete the extensive equipment design and fabrication efforts underway to support the channel processing and shipping portion. As such, the project began with the removal of the spud ends and Stellite rollers of the CRBs in the fuel pool using the UVLS and the CRP.

3.1.1 CRB and Incore Processing in Fuel Pool

CRB processing was a four-step process that began in the SFP. The four upper stellite rollers were removed using the CRP and deposited into a stellite roller container that hung from the side of the CRP. After the four (4) stellite rollers were removed from the top portion of the CRB, the CRB was moved to a low dose area in the SFP and dose profiling. The dose profile consisted of obtaining dose measurements every 12" along the CRB axis at a distance of 6". An Eberline R07 meter with the appropriate underwater probe and housing was used. Documented results of the CRB dose profile survey were used to complete the waste characterization of each component and support final packaging plans. The last processing step in the SFP was the removal of the lower drive connection (spud end) using the UVLS. After its removal from the CRB, the spud end was moved from the UVLS and stored in the SFP. The CRB was either returned to its original storage location or placed in a temporary storage location until the plan called for transferring it to the RV for final processing and packaging. All components moved within the SFP were tracked using Move Sheets that were generated by BRP to trace component accountability.

The other processing evolution that occurred in the SFP was the sectioning of the incore detectors. The incore detectors were comprised of three sections that we identified as the hot end, mid section, and the cold ends. The incore detectors were processed using the mobile rod cutter (MRC) that was suspended by a cable from the refueling bridge hoist. The hot end was removed first to make handling the detector easier. Once the hot end was removed the remaining sections were cut into appropriate lengths for handling and packaging.

3.1.2 CRB Processing and Packaging in the Reactor Vessel

While the CRB preparation and incore processing work was underway in the fuel pool, preparations were being made in the RV for the balance of the processing and packaging of the CRBs. This included the installation of a specially designed three-piece base plate that provided supports for the Fort St. Vrain FSV-1 cask (see Figure 1), liner, and liner basket. This base plate was installed on the "mushrooms" (fuel support pieces) providing a flat working platform with uniform load distribution. Once this was completed the FSV-1 liner and CRB compactor was installed. Completion of this process was coordinated with the on-going fuel pool work such that the CRBs that had previously been suspended from the equipment rail in the fuel pool could be transferred directly from the preparation process to the RV for processing and packaging in the FSV-1 liner(s).

The CRBs were transferred to the RV in support of the packaging plan. BRP Operators transferred the CRBs one at a time from the SFP to the RV using the BRP transfer cask. After each CRB was transferred to the RV, it was lowered into the CRB compactor and released from the transfer cask hook (see Figure 2). The transfer cask was returned to the fuel pool to retrieve another CRB while the CRB in the CRB compactor was flattened and moved to either the FSV-1 waste liner or the CRB basket for later placement in the FSV-1 liner per the packaging plan (see Figure 4). The packaging plan developed by NUKEM was fine-tuned after the waste characterization was completed and before the components were actually packaged.

The RV work began with the processing and packaging of the CRBs in the FSV-1 liners. This required a specific processing and packaging sequence to maintain ALARA due to the overall length of the FSV-1 and its liner and the limited depth of water in the RV. With the FSV-1 liner being roughly fifteen feet long and the cask approximately 18 feet, the stack up of the liner over the cask brought the upper portion of the liner out of the water. Fortunately the BRP fuel assembly design is approximately half as long as standard BWR fuel design thus the BRP CRBs are correspondingly half as long. This allowed us to package the liner in two layers consistent with the plan that dictated that only the bottom portion of the liner be loaded prior to placement in the cask.

Once all of the components identified for the shipment were transferred to the RV, the BRP transfer cask was disconnected from the 75 ton crane and the CRB compactor was removed from the RV. The open area of the lower process support plate was cleaned using the Tri-Nuclear V-260 vacuum system in preparation for the FSV-1 cask handling.

3.1.3 Shipment of CRBs Using FSV-1 Cask

The FSV cask arrived at BRP and a receipt survey was performed. The cask was moved into the Turbine Building where it was removed from its transport trailer and placed on a transfer trailer. The transfer trailer was moved around to the cask loading dock area where the cask was removed from the transfer trailer and placed onto the BRP transfer cart. The transfer cart was moved into the Reactor Sphere via the equipment airlock using normal means.

Once inside the sphere, the cask was removed from the transfer cart and placed on timbers in the horizontal position where the purge cover was removed and the cask closure head bolts loosened. The cask was then moved to the vertical position using the hydraulic yoke and prepared for submersion in the RV by changing the transport sockets with the in-pool sockets, installing the yoke guide pins, and removing the cask closure head. The cask was then moved to the reactor deck where the center plug was removed. The cask was moved into place in the RV where it was placed on the lower process support plate with the lifting rig disconnected and removed from the RV (see Figure 3).

Again due to the height of the FSV-1 cask in combination with the depth of the water in the RV, the cask could not be loaded by simply placing the liner in the cask. In order to avoid bringing irradiated material too close to the water surface during loading, each liner and cask was loaded in a multiple step fashion to ensure that dose rates in the general work area did not increase during the evolution. This was identified in the planning stages and resulted in the design and use of the CRB basket. The plan called for the burial liner loaded with seven CRBs and the stellite roller container to be dose profiled and loaded into the cask. After this was accomplished, the CRB basket containing six CRBs was loaded into the liner on top of the stellite container. Any void space at the top of the liner was then filled with spud ends and other miscellaneous items. Once the liner was filled, the liner lid was installed and the liner was raised out of the cask sufficiently to dose profile the remaining upper half of the liner.

Once the cask was loaded, the cask closure head was installed along with the lifting rig and the cask was raised to the RV surface (see Figure 4). The lid of the cask was raised above the water surface to verify that a seal had been obtained. Upon a satisfactory seal, the cask was raised out of the water and the draining process started. Service air was introduced through the purge fitting to speed the draining. Once drained, the center plug was reinstalled and the cask was moved to the sphere laydown area where it was decontaminated and configured for shipping. The cask was then placed in the horizontal position where all of the closure bolts were torqued and the leak test performed. Upon completion of the leak test, the cask was moved back to the Turbine Building where it was placed back on the transport trailer and turned over to the Radwaste Department for final survey and release from the site.

3.2 Channel Support Tube Processing and Packaging

After the completion of CRB disposal, the next operation was the processing of the CST components. It was determined that the 8-120B cask would be cost affective to ship the GST because its ample available internal volume combined with its adequate shielding capacity (see Figure 5). The CST consisted of three pieces: a support tube, transition piece, and a channel section, with an overall length of approximately 13 feet. First the RV was prepared by removing the two FSV liners and liner basket receptacles from the support plate to make room for the 8-120 burial liner. The 8-120 burial liner without the lid was placed on the support plate in the RV and the NUKEM Underwater Shear/Crusher (USC) was placed on top of the liner (see Figure 6).

The BRP operators transferred the CSTs one at a time from the SFP to the RV using the BRP transfer cask. Before transferring a CST to the RV, it was dose profiled then staged in the channel rack. Once the CST was in the RV it was lowered into the USC support tube end first where it was crushed and sheared in predetermined increments up to the transition piece.

Early on in the processing of the GSTs it was discovered that depending on their exposure history, they reacted differently during the volume reduction process. Often times the more highly activated the components were the more stress corrosion fracturing was present and the more brittle they were. The result was that we didn't know until we started volume reduction of each component whether or not it would maintain its structural integrity throughout the volume reduction process. Ultimately we developed two processing strategies that combined provided adequate flexibility to addressed any condition the component may be in.

If the channel section separated from the transition piece during the last crush and shear on the support tube and transition piece was placed in a transition piece basket and the channel section was crushed and sheared in predetermined increments from bottom to top (see Figure 7). If the channel section did not separate from the transition piece at the last crush and shear on the support tube and transition piece the channel section was inverted and processed from the top end down to the transition piece. Then when the transition piece separated from the channel section it was placed into a transition piece basket. This process was repeated until the liner was full and any void space at the top of the liner was filled with CRB spud ends and other miscellaneous items.

Once all of the components identified for the shipment were transferred to the RV the BRP transfer cask was disconnected from the 75 ton crane and the UCS was removed from the RV and stored on the reactor deck. In preparation for cask loading, the liner lid was installed on the liner and the liner dose profiled.

The 8-120B cask was receipt surveyed by BRP and moved into the Turbine Building where the upper impact limiter was removed and the cask tie down cables disconnected. The cask moved around to the cask loading dock area where the cask was removed from the trailer and placed onto the new BRP transfer cart. The transfer cart was moved into the Reactor Sphere through the equipment airlock by opening and closing the airlock doors. This was accomplished without jeopardizing the sphere containment boundary.

Once inside containment, the cask was prepared for loading by removing the primary lid, wrapping the cask in plastic, installing the wood cribbing, and installing a plastic bag inside the cask to protect it against gross contamination. Once the cask was prepared, the liner was raised from the RV using the transfer bell/grapple (see Figure 7) and allowed to drain. Once drained, the snap on liner bottom tray was installed and the liner was placed in the cask (see Figure 8). The grapple was disengaged from the liner lid and the transfer bell/grapple moved to the reactor deck for storage. The cask lid was reinstalled, lid bolts torqued, and a leak test performed. Upon a satisfactory leak test, the cask was moved back to the Turbine Building where it was prepared for final survey and shipping by the Radwaste Department.

After dose profiling all of the CSTs there were two stainless steel CSTs that the channel section and transition piece characterized as Greater Than Class C (GTCC) material which meant they could not be shipped for off-site disposal. These support tubes were within shipping limits but had to be removed from the transition piece before they could be transferred to the RV and processed. At the request of BRP, NUKEM cut the support tube from each of the stainless steel CSTs using a specialized underwater hack saw. The support tubes sections were transferred to the RV and processed into the last 8-120 burial liner.

During CST processing the transition piece from each CST was packaged into a transition piece basket that was stored in the RV. Once the CST processing was complete, there were 12 baskets containing all of the transition pieces. These baskets were packaged into three FSV-1 liners and ultimately shipped off site in three FSV shipments in the same manor as the CRB shipments.

After all of the primary components were either moved or packaged and shipped from the SFP, the fuel pool had adequate space to determine the LSA and miscellaneous item requirements. Items such as the channel rack, periscope, sipper cans, and miscellaneous containers were pressure washed, removed from the SFP, wrapped in plastic then placed in a storage box or moved to a storage location to await final disposition.

4. SUMMARY

The Big Rock Point fuel pool cleanout project is said to be the largest project of its kind to date. Its successful completion was the first milestone in the BRP decommissioning program and remains a source of pride for both the plant and NUKEM. The applied innovative methodology and equipment proved key to the success of the project by eliminating potential problems through engineering controls. Additionally, the team effort put forth by Big Rock Point and NUKEM staff ensured that the project realized the best ideas and resolutions to issues always employing ALARA principals and cost controls.

In the end, the project was completed within the ALARA budget, the project budget, the plant schedule, and with no reportable incidents. Big Rock Point is currently upgrading the containment crane system as well as other work in preparation of the dry fuel storage of the spent fuel.

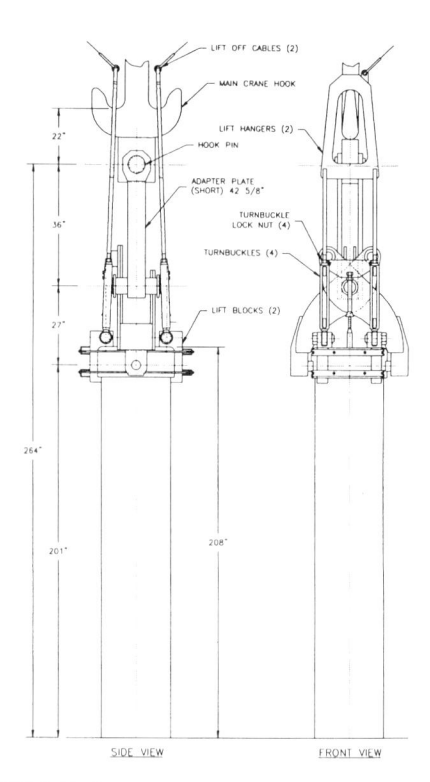

Figure 1   FSV-1 Cask

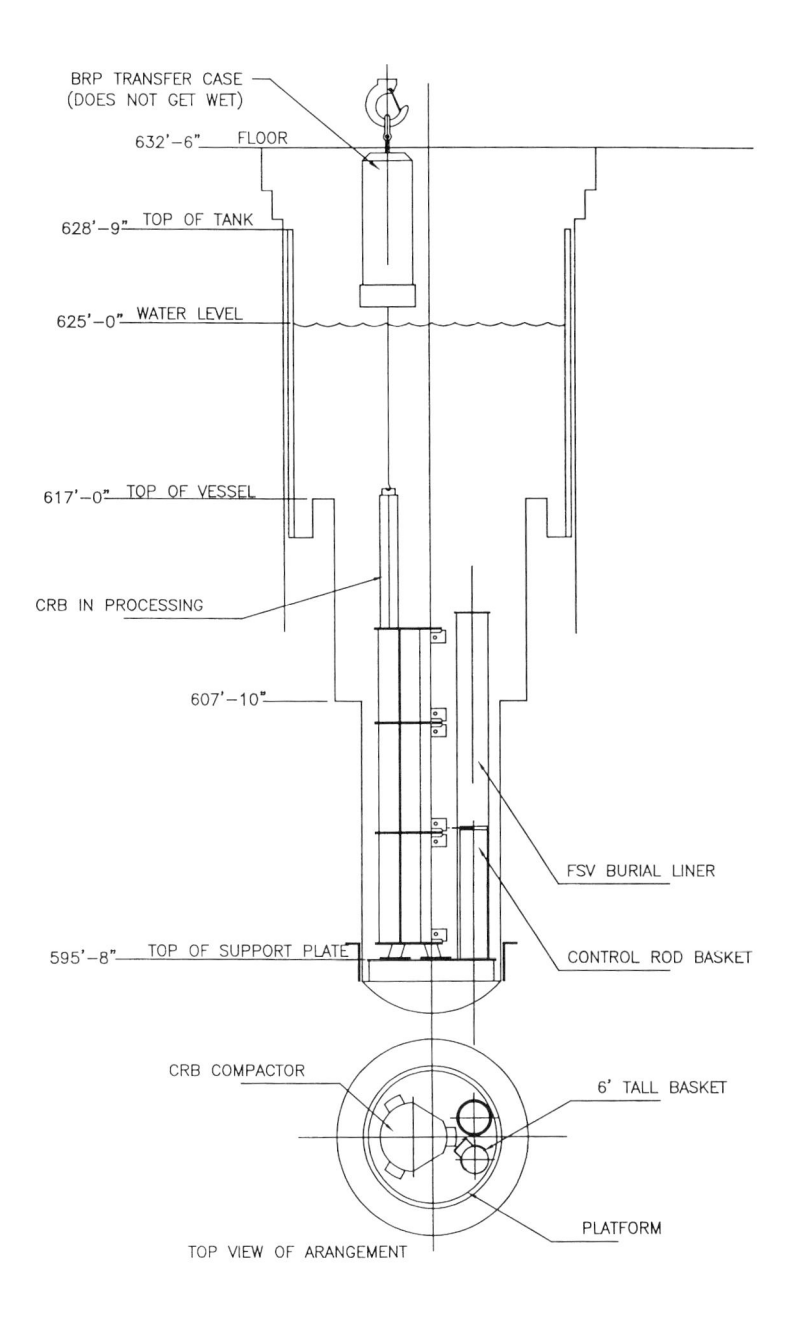

Figure 2   **CRB Processing Equipment Layout in RV**

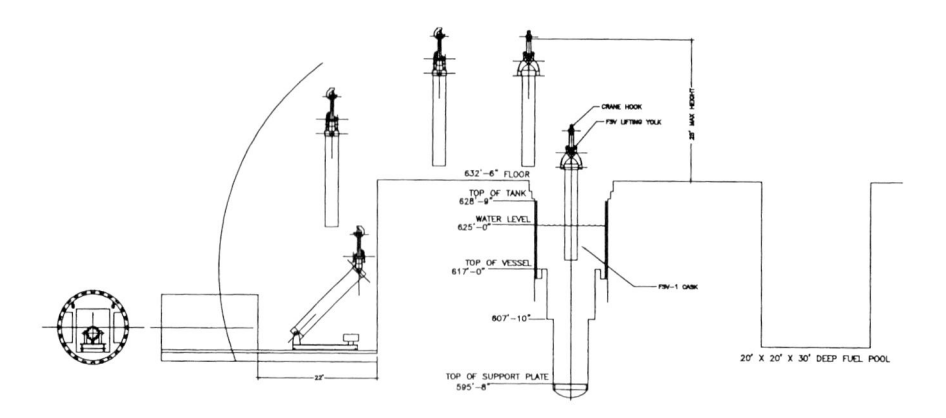

Figure 3   FSV-1 Cask handling in Containment

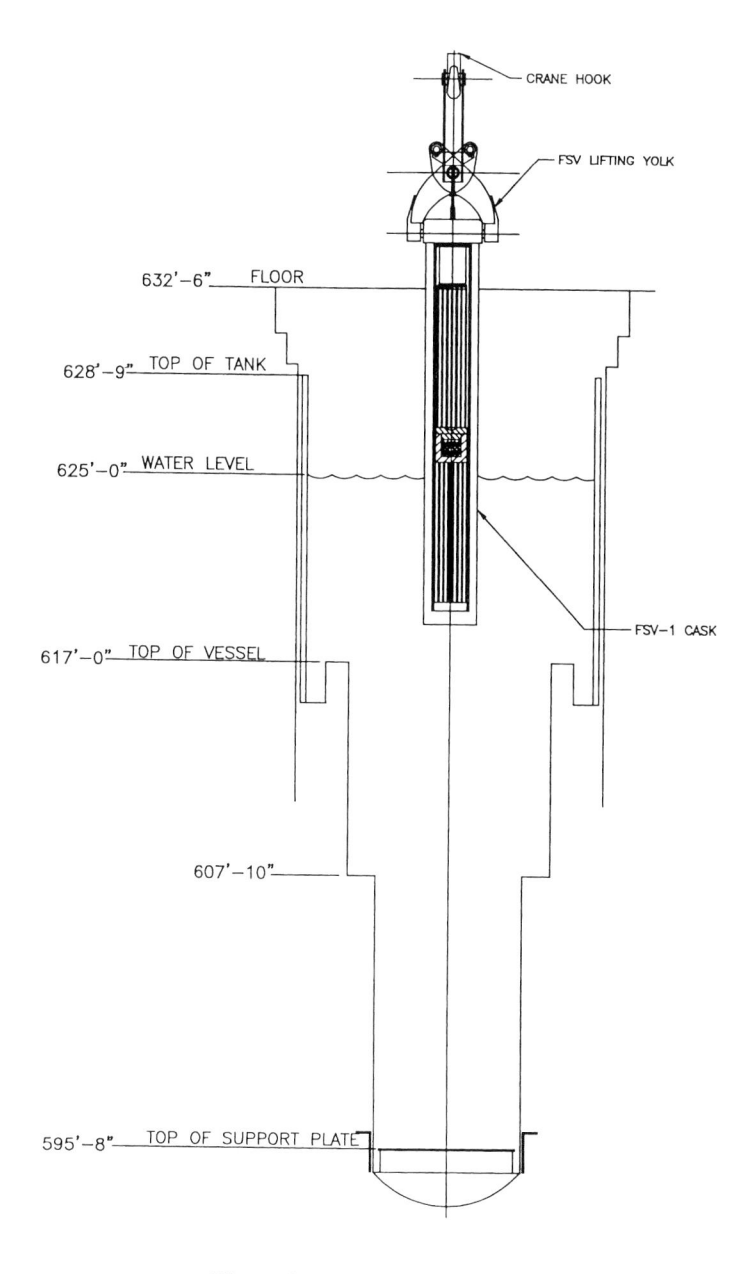

Figure 4   Loaded Liner and FSV-1 Cask

Figure 5   8-120B Cask

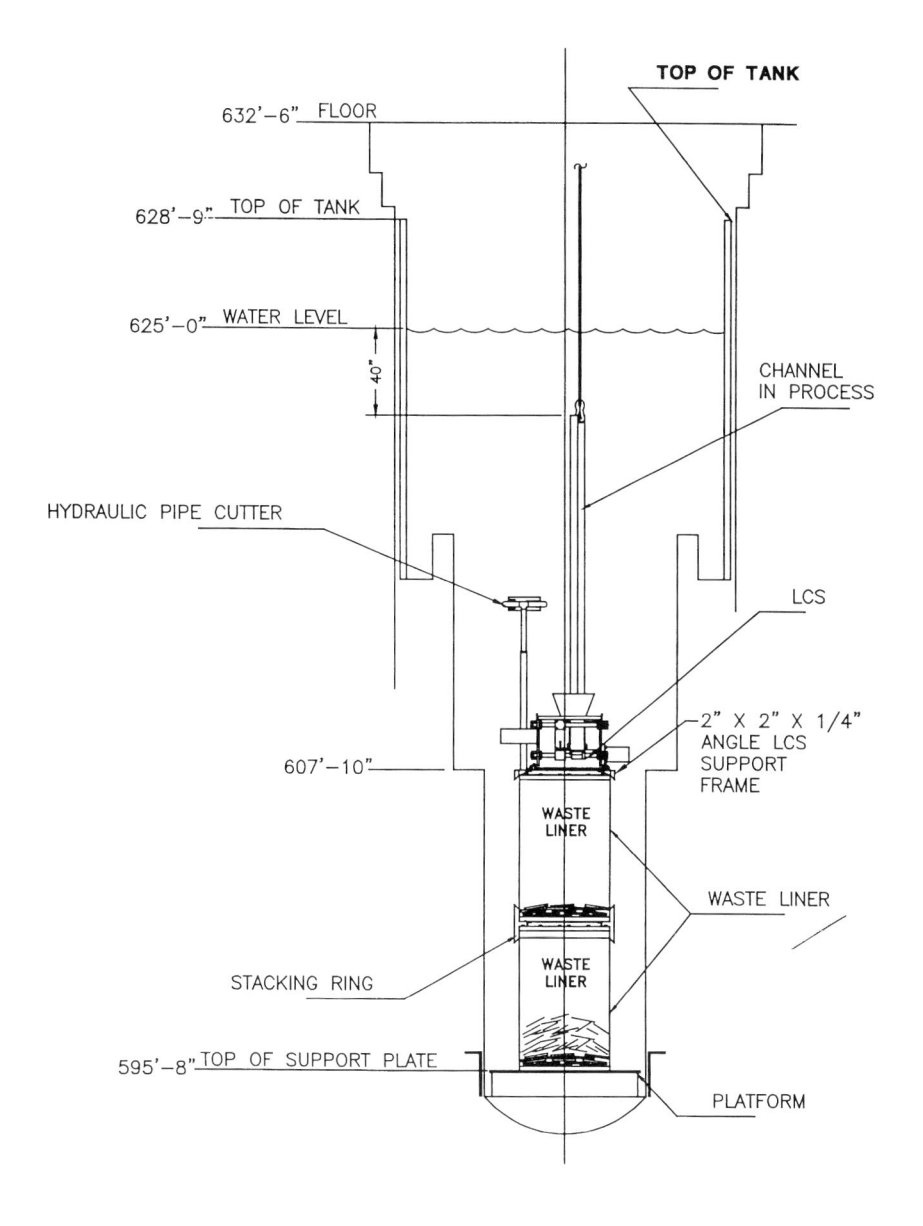

Figure 6   USC & Pipe Cutter Mounted on Waste Liner

Figure 7   Transfer Bell and Grapple Arrangement

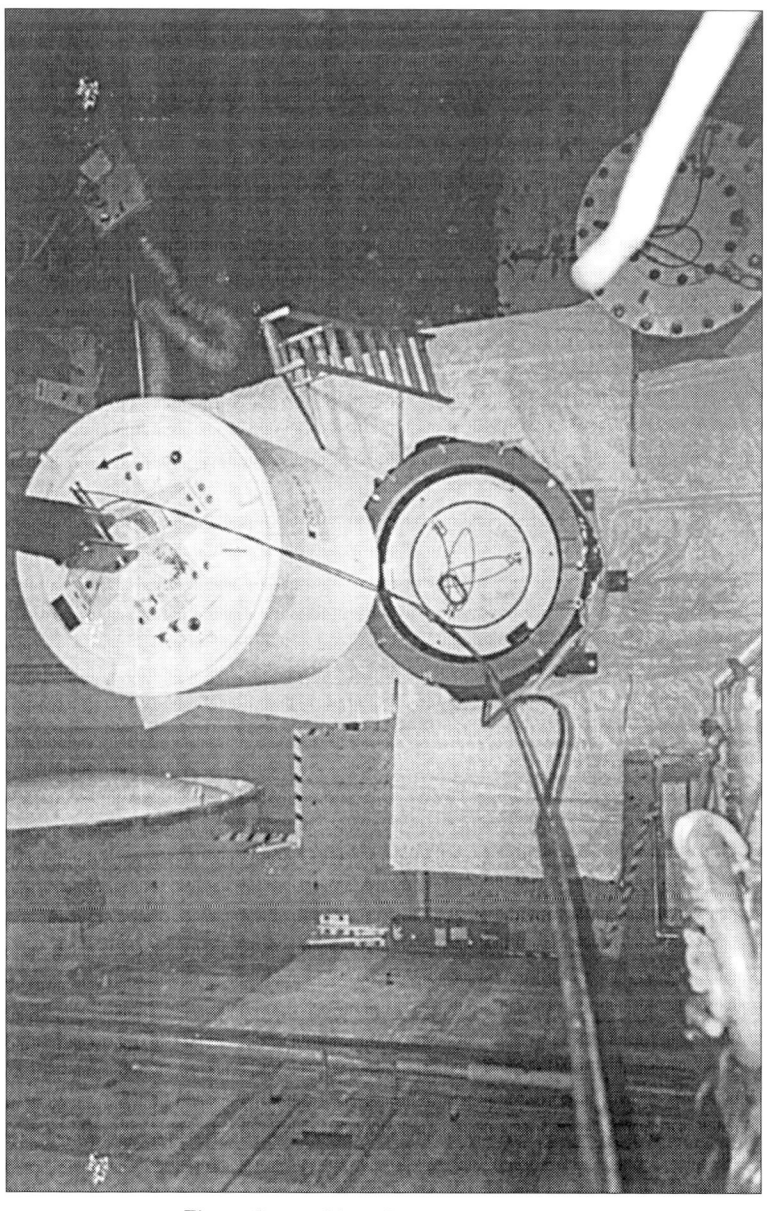

Figure 8   Liner Loaded in 8-120B Cask

© With Author 2001

C596/040/2001

Aspects of decommissioning the UTR-300 reactor at the Scottish Universities Research and Reactor Centre

H M BANFORD, R D SCOTT, I ROBERTSON, and **J D ALLYSON**
Scottish Universities Research and Reactor Centre, Glasgow, UK
T M McCOOL
Nuclear Decommissioning and Cleanup Business Group, British Nuclear Fuels plc, Sellafield, UK

ABSTRACT

The decommissioning of a 300 kW Argonaut type universities research reactor is in progress at the Scottish Universities Research and Reactor Centre. The strategy adopted for the demolition of the reactor and its ancillary plant is discussed. The main focus of the presentation, however, is the effort expended on the determination of the radioisotope abundance of the reactor structure and thereby the waste characterisation. Emphasis is placed on the monitoring of tritium and the mapping of the boundary between free-release and radwaste material within the concrete bio-shield.

1. INTRODUCTION

A 300 kW research reactor (UTR-300) was operated at the Scottish Universities Research and Reactor Centre (SURRC) in East Kilbride near Glasgow between 1963 and 1995. It was used for university teaching and research, with its main activities centred upon basic reactor physics experiments, activation analysis and the provision of short-lived radioisotopes. The reactor is of the Argonaut type and comprised a core of highly-enriched uranium that was moderated and cooled by light water, reflected by graphite and shielded by concrete. Following permanent shutdown in September 1995, the fuel was removed in January 1996 and work then commenced in decommissioning peripheral ancillary plant. Ultimately funds were acquired for full-scale decommissioning in 1999. Accordingly, since early 2000 the licensee, the University of Glasgow, has been involved in a cooperation agreement with the Nuclear Decommissioning and Cleanup Business Group of British Nuclear Fuels plc to demolish and remove from the nuclear site the UTR-300 reactor and all associated ancillary plant plus the building in which the reactor is housed. The ultimate aim of the work is to end the licensee's period of responsibility for the nuclear site.

2. THE REACTOR AND ASSOCIATED PLANT

Associated with the reactor are a plant room, a process pit, fuel pits and an active drain system.

2.1 Reactor

A Side Elevation and a Plan of the reactor are given in Figures 1 and 2 respectively and a diagrammatic view of the reactor core is given in Figure 3. There are two core tanks with six slots apiece that accommodated one fuel element per slot. The tanks are held in a matrix of graphite, which constitutes the neutron reflector, and are connected to pipelines that run from a process pit set into the floor of the reactor hall. Light water was pumped from a holding tank in the process pit into the bottom of the core tanks and then flowed through the tanks and back to the process pit. Here the thermal energy of the primary coolant was transferred to a secondary coolant circuit via a heat exchanger. Reactor power was regulated by four control absorber assemblies of the "window-shade" pattern. Two graphite thermal columns are coupled to the reflector. The large one was used for irradiating samples and the small to conduct neutrons to a shield tank. The reactor shielding is reinforced concrete. Various penetrations and support assemblies are mounted within the shielding to act as beam tubes, ion chamber carriers, ventilation lines, control absorber drives and experimental facilities. There are three removable closure blocks above the core void and a door that gives access to the thermal column.

The reactor was constructed from a variety of materials. Aluminium alloy was used for the reactor primary circuit, part of the structure of the control absorber assemblies and some irradiation facilities, but generally the metal elsewhere, such as the core support plate, core restraints and the reinforcing bar within the concrete, is mainly mild steel. However, there are also some stainless steel components within the reactor core. The reflector and thermal columns are graphite and the concrete is ordinary concrete except in the two closure blocks immediately above the core. In this case the aggregate is barytes.

The reactor sits on a raft of reinforced concrete 45 cm (18 inches) thick which was laid over material denoted as "suitable fill". This "fill" is an unknown quantity, but it is probably whinstone, which is a common bottoming material used in the West of Scotland. The subsoil is clay.

2.2 Plant room

The equipment in the plant room comprises a ventilation system, including the reactor exhaust stack, various pumps and fans, cooling tower, heating plant, fumehood exhaust systems, a water treatment system for primary circuit make-up feed, a pneumatic transfer system to blow sample containers into and out of the reactor, batteries and a variety of electrical switchgear panels.

2.3 Process pit

The equipment within the process pit comprised the bulk of the components of the reactor primary water circuit. Principal items included the primary pump, a dump tank where the moderator/coolant was kept during periods of reactor shutdown, an ion exchange resin assembly, various control valves and associated pipework and a heat exchanger to thermally couple the reactor primary and secondary coolant circuits.

2.4 Active drain system
The active drain allowed radioactive liquid effluent from the reactor and several laboratories within the reactor building to be collected in delay tanks. Here the effluent could be tested for levels for radioactivity prior to discharge under authorisation to the public foul drain. It comprises various runs of piping, sinks, shower trays, pumps and two delay tanks located in a pit external to the reactor building.

2.5 Fuel pits
There are two fuel pits set into the floor of the reactor hall. On contained a spare core of fresh fuel and the other was used to store the active core when it was required to be removed during maintenance shutdown periods.

3. DECOMMISSIONING STRATEGY

The strategy adopted is to remove all non-radioactive material from the reactor structure before removing activated, thereby avoiding cross contamination of clean material.

3.1 Containment
Primary containment of the reactor consists of a PVC tent draped from a platform suspended from the beams of the hall roof as illustrated in Figure 4. The walls of the tent are fixed to the floor with battens. There is no support structure inside the tent. The tent covers the reactor itself, the process pit and the two fuel pits, thereby encompassing a floor area of 250 m^2. There is a single door at one end for personnel access and a large door at the opposite end for equipment access. A ventilation system has been installed to give a slight pressure depression in the tent with a designed extract rate of 5690 m^3/hour via pre-filters, intermediate filters and finally HEPA filters. The exhaust from the tent couples to the original building ventilation system that then exhausts through the reactor stack as normal. Radiological protection is afforded by two fixed gamma monitors and three fixed beta-in-air monitors, one of which is connected into the tent exhaust downstream of the HEPA filters. A further beta-in-air unit is available for mobile use as required.

The reactor is being demolished primarily using remotely operated vehicles in the form of two BROKK machines. The main machine is a BROKK 330 along with a much smaller BROKK MiniCut for light work and for accessing places that are difficult for the larger machine. A system of cameras has also been installed in the tent. There are two mounted on poles at diametrically opposite corners, one on the tent ceiling immediately above the core void and one on the BROKK 330. These cameras are coupled to CCTV monitors and VCRs in an operations control room and in conjunction with another camera and time-delay recording system are being used to make a video film of the complete decommissioning job. The BROKK machines can be operated from the control room using the CCTV or from the hall floor, but outside the tent.

3.2 Waste strategy
3.2.1 *Free-release*
Free-release material (1) is being disposed of by normal dumping routes. All free-release components from the biological shield have now been consigned.

3.2.2 *Low Level Waste (LLW)*
LLW is being packed into ISO containers for disposal to Drigg.

3.2.3 *Intermediate Level Waste (ILW)*
ILW is being packed into Chapelcross flasks and shall be sent to the MBGWS at Sellafield to await ultimate disposal in the fullness of time.

4. ESTIMATION OF THE RADIONUCLIDE INVENTORY

A reasonably accurate estimate of the inventory of radionuclides remaining after removal of the fuel is important for a number of reasons. First, it is essential in the overall planning of the operation and production of the safety cases; secondly it gives a basis for the costs of, and strategy for, waste disposal based upon the predicted quantities of LLW, ILW and free-release material; thirdly, there are legislative requirements for such an estimate and, finally, the appropriate Environment Agency will not grant an authorisation for removal of waste from the licensed site without a justifiable estimate of the quantities involved. It is particularly important to establish accurately the categorisation of material close to the LLW/ILW boundary (for reasons of cost) and the LLW /free release boundary (to avoid needless use of storage space at Drigg) and this requires accurate measurements on appropriate samples.

In order to calculate activities it is necessary to have values for the neutron flux, the material composition and the irradiation time, of which only the last is well known. The following assumptions were made in formulating a model for the calculation of activities at shutdown (11.9.95).

i) the operating history was 18 hours/week for 11 months each year;
ii) the reactor operated at 100kW for 7y followed by 25y at 300kW;
iii) a simple analytical form could be assumed for the flux distribution, namely a product of cosine functions in the core and an exponential decrease along the length of the thermal columns with parameters deduced from flux contours in the core and measurements in the large thermal column;
iv) as the period between shutdown and dismantling of the structure is more than 5y, activation products with half-lives of less than a year or so can be ignored;
v) in the calculation of activation, fluxes derived from assumption (iii) were multiplied by a factor corresponding to the 'on' portion of the duty cycle (approximately 0.1), but it should be noted that this procedure would be incorrect for short-lived products (ie having half-lives << the 'on' period) which saturate at a value corresponding to the true flux;
vi) only thermal neutron activation was important.

4.1 Graphite
The obvious product here is ^{14}C, which can be produced in the following ways:

$$^{13}C (n,\gamma) \,^{14}C \qquad (1.37 \times 10^{-3} \text{ b})$$
$$^{17}O (n,\alpha) \qquad (0.24 \text{ b})$$
$$^{14}N (n, p) \qquad (1.8 \text{ b})$$

The activity arising from the first of these was calculated as an average over the core and each of the two thermal columns separately. This gave 0.9 GBq, 0.08 GBq and 0.03 GBq for the core, large thermal column and small thermal column respectively with a total mass of 7.3 te in a volume of 4.6 m^3. This 1 GBq is well inside the upper LLW limit of 12 GBq te^{-1}.

The contribution from the other two reactions is much more difficult to assess since it would arise from activation of the air passing over and through the graphite (and therefore of unknown residence time) and of any moisture in the graphite. Also, there might be air in the structure of the graphite and it is clear that the calculated value must be a lower limit. Measurements of ^{14}C in graphite samples from the large thermal column were therefore carried out and the results are shown in Table 1.

Table 1: Comparison of measurements and calculation of ^{14}C. Distances are from the inner end of the central stringer in the thermal column and the off-centre samples were displaced 25 cm horizontally and 45 cm vertically from the centre line. The number of measurements at each position is in brackets.

Sample Position	^{14}C Bq g^{-1} (measured)	^{14}C Bq g^{-1} (calculated)
Core Tank	810 (3)	650
Inner End (off centre)	140	110
30 cm	190 (2)	80
60 cm	66	30
75 cm (off centre)	<30	7
90 cm	35 (2)	9
120 cm	9	3
150 cm	1	1

The trace element content of the graphite is unknown, but, given the date of construction, it seemed reasonable to adopt the specification for Magnox graphite (2) in calculations of other possible radionuclides. Production of ^{3}H turned out to be by far the largest through the reaction ^{6}Li (n,α) ^{3}H which has a cross section of 940b, sufficient to generate appreciable amounts of ^{3}H although the Li content is given as only 50 ppb. The calculation gave 8 x 10^{10}Bq, 8.3 x 10^9 Bq and 2.8 x 10^9 Bq at the shutdown in the core, large and small thermal columns respectively and the theoretical values are compared in Table 2 with measurements performed on the same samples as were used for ^{14}C analysis.

The ratio of ^{3}H to ^{14}C activity at shutdown would be 100:1 at the assumed Li content of the graphite and considering only the ^{13}C (n,γ) ^{14}C reaction. The experimental values were generally less than this (taking ^{3}H decay into account) but this is not surprising given the mobility of ^{3}H and the possibility of the other mechanisms for ^{14}C production. γ-ray spectrometry of a small sample showed only ^{60}Co and ^{152}Eu, both at around 3000 Bq g^{-1} in a region where the flux is 2.6 x 10^{15} m^{-2} s^{-1} (including the duty factor). Comparison with calculation indicated that the quoted Co content (0.02 ppm) was rather low and that of Eu (4 ppb) rather high. The activity of these two nuclides is some 5 GBq each and the specific activity of the graphite is well within the LLW limit.

Table 2: Comparison of measurements and calculation of ^{3}H at various times

Sample Position	^{3}H (Bq g^{-1}) Measurements (Calculation)		
	12/96	7/97	3/99
Core Tank	1.2 x 10^5 (5.7 x 10^4)	9.9 x 10^4 (5.5 x 10^4)	1.2 x 10^5 (5.1 x 10^4)
Inner End (off centre)			6500 (9050)
30 cm		5450 (7100)	3780 (6400)
60 cm		2130 (2350)	
75 cm			610 (570)
90 cm		1070 (770)	770 (700)
120 cm		1100 (250)	
150 cm		190 (80)	

4.2 Concrete

The biological shield is some 2.8m thick and, as it was expected that a considerable quantity towards the outside would qualify as free release material (<0.4 Bq g^{-1}), a series of concrete cores was extracted by diamond drilling in order to establish, by γ-ray spectrometry, the radionuclide profile. Most of these cores were 2m long, but a few were drilled right through to the inside face. The only γ-emitting nuclides revealed by the γ-ray spectrometry were ^{60}Co, ^{152}Eu and ^{154}Eu. The ratio of the ^{152}Eu to ^{60}Co activity averaged 2.8 and this implies a trace concentration ratio of Co to Eu of around 12:1, close to typical literature values of 6 ppm Co and 0.6 ppm Eu. The results implied that the induced activity fell to below the free release value at a depth of about 1.5m from the outside face, but a search of the literature revealed that trace concentrations of Li in the aggregate might lead to a ^{3}H activity tens of times greater than that of ^{152}Eu. It was therefore considered necessary to assay for ^{3}H in the concrete to check whether or not its presence altered the position of the free release boundary.

Samples were initially sent to a commercial organisation for assay by prolonged heating, gas collection in a bubbler and liquid scintillation counting. However, a simple leaching procedure has been developed in-house which gives very similar results and is now used routinely in conjunction with liquid scintillation.

Results for a sample taken from the side elevation of the concrete shield at a position 72 cm vertically from floor level and 36 cm horizontally from the centre line of the reactor core are shown in Fig. 5. It is immediately obvious that the presence of ^{3}H shifts the free release boundary outwards by about 0.5m and that the halving distance for ^{3}H is considerably greater (some 9 cm) than that of the other induced activities (some 6.5 cm). Also, the ^{3}H does not increase continuously to the inside face of the concrete shield but shows a peak at a distance of about 30 cm from the inside. These results indicate that ^{3}H, most likely in the form of tritiated water, is mobile in the concrete and the general behaviour is consistent with a thermal gradient in the concrete coupled with escape from the inside face (the maximum graphite temperature under operating conditions was 140° C). Co and Eu inventories were calculated by extrapolating the measured results from this core to the inside face (the inner samples were too active for γ-ray spectrometry) and comparing these with the predictions of the flux model and the assumed typical Co and Eu trace concentrations. This gave a scaling factor by which the model results obtained by using volume average fluxes (over the faces of the reactor core and thermal columns and into the concrete with the measured half distance) could be adjusted. It is clear that the fact that ^{3}H results exhibit a turnover has a large effect on the inventory because of the exponential behaviour of the concentration. In fact the inventory of ^{3}H for the concrete core shown in Fig. 5 is only one quarter of that which would be obtained by extrapolating the outer measurements to the inner face. The inventory in this core was used to estimate the total activity in the concrete and it was concluded that, in some 130te of concrete (including the raft underneath) there will be, at a generous estimate, 330 GBq of ^{152}E, 110 GBq of ^{60}Co and 90 GBq of ^{3}H with much smaller quantities of ^{55}Fe and ^{154}Eu.

4.3 Mild Steel

The main components here are the core and thermal column support plates, the inner ends of the beam plugs and closure block plugs, the core and γ–curtain restraints, the inner re-bar and some support structure in the concrete. There are also some small volume stainless steel components associated with the control blades (boral clad in aluminium) and all the above have to be classed as ILW.

The model was used to obtain average values of flux over these components and activities calculated for ^{55}Fe, ^{60}Co and the long-lived nickel isotopes ^{59,63}Ni. Co and Ni are trace constituents of mild steel and their concentrations vary from type to type, so these were measured by inductively-coupled mass spectrometry for (inactive) samples from the top closure block and the outer re-bar. The concentrations were slightly different with the Co concentration at 160 and 135 ppm and the Ni at 700 and 1050 ppm for the top closure and re-bar respectively. This means that the ^{55}Fe to ^{60}Co ratio is in the range 28 - 23:1 at shutdown, whereas the ^{63}Ni activity is around 2% of that of ^{60}Co and ^{59}Ni is down by another two orders of magnitude. It is important to know these ratios as ^{55}Fe and ^{59}Ni decay by electron capture and ^{63}Ni is a weak β emitter and their activity is most readily ascertained by comparison with the easily measured ^{60}Co. It was found that the specific activities of ^{55}Fe and ^{60}Co were in the range 10^7 - 10^6 and 10^6 - 10^5 Bq g^{-1} respectively. Thus we estimate that mild steel components containing some 5 x 10^{12} Bq ^{55}Fe and 2 x 10^{11} Bq ^{60}Co at shutdown in a volume of around 0.2m^3 will be consigned for storage as ILW.

5. CONCLUSIONS

Given the uncertainties in the trace element contents of the various components and the sampling problems following from the distribution of these elements within the bulk matrix, the combination of which exceeds the purely analytical uncertainties arising in the measurements, it appears that the predictions of a fairly crude flux model backed by a judicious sampling and measurement strategy can produce estimates of total activity which are reliable enough both to delineate the various waste boundaries and set appropriately conservative limits to ensure authorisations are not breached.

6. REFERENCES

(1) The Radioactive Substances (Substances of Low Activity) Exemption Order 1986, HMSO London (1986).
(2) Handbook on decommissioning of nuclear installations, EUR 16211 EN, p.38, (1995).

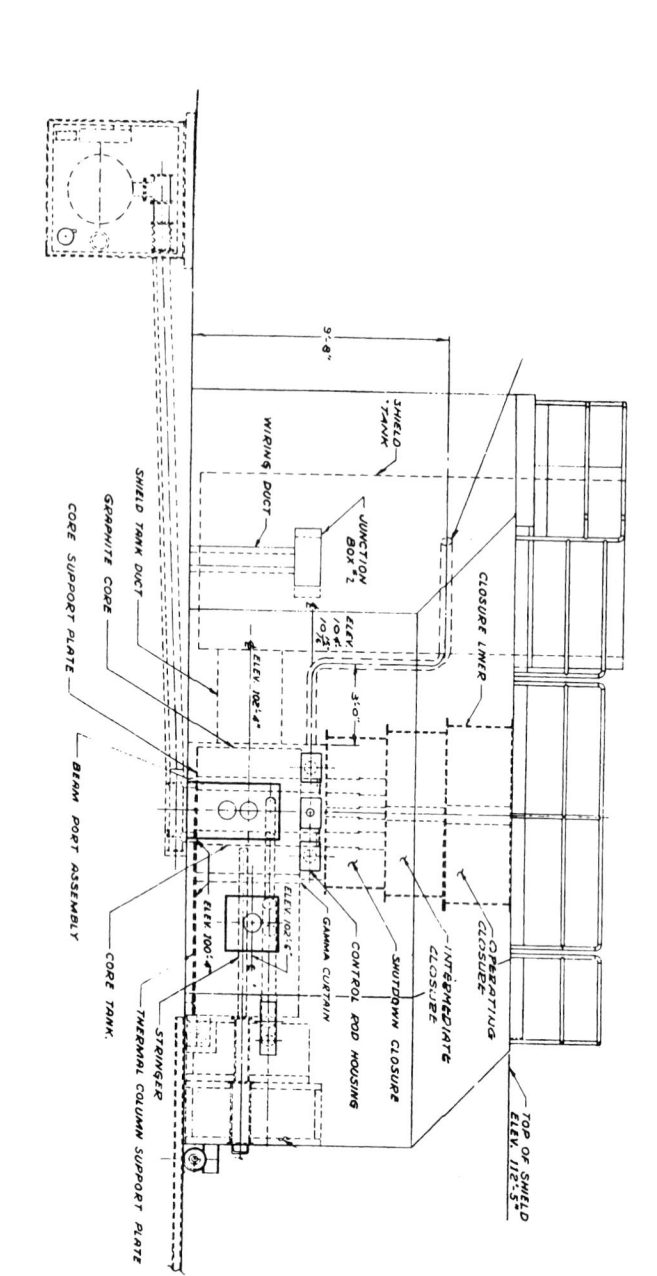

Figure 1. Side elevation view of UTR-300 reactor

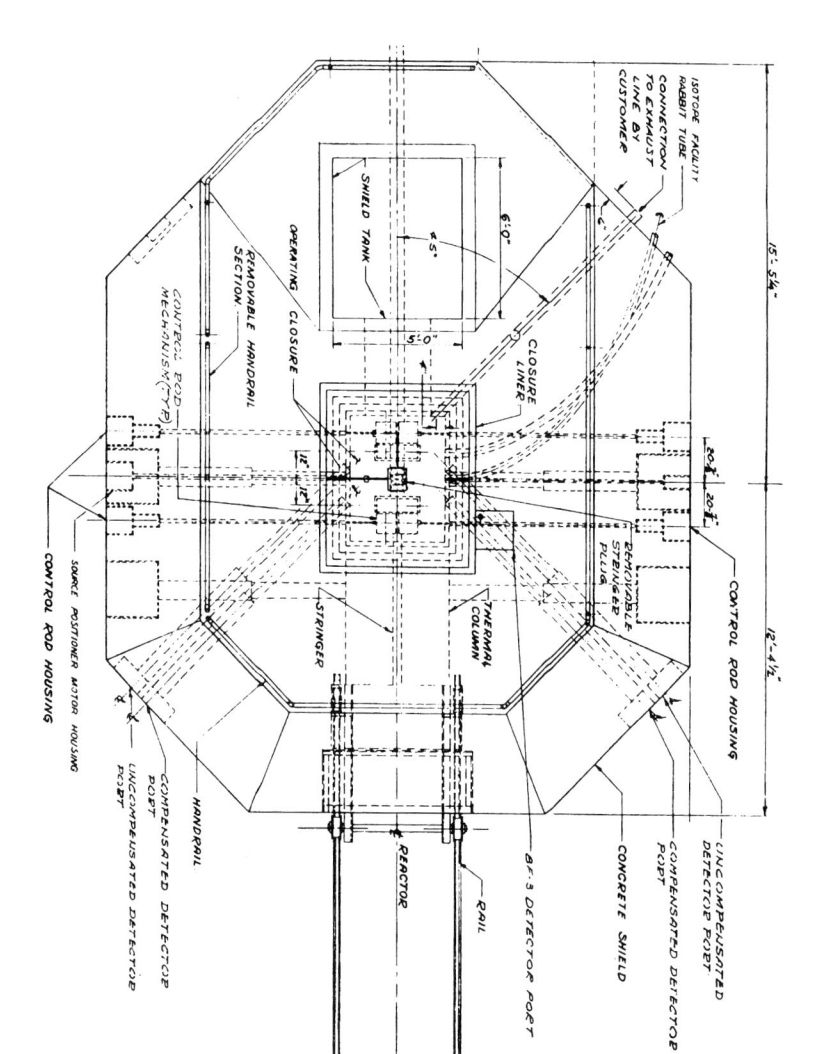

Figure 2. Plan view of UTR-300 reactor

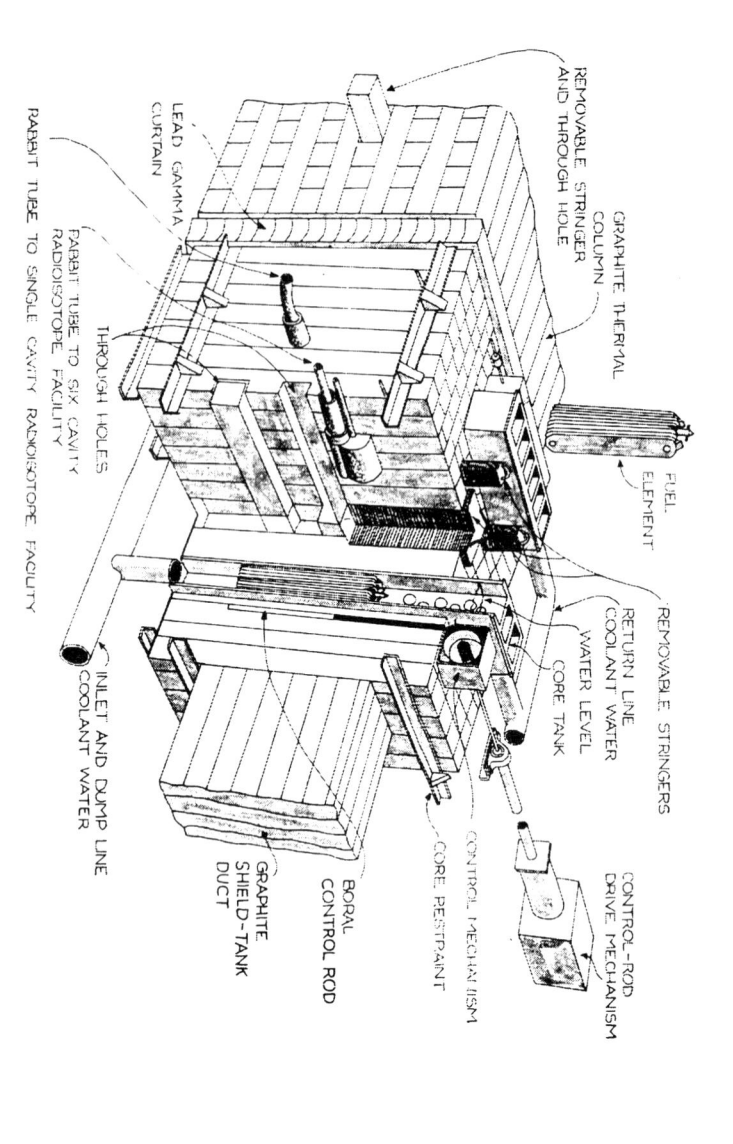

Figure 3. Diagrammatic view of UTR-300 core

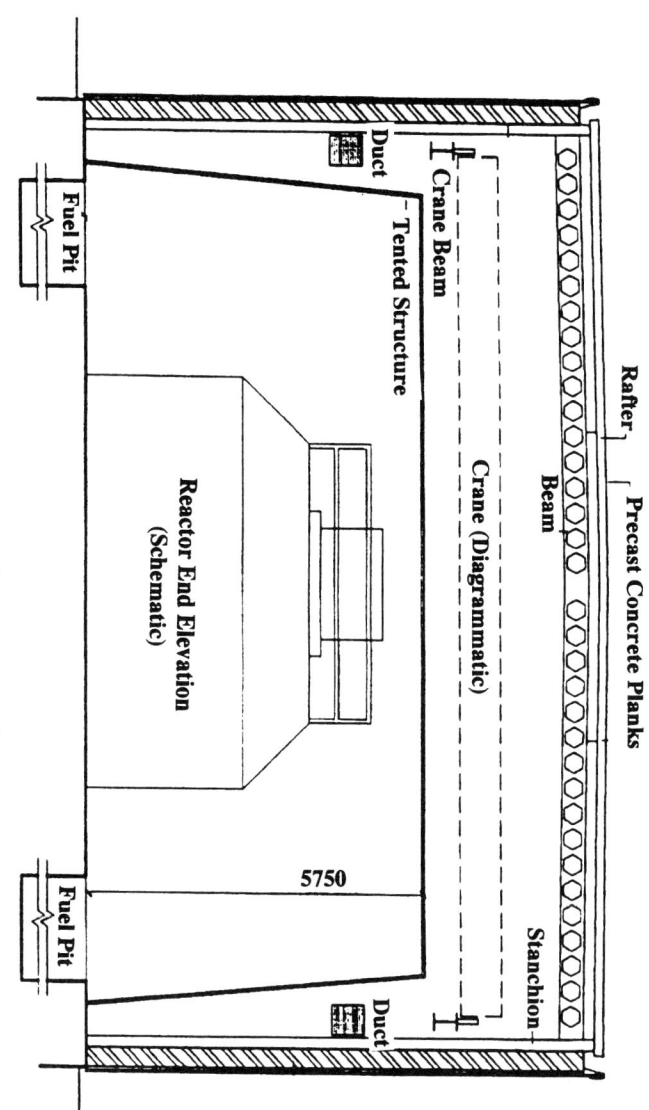

Figure 4. Schematic of primary containment

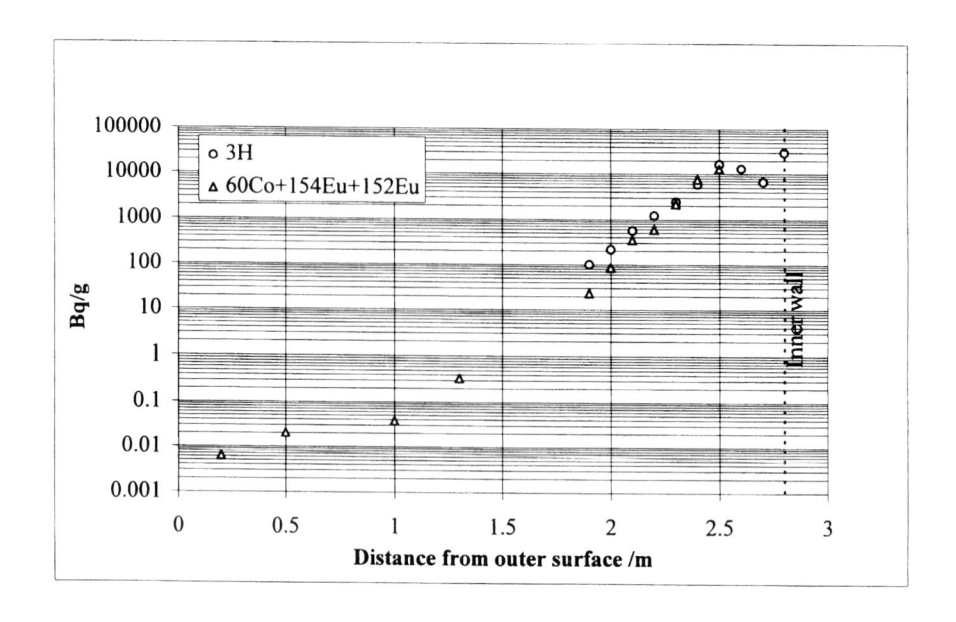

Figure 5. Profile of radionuclides along a core of concrete from the biological shield

C596/022/2001

Assessment of the costs of decommissioning the nuclear facilities at Risø national laboratory

K LAURIDSEN
Risø National Laboratory, Roskilde, Denmark

1 INTRODUCTION

Risø National Laboratory (RNL) was established in the late 1950'es as a Danish research centre for preparing the introduction of nuclear energy. Three research reactors and a number of supporting laboratories were built. However, Denmark has not yet built any nuclear power plants. In 1980 the Danish Parliament decided that nuclear power should no longer be an option in the national energy planning. Subsequent to that decision the research at RNL related to nuclear power was reduced and the utilisation of the facilities concentrated on other applications, such as basic materials research, isotope production and silicon transmutation doting. Already in 1975 one of the reactors had been taken out of service for economical reasons and the activities moved to the 100 MW materials test reactor, DR 3. Furthermore, in 1989 the hot cell facility was closed, and over the next four years it was partly decommissioned.

As part of Risø's strategic planning in 2000 it was taken into account that the largest research reactor, DR 3, was approaching the end of its useful life, and that the decommissioning question was becoming relevant. As most of the other nuclear activities at Risø depended on the running of DR 3, it was decided to decommission all nuclear facilities at Risø National Laboratory once the reactor had been closed. Therefore, a project was started with the aim to produce a survey of the technical and economical aspects of the decommissioning of the nuclear facilities. The survey should comprise the entire process from termination of operation to the establishment of a "green field", giving an assessment of the manpower and economical resources necessary and an estimate of the amounts of radioactive waste that must be disposed of. The planning and cost assessment for a final repository for radioactive waste was not part of the project. Such a repository was considered a national question, because it will have to accommodate waste from other applications of radioactive isotopes, e.g. medical or industrial.

In September 2000 Risø's Board of governors decided that DR 3 should not be restarted after an extended outage. The outage was caused by the suspicion of a leak in the primary system of the reactor, and followed after the successful repair of a leak in a drainpipe earlier in the year. Extensive inspection of the reactor tank and primary system during the outage showed that there was not any leak in the primary system, but at the same time some corrosion was revealed in the aluminium tank. According to the inspection consultant the corrosion called for a more frequent inspection of the tank. Therefore, the management judged that the costs of bringing the reactor back in operation and running it would outweigh the benefits from continued operation in the remaining few years of its expected lifetime.

The closure of DR 3, of course, accentuated the need for decommissioning planning and the results of the project. But the project was allowed to finish according to the original schedule, delivering a draft report by the end of 2000 and a final report by the end of February 2001. The main report (1) was written in English, but a summary report in Danish (2) was produced, as well.

2   PROJECT ORGANISATION

The project was run by a small project group of four persons with the right to draw on relevant experts at Risø and externally. A steering group with participation by the Risø management and relevant technical experts followed the work closely, and most members of the steering group participated in the writing of the report.

In the early phase of the project visits were paid to laboratories in other countries with ongoing decommissioning programmes in order to learn from their experience. The laboratories visited were the Paul Scherrer Institute in Switzerland, the Forschungszentrum Karlsruhe in Germany and the UKAEA's Harwell laboratories in England.

3   DESCRIPTION OF THE NUCLEAR FACILITIES

The nuclear facilities at Risø National Laboratory to be decommissioned are three reactors, a hot cell facility, an isotope laboratory, a fuel fabrication plant and a radioactive waste treatment plant and its associated storages.

DR 1 ("Danish Reactor 1") is a 2 kW thermal homogeneous, solution-type research reactor which uses 20 % enriched uranium as fuel and light water as moderator. The first criticality was obtained on August 15th, 1957. During the first 10 years of operation the reactor was heavily used for making neutron experiments, which later were transferred to the DR 2 and DR 3 reactors. Since then the reactor has been used mainly for educational purposes. The reactor is still operational and may be used for materials investigations in the preparations for decommissioning of the other facilities. Afterwards it will be closed and decommissioned. Figure 1 shows a sketch of the reactor.

DR 2 was a "swimming pool" type, light-water moderated and cooled reactor with a power level of 5 MW$_t$. It went critical on December 19th, 1958. The reactor was finally closed down on October 31st, 1975 and later partially decommissioned. At present the reactor block and part of the coolant circuit remain. The reactor tank is made of aluminium. It has a wall thickness of 9.5 mm, a diameter of 2 metres and a height of 8 metres.

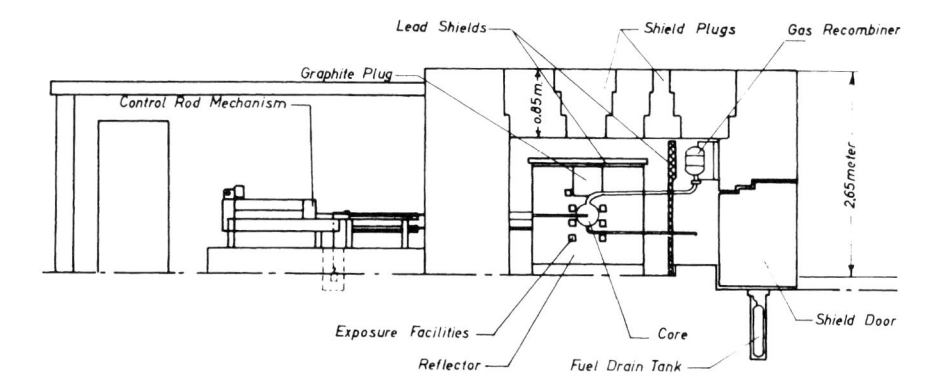

Figure 1 Sketch of the DR 1

The largest reactor, DR 3, was a 10 MW materials test reactor of the same type as the PLUTO reactor at Harwell and the DMTR at Dounreay. It was cooled and moderated by heavy water. The reactor went into operation in January 1960, using highly enriched fuel for the first 30 years, but in 1990 a change was made to low enrichment (20%). It had a number of in-core irradiation facilities and four tangential beam tubes used for neutron scattering experiments and silicon transmutation doting. As a side-benefit the reactor contributed to the central heating system for the laboratory. DR 3 will be by far the largest source of activity and the one that will be most difficult to dismantle. Figure 2 shows a sketch of the reactor block.

The Hot Cell facility was in active use during the period 1964 - 1989. Its six concrete cells have been used for post-irradiation examination of fuel from test reactor fuel pins irradiated in the DR 3 reactor, the Halden reactor in Norway etc. Power reactor fuel pins, including plutonium-enriched pins, from several foreign reactors have been examined. Also HTGR fuel from the English Dragon reactor has been examined. All kinds of non-destructive and destructive physical and chemical examinations have been performed. In addition, various radiotherapy sources - mainly ^{60}Co sources - have been produced from cobalt pellets irradiated in DR 3. Following a partial decommissioning in 1990-94 a row of six concrete cells now remains in the building, which is now being used for other purposes.

The Isotope Laboratory was connected to DR 3 via two pneumatic tube systems, which were used for sending samples to be irradiated in the reactor and receiving them back to be processed and shipped from the laboratory. Some contamination remains in one laboratory.

In the fuel fabrication facility fuel elements for the DR 3 reactor have been produced and handled for more than 35 years. Up to 1988 the fabrication was based on highly enriched (93% ^{235}U) metallic uranium; but from then on the elements have been made from low enriched (< 20% ^{235}U) U_3Si_2 powder. Due to the pyroforic nature of the U_3Si_2 powder it has been handled only under argon atmosphere in closed containers or in gloveboxes. This limits the necessary cleanup mainly to the equipment in this room and the connected ventilation channels.

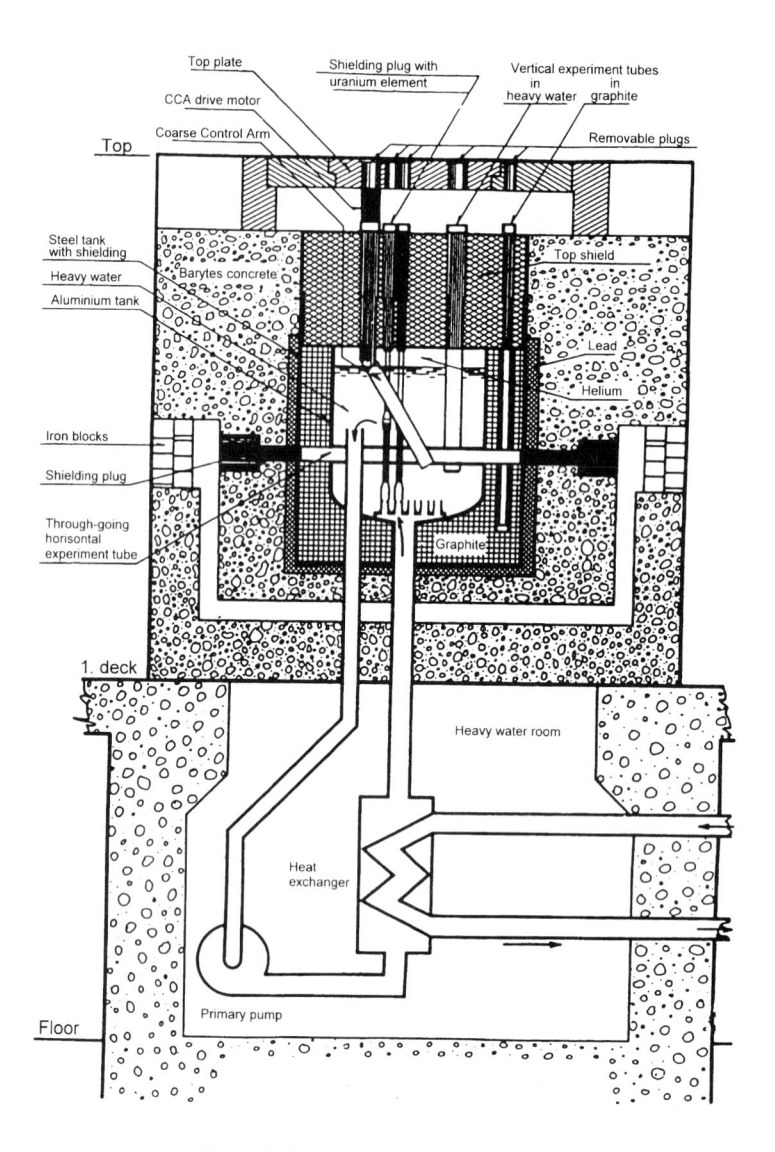

Figure 2 Sketch of the DR 3 reactor block

The waste management plant is responsible for the collection, conditioning and storage of radioactive waste from the laboratories and the nuclear facilities at Risø and from other Danish users of radioactive materials. The latter is a commercial service provided since 1972. Decommissioning of the waste storage facilities is primarily a question of providing final disposal for the waste units, while the following decommissioning of the empty facilities should be fairly simple. The process equipment used in treatment of the waste will only be slightly contaminated, but the decommissioning will have to be postponed until there is no more use for the facilities or suitable substitutes have been provided.

After decommissioning of the nuclear facilities at Risø there will still be a need for a treatment system for radioactive waste in Denmark, because radioactive isotopes will continue to be used in medicine, industry and research. Facilities for handling, conditioning and temporary storage of waste from such activities must be available, but could of course be situated elsewhere.

4   APPROACH TO COST ASSESSMENT

The facilities at Risø National Laboratory vary much in size and complexity. Therefore, the approach to cost assessment has been different for different facilities. Although a standardised approach has its advantages it has not been considered sensible to use the same method for e.g. DR 3 and the fuel fabrication laboratory.

Basically, for each facility the decommissioning tasks and their associated needs for manpower and other expenses have been identified and the costs assessed in fixed prices ("Year 2000 prices"). Those tasks where changes in radiation levels result in different needs for protection between the scenarios, described below, have been marked during the task identification, so that differences in cost estimates arising from this reason can be identified. Similarly, differences in costs arising from the different demands to care and maintenance between the scenarios have been identified.

4.1   Methods used for cost assessment

As indicated above, the decommissioning work required for some facilities, e.g. the Isotope Laboratory, could easily be identified, whereas for others a systematic approach was necessary. In particular for DR 3 a standard list of costing items (3) was used as a template for specifying the costs of decommissioning operations. This list has been established in collaboration between the OECD/NEA, the IAEA and the EC. It is aimed at nuclear power plants, but most of the items listed are valid for a research reactor, as well. Also for other facilities than DR 3 the list has been used as a checklist.

For each of the items addressed the required labour effort was estimated - either by Risø staff, where we felt we had sufficient insight, or with the help of consultants or the PRICE programme (4), developed by the UKAEA. A standard rate of 231 DKK/hour was used to calculate the labour cost. This labour rate was obtained by calculating a suitable average of the costs of the staff categories foreseen for the decommissioning organisation.

In PRICE a facility is broken down into simple building blocks or "Components". For each component data are stored from previous projects on the resources (man-hours) required to remove unit quantity of that component. This is termed the "Norm", which varies depending on "complexity" and "radiological condition" classifications attributed to the component. PRICE produces an estimate of the man-hours needed to decommission the component in question. The associated cost is then calculated by multiplication with the standard rate.

For DR 3 the costs were entered into an Excel sheet, based on the costing items in the above-mentioned standard list. For DR 1, DR 2 and the Hot Cell facility decommissioning operations were identified by Risø staff, and PRICE was used to calculate the cost.

One point where we deviated from the standard list was in the assessment of the health physics assistance needed. Here the list prescribes the specification of health physics effort for each task. However, we found that - at this preliminary stage - the necessary health physics staff and the required equipment can be assessed on an overall basis, taking into consideration more broadly the tasks that are to be performed.

4.2 Scenarios considered

Three different temporal scenarios were considered. The deciding difference between the scenarios is the cooling time foreseen for reactor DR 3 from termination of operation to final dismantling. Cooling times of 10, 25 and 40 years were considered. The total durations of the scenarios were estimated to be 20, 35 and 50 years, respectively. The shortest scenario is sketched in Figure 3.

Construction of the final repository for radioactive waste is foreseen to take place in stages as the demand arises. The stages shown in Figure 3 should be taken as indicative only and not as an actual plan for the repository - which was outside the scope of the project.

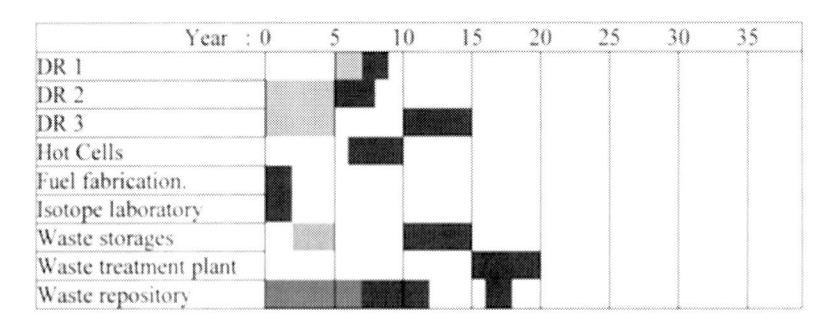

[Figure showing Gantt chart with rows: DR 1, DR 2, DR 3, Hot Cells, Fuel fabrication, Isotope laboratory, Waste storages, Waste treatment plant, Waste repository, across years 0 to 35]

: Dismantling of external circuits etc.
: Final dismantling of facility, transfer of waste to repository
: Establishment of intermediate storage facility and/or handling facility
: Identification of a disposal site and acceptance and licensing of the construction
: Construction of waste repository (in several stages)

Figure 3 Scenario 1 - "20 years scenario" (10 years cooling time for DR 3)

In all three scenarios the two small facilities, the fuel fabrication laboratory and the Isotope laboratory, are decommissioned first. Both are considered as being only lightly contaminated, and the buildings can be used for other purposes. Furthermore, it is assumed that Hot Cells, DR 1 and DR 2 are decommissioned during the first ten years in all scenarios. The transfer of waste from the storages at Risø to the final repository can - more or less - be carried out at any time after the repository has been constructed. Irrespective of the scenario, after the end of decommissioning some activity will still take place either at the waste treatment plant or at the repository, receiving radioactive waste from other activities than Risø's.

A more detailed planning may show that the whole process can be carried through in an even shorter time than the 20 years foreseen here. However, in the short scenario in particular it may turn out to be necessary to establish an intermediate storage facility for waste arising from the dismantling of active parts. The necessity of this will depend on the time required for selecting a site for the final repository and going through the safety case for this site. Needless to say, this process may be difficult due to the political overtones.

5 ESTIMATED COSTS

During the project it was realised that there will probably not be big differences between the three scenarios considered with respect to the protective measures needed for the personnel carrying out demolition work. Even the long scenario with a cooling time of 40 years for DR 3 does not result in a reduction of the ^{60}Co activity in the reactor internals large enough to permit manual manipulation. However, it must be expected that personnel doses from the demolition work will be lower.

Thus, the costs for the three scenarios will be equal in fixed prices, apart from the differences due to expenses for keeping the organisation running for different periods of time and for keeping some facilities in safe storage for the longer scenarios.

Table 1 gives the cost distributions over the "20 years scenario" in five-year periods. As can be seen, the dominant contributor to the total costs is DR 3. A further refinement of the present cost assessment, therefore, should look at this facility first.

Table 1 Staffing and costs at 20 years scenario, millions DKK

5-year Period	1	2	3	4	5-10	Total
Year	1-5	6-10	11-15	16-20	21-50	
Staff pr year	*70*	*70*	*70*	*35*		
General Costs	59	59	59	30	0	207
Administration	20	20	20	10	0	70
AHP*+QM*	21	21	21	10	0	73
DR 1	0	6	0	0	0	6
DR 2	0	28	0	0	0	28
DR 3	181	0	238	0	0	419
Hot Cells	0	25	0	0	0	25
Fuel fab. facility	~0	0	0	0	0	~0
Isotope Lab.	~0	0	0	0	0	~0
Waste Plant	44	43	44	32	90	253
Total costs	**325**	**202**	**382**	**82**	**90**	**1081**
Salary share	125	125	125	63	0	438

* Applied Health Physics   **Quality Management

The item "General costs" is made up by operational costs, site security, inspection and maintenance of buildings and other services from Risø's technical department. The operational costs cover computer licence and maintenance, stationary, travel costs, etc. The costs listed for the Waste Plant include operational costs during the decommissioning of the other facilities, in addition to the costs of decommissioning the plant itself and moving the contents of its storages to a permanent waste repository. Furthermore, even though the decommissioning of all facilities will be complete after 20 years in the scenario shown, an expense of 3 million DKK per year until year 50 has been included for the continued operation of the waste plant. This is also the case in the other scenarios, and has been done in order to underline the fact that an expense will remain to treatment and disposal of waste from other Danish sources. This treatment may take place in the present plant or in a new one, should the final repository be placed somewhere else.

Total costs - in fixed prices - for the three scenarios considered range from about 1080 to about 1180 million DKK, i.e. on the average about 55 million DKK per year in the years where substantial work is taking place. The difference in cost between the scenarios is solely due to the expenses for keeping the organisation running for different periods of time and for keeping some facilities in safe storage for the longer scenarios.

6   UNCERTAINTIES

The study described in this paper was the first attempt to go into detail in the assessment of costs of operations to be performed when decommissioning Risø's nuclear facilities. Therefore, there are many tasks for which no prior experience exists concerning the manpower needed. As far as possible, experience from other countries has been taken as a guideline; but it must be anticipated that the cost estimates produced will change as experience grows and more detailed studies are performed. Also the total time necessary to perform the decommissioning may turn out to be different from the 20 years used in the scenario shown here.

The dominant contributor to the total costs in all three scenarios is DR 3. In particular concerning a number of procurements in the DR 3 figures, only a rough estimate has been possible during the project. It is believed that the cost assessments in these cases have been conservative and that a more detailed study will lead to lower estimates.

7   FURTHER WORK ON DECOMMISSIONING

After the decision had been taken to close DR 3 the Ministry of Research initiated the preparations for the establishment of a separate organisation to carry out the decommissioning. The primary purpose for setting up a separate organisation was to keep the budget for decommissioning and the one for Risø's continued research entirely independent of each other. The new organisation, Danish Decommissioning (DD) has been established with a managing director, and the plan at the moment is that the responsibility for all the nuclear facilities should be transferred from Risø to DD early in 2002, subject to permission from the nuclear authorities and funding by the government. Meanwhile, the staff in the Nuclear Facilities Department has started the dismantling of non-active equipment and the planning of further decommissioning work. Also a refinement of the cost estimates described here is going to be performed during 2001.

8 ACKNOWLEDGEMENTS

The results described in the present paper have been produced in close collaboration with members of the project group, Nils Hegaard, Klaus Iversen and Erik Nonbøl, and other members of the "writing group", Knud Brodersen, Per Hedemann Jensen and Povl L. Ølgaard. Their contributions are gratefully appreciated.

9 CONCLUSION

The study described here was the first attempt to go into detail in the assessment of the costs of decommissioning Risø's nuclear facilities. It has comprised a quite detailed identification of tasks to be carried out for the individual facilities and an estimate of the number of manhours and possible special equipment needed for each task. The total cost arrived at for all facilities is around 1100 million DKK, but this figure still bears in it substantial uncertainties.

Contrary to the work group's expectations at the outset, it has been realised during the project that a postponement by 40 years of final dismantling of DR 3 will probably not alleviate the need for remote handling operations. Therefore, a fast decommissioning seems to be preferable, enabling the use of existing knowledge within the staff about the facilities. Unless the political climate with respect to nuclear power changes in Denmark, there will not be any Danish expertise on the subject area in 40 years.

10 REFERENCES

1) Lauridsen, K. (Editor), Decommissioning of the nuclear facilities at Risø National Laboratory. Descriptions and cost assessment. Risø-R-1250(EN). ISBN 87-550-2844-6. Risø National Laboratory, February 2001. Available as PDF-file at the Internet address: http://www.risoe.dk/rispubl/SYS/ris-r-1250.htm

2) Lauridsen, K., Dekommissionering af Risøs nukleare anlæg - vurdering af opgaver og omkostninger. Risø-R-1251(DA). ISBN 87-550-2845-4. Risø National Laboratory, March 2001. Available as PDF-file at: http://www.risoe.dk/rispubl/SYS/ris-r-1251.htm

3) A proposed standardised list of items for costing purposes in the decommissioning of nuclear installations. Interim technical document. OECD/NEA 1999.

4) Bayliss, C., Practical Application of the UKAEA's Decommissioning and Liability Management Toolbox. 7th IBC Conference - Decommissioning Nuclear Facilities, October 30-31, 2000.

C596/045/2001

Robotic decommissioning of a Caesium-137 facility

A MURRAY
AEA Technology plc, Amersham, UK

Synopsis

Amersham owns a former Caesium-137 sealed source production facility. They commissioned AEA Technology to carry out an Option Study to determine a future strategy for the management of this facility and the subsequent decommissioning of it.

This paper describes the remote equipment, the operations and the Safety Management arrangements.

Using this equipment the facility has been dismantled and decontaminated remotely. Some 2300kg of waste containing in the order of 150 TBq have been removed roboticaly from the facility. Contact dose rates have been reduced from 10's of Sieverts per hour to 100's of micro Sieverts per hour.

INTRODUCTION

AEA Technology has accepted on a turnkey basis a contract to decommission a Caesium-137 sealed source production facility owned by Amersham. The facility consisted of eight glass-fibre containment boxes and a single fabricated steel material entry/exit box (see Figure 1). Up to 175 mm of lead shielding was provided along the working face of the facility to minimise operator dose uptake. Materials and equipment inside the cell line were manipulated with tongs that passed through the front working face. Each cell was connected to its immediate neighbour by a 275 mm diameter tunnel that was used for inter box transfers. The boxes were supported on a steel and concrete plinth.

The cell line was used to manufacture encapsulated ^{137}Cs pellets for approximately ten years. The caesium was supplied either as a solution or a powder and subjected to various processes including wet chemistry before conversion into small ceramic pellets. The pellets were then encapsulated in stainless steel capsules and shipped to customers. The finished capsules could be as small as 4 mm diameter and 8 mm long containing 75GBq of activity. The pellets were smaller often 2mm diameter and 2mm long. Given the need to manipulate the in cell equipment using tongs alone, spillage of capsules and/or pellets was a significant possibility.

Once spilt they were difficult to recover and could pass into areas that could not be reached by the tongs.

When manufacture ceased, the facility was put on a programme of care and maintenance and the shielded cells used for the temporary storage of waste. With the passage of time the high radiation dose rates caused the fibreglass, used in the construction of the inner cells, and the plastic ventilation system to deteriorate.

Amersham therefore invited AEA Technology to undertake an Option Study in 1993 to determine the optimum strategy for the future management of the facility. This paper describes the decommissioning so far carried out.

Choice of Decommissioning Method
At the time the Option Study was carried out the following facts were clear.

- The facility was located adjacent to other working plant and the need to keep this plant operational would constrain the lay out and size of any equipment provided to facilitate the decommissioning.

- There was a large volume of caesium-contaminated waste and part finished sources inside the facility. Dose rates of the order of 30 Sv/h have been measured on some of these items and to date significant quantities of activity of the order of 150 TBq have been removed from the facility.

- The associated radiation levels had damaged the facility local ventilation system and the box containment.

- The radiation levels inside the maintenance corridor were such that man entry was impossible and consequently it was impossible to repair the ventilation system.

- The radiation had also damaged the internal seals on the tongs and crazed the viewing panels on the boxes. This made it impossible to use the tongs to manipulate the waste and made it difficult to see into many of the boxes.

- Although access ports existed in three of the cells the high radiation levels precluded their use for waste removal operations.

- Whilst the "plug" door was effective as a shield door it could not be used as a means of access during the initial decommissioning operations.

Four options were considered for the resolving the problems.

1. Carrying out the minimum work necessary to enable an appropriate care and maintenance regime to be instituted.

2. To undertake early post operational clean out but delay final (remote) dismantling of the facility.

3. To refurbish the existing equipment and then use it to undertake the clean out and to follow this by completing the waste removal and the dismantling of the facility using advanced remote handling techniques.

4. To undertake post operational clean out and the dismantling of the facility using advanced remote handling techniques.

Option 4 was the preferred approach.

The overall objective of the work is to decommission the facility in a safe and cost effective manner. The physical end point involves the complete removal of the facility leaving only the decontaminated areas that form part of the building structure. The target Radiological End Point is 40 Bq/cm^2. The actual decommissioning itself involves two basic phases, a remote phase followed by a manual one. The aim of the remote phase being to reduce the local dose rates to the lowest practically achievable level and to minimise the amount of manual work that has to be carried out. At the time of writing all the robotic decommissioning was complete including size reducing and removing all of the contents of the boxes and the boxes themselves. Some of the steel work was removed and the remaining steel and concrete structure was then thoroughly decontaminated remotely.

FACILITY MODIFICATION TO PERMIT DECOMMISSIONING

The Decommissioning Plan followed directly from the Option Study. It confirmed that to complete the work it would be necessary to provide an extended primary containment together with a remote waste posting facility, a new ventilation system, a new secondary containment and a Telerobotic System to undertake the actual work. This work started in 1995 with detailed design and commissioning was completed by 1997.

The primary containment and its ventilation plant

The primary containment and shielding was provided, with the exception of the west wall, by parts of the original structure of the building in which the facility is located. The Maintenance Bay and the Waste Posting Facility completed the primary containment, as shown in Figure 1. The basic structure and purpose of these two new items of plant are essentially similar. Both maintain containment and shielding for waste as it is transferred from the cell line into the Waste Posting Facility and hence out of the facility. In addition in comparison with the maintenance corridor, the Maintenance Bay provides a relatively low radiation dose storage and maintenance area for the Telerobotic System.

The Maintenance Bay and Waste Posting Facility are constructed from a combination of lead, Perspex and mild steel assembled on steel frames to form a single shielded enclosure.

The Maintenance Bay straddles the access into the facility maintenance corridor, from which it can be shielded by the new sliding door that replaced the original plug door. A removable wall allows essential man entry to maintain the Telerobotic System if remote repair is impossible.

The original ventilation system that served the primary containment has been isolated, sealed and much of it removed as part of the initial decommissioning operations.

The New Ventilation System produces an air velocity of at least 0.5m/s through all known openings in the facility and a small depression relative to the secondary containment.

It protects the operators by drawing air from the secondary containment into the facility through filtered air inlets in the Waste Posting Facility and cells 1-3. This air flows through the facility, is collected into a pipe in cell 9, flows through a pre filter and then exhausts through a disused posting port in the front face of the cell. The air then passes through two stages of High Efficiency Particulate Air (HEPA) filtration to remove particulate before passing to the Amersham extract system, where it passes through a further HEPA filter. The lay out of the new system is shown in Figure 1. Only one primary filter is on line at any one time. All the exhaust filters are of a safe change design and consequently a primary filter can be changed whilst the system is operating.

The New Ventilation System was designed to remove 50 litres/second of air from the cell line (equivalent to four air changes per hour). The system has been tested and shown to perform satisfactorily.

The Secondary Containment
The general layout of the secondary containment is shown in Figure 1. It is built up from units using AEA Technology's Modular Containment System. The containment is some 165m^3 in volume and is attached to the caesium facility by standard engineering fasteners.

The secondary containment is basically divided into an entry/exit, monitoring and decontamination area and an inner operational area. It comprises the following essential features:

- An entry/exit tunnel incorporating a monitoring area and emergency shower;
- Service ports;
- Viewing windows and lighting panels;
- An airlock for waste transfers.

Personnel enter into, and exit from, the inner operational area via the entry/exit shower tunnel, which in turn is accessed from the general laboratory area. Materials are taken into the secondary containment via the airlock.

The secondary containment is normally a free breathing area and has its own dedicated ventilation system. This draws air into the containment from the surrounding areas and discharges it to the atmosphere via a dedicated primary HEPA filter; a mobile filtration unit and the building filtered space extract system. A small fraction of this air is sucked from the secondary containment by the New Ventilation System and forms the supply to the extended primary containment. The mobile filtration unit is located in a laboratory area adjacent to the secondary containment. The two ventilation systems do not interfere and neither interferes with the Amersham systems. The mobile filtration unit is fitted with two stages of HEPA filtration and a two stage axial fan. The system is designed to ensure that the flow velocity in any opening will not fall below 0.5m/s. It is not intended that there should be a significant depression in the secondary containment relative to the surrounding laboratory areas. The secondary system is interlocked with the New Ventilation System such that if the latter system fails the mobile filtration unit cuts out.

The Telerobotic System
The Telerobotic System basically comprises of a NEATER N670SB telerobot, including auto-tool change equipment, and a supporting trolley assembly that includes tool and waste storage locations, viewing systems and a cable deployment system.

The robot had a maximum 25 kg lifting capacity and was been designed to withstand a total integrated dose of up to 10^6 Gy. An operator controled it manually using twin joy-sticks. The operator was normally seated at a control console situated in the secondary containment but, as an alternative, could control the robot from anywhere within the containment using a portable panel that is connected to the main control desk by a substantial umbilical cable.

The robot was supplied with a selection of tools. Saw blades and the router or drill's bits were changed remotely. It is unnecessary to breach either the containment or the shielding.

As remote decommissioning progressed the tooling was developed. The most significant development was the introduction of a grinder with vacuum recovery for size reducing the boxes. This allowed the contaminated dust generated from cutting to be collected at source reducing the need for operational vacuuming. Several TBq's of activity were collected using this system.

The robot is mounted on a trolley to allow deployment and positioning of the equipment in the maintenance corridor. The trolley consisted of a basic frame mounted on four wheels. An electric motor powered all four wheels moving the trolley at a maximum speed of 40 mm/s. The robot was positioned on the trolley so that it can retrieve dropped items in front and behind the trolley.

Electrical power, compressed air for the pneumatic tools and equipment, and all control signals were supplied via cables and pipes that are bundled together to form a single umbilical cable. A cable engine and cable guide, located in the Maintenance Bay, automatically payed out and retrieved this umbilical as the trolley moves along the corridor. One of the power cables failed and was by-passed by a parallel supply cable on a separate reel mounted in a shielded box just outside the Maintenance Bay. In the event of a breakdown, the Telerobotic System could be recovered into the Bay using a manual winch operating on a wire rope attached to the rear of the trolley.

The viewing system consisted of two 2D black and white cameras. Illumination was originally supplied by lamps mounted adjacent to the cameras. Each camera was mounted on a remotely operated pan and tilt unit that allows the camera to be orientated to suit the operation in progress. One of the cameras was mounted on an independent trolley attached to the Telerobotic System by an umbilical. The other camera, lighting unit and pan and tilt unit is attached to the trolley. Towards the end of the remote decommissioning phase the lamps and cameras failed and additional cameras were deployed through the service holes with the lights.

The viewing station that supported the control system was located in the secondary containment. The station comprises two black and white monitors and a video recorder. These were positioned close to the operator console with controls for operating the camera zoom, camera movements and lighting.

DECOMMISSIONING OPERATIONS

The remote decommissioning operations began in 1997 and involved the use of the new plant to remotely decommission the cell line to a condition where man entry and manual decommissioning were justified on ALARP grounds.

The overall strategy applied to the remote phase involved removing the back from one cell and then removing the majority of its contents before breaching the next cell. Work started with the least active cell first. This arrangement kept the contamination levels inside the facility as low as possible and also allowed the robot a "shake down" period after it was moved from the mock up to the facility proper. Once in the facility the Telerobotic System could not be run without exposing it to activity. If, as happened, the move induced faults in the robot the lower initial activity meant that the contamination levels in the Maintenance Bay were initially fairly low, thus making any necessary repairs easier.

The main phases of the work are summarised below.

- Removal of the accumulated waste from the boxes. All the box contents were removed during this part of the work. Dose rates as high as 30Sv/hγ were observed on some of the waste cans and some 90 TBq of ^{137}Cs activity were removed. Despite the removal of the waste, contact dose rates as high as 25Sv/hγ remained in some areas. This was thought to be due to the presence of dust and small pellets.

- Vacuuming the boxes to remove loose contamination. Vacuuming the accessible areas removed a further 40 TBq of ^{137}Cs. The removal of the waste and the initial vacuuming typically reduced the original contact dose rates by an order of magnitude and the new contact dose rates lay in the range 500mSv/hγ – 2Sv/hγ.

- Removal of the fibre glass boxes removed a further 20 TBq of ^{137}Cs. This typically reduced the contact dose rates by a further order of magnitude with the exception of one area where there was known to have been liquid spills during operations. The dose rates here were largely unchanged.

- Scabbling of accessible concrete area reduced general contact dosreates to 100'sμSv/hγ.

- Steel plates were removed to remove contamination where liquid had seeped into the concrete below. This allowed further scabbling to depths of 150mm to reduce doserates down to levels acceptable for man entry. Remote decontamination of the plinths and the maintenance corridor floor removed a further 4TBq of ^{137}Cs.

- Remote decontamination of the installed decommissioning equipment reduced contact doserates to a few mSv/hγ as anticipated. Despite this further manual decommissioning of the equipment proved difficult and a higher than expected dose burden was required to remove the equipment.

OPERATIONAL EXPERIENCE

There have been a number of failures of the remote equipment but all have been associated with the peripherals rather than the robotic arm itself. The umbilical cable has been a particular point of weakness. In addition, the lights and cameras originally supplied proved very difficult to replace after extensive use and an alternative arrangement was substituted.

The sizes of the Maintenance Bay and to a lesser extent that of the Waste Posting Facility are dictated by the limited space available to house them. In the case of the Maintenance Bay the space is both the maximum that realistically could be made available and near the minimum that is acceptable. The small size has little effect on day to day operations. However it is believed to have contributed to the umbilical failures because it is necessary to store the cable in a locker with many tight bends rather than wound around a drum of relatively large diameter. The gentler curves associated with the latter arrangement would have fatigued the cable less.

The congestion also posed problems when it proved necessary to enter the bay in order to effect a repair. It was difficult to decontaminate the area to a radiation level where man entry became possible and the work was further slowed down because the operator had to cope with the congestion. In practical terms there was little that could have been done to alleviate the situation. If more space had been available a larger facility and more remote decontamination facilities would have been desirable.

In line with the rest of the facility the Waste Posting Facility is fitted with tongs. Their limited ability of manoeuvre is well known. In future work it would be worth considering the possibility of supplying one or more mini Master Slave Manipulators. Their greater flexibility could justify their greater cost. The currently perceived wisdom in decommissioning circles appears to be that if manipulators are not already present they are not usually worth installing. Our experience would suggest that this conclusion is not necessarily axiomatic.

Minimising the volume of waste during decommissioning can significantly reduce the overall cost of a project. This was particularly important for this project as all waste was expected to be Intermediate Level Waste. During the pre-works a can crusher was installed in the Waste Posting Facility to reduce the volume of 2.2litre waste cans that were known to be in the facility. Due to the limited space no other equipment could be installed at the time. After the first cell was decommissioned the space was used to install further size reduction equipment. The cell face tongs were re-furbished to operate the equipment. This allowed the Telerobotic System to continue as planned, whilst at the same time the waste was size reduced further. This gave better packing factors and thus less volume of Intermediate Level Waste. A waste transfer arm was also installed in the cell face allowing the waste to be transferred to the Waste Posting Facility without interfering with the Telerobotic System.

Radiation tolerant cameras can be expensive. They are however the eyes of a job such as the present. Operators spend many hours looking at the images on the screen. It is therefore cost effective to provide them with the best possible image quality. A good sharp image makes it easier for operators to deliver the required high levels of manual skill and also reduces the strain in so doing. A fundamental weakness of the robotic viewing system is that it is one sided in that it can only view the work from the maintenance corridor. This limitation means that only highly skilled technicians could carry out the work.

In some cases, where the cell front viewing windows were not too seriously crazed, it was possible to look into the cells from the front. In these cases the operator could gain an all round view of the object on which he was working and significantly faster progress was possible.

All the tools supplied with the robot have proved of considerable value. The vacuum cleaner was however of particular value because of two reasons. There were hundreds of small pellets and capsules that were dropped during manufacture which had to be recovered. Also cutting the glass-fibre containment boxes produced significant amounts of contaminated dust that had to be recovered. In total approximately 40TBq of loose activity was recovered using the vacuum.

The cleaner is wholly contained within the primary containment so that its operation poses no radiological hazards. Considerable effort has gone into developing the optimal configuration. For operational reasons its construction is divided into two parts, a dust collection and a fan unit. Dust and small particles are picked up using a flying hose and the two units are connected together using a second hose. The hoses are sized such that it is impossible to assemble the unit the wrong way round. Such a misconnection would have no safety connotations but would be operationally inconvenient. The dust collection unit is made from a standard waste can and thus disposable. When full it is posted out from the facility in the same way as other waste cans.

Prior to its insertion into the facility the Telerobotic System was extensively tested in a mock up of the real system. The tests carried out were based upon information about the difficulties that were expected to arise. This information was gleaned from a range of historic documents.

As with many an old plant the documentation was somewhat patchy. The decommissioning arrangements were therefore designed to be flexible. This flexibility has proven to be an asset and in particular has meant that unforeseen difficulties could be tackled in a safe manner. Experience however confirms the basic fact that for overall efficiency there is no substitute for prior knowledge.

An in house team of decommissioning technicians was used to decommission the facility. Each one had completed a mechanical apprenticeship and supplemented this with several years of decommissioning experience. Their knowledge of tooling and size reduction in a nuclear environment gave longevity to equipment and allowed equipment and techniques to be adapted when unforeseen tasks arose.

Dose to operators has been kept within 5mSv per year for all of the decommissioning team. The 5mSv/y figure is now commonly used as a Dose Restraint Objective in the British Nuclear Decommissioning industry. This may appear to be easily achievable for remote decommissioning projects. However in the circumstances that we encountered this was not the case. With operational plant close by and the gradual increase in doserate from the primary vent system pipework between the pre-filter and Primary HEPA filter the background doserate was just over 2µSv/h. Working a nine hour day over 220 working days this equates to approximately 4mSv. This only leaves a small allowance for other work such as changing HEPA filters, handling waste packages and non-routine maintenance which are significant when dealing with large amounts of activity such as this.

SAFETY MANAGEMENT AND RELATED DOCUMENTATION

The Nuclear Installation Inspectorate licenses Amersham. One consequence of this is that Amersham retains overall responsibility for the safety of the operations carried out by its contractors. All safety related documents are therefore written to conform with and cleared through the Amersham system. The systems in place there allow:

- Contractors sufficient freedom to utilise their own skills.

- Amersham to process documents in an effective manner contributing their skills not only to the management of safety but also to the effective execution of the project.

- Provide the opportunity for contractors to manage safety in their own way whilst at the same time ensuring that Amersham's requirements are met.

A schematic diagram indicating the relationship and extent of the safety documentation that has been produced for this project is contained in Figure 2. The Stage 1/2 Safety report was the original overarching Safety Document. It set out the overall objective, defined the radiological end point and justified the general approach that was to be deployed. In turn it identified the additional documentation that would be required. The document was peer reviewed and endorsed by the Amersham Safety Committee.

The report also confirmed the need for the additional equipment described above and concluded that each new item of plant would need its own Stage 2 (New Plant) Safety Report and a corresponding Stage 3 (New Plant, Commissioning) Safety Report. The Stage 2 Reports basically described the plant, justified the approach and set out how it would be commissioned. The corresponding Stage 3 Reports explained how the conditions set out in the Stage 2 Report had been met.

Finally, the Stage 1/2 Report concluded that a Stage 3 (Pre Commencement) Safety Report for the complete project would need to be prepared, peer reviewed and then approved by the Safety Committee, prior to the issue of an Authority to Proceed (ATP).

By mutual agreement the operations were divided into two phases:

- Phase A, embracing the remote decommissioning and decontamination operations which are currently being carried out;
- Phase B, covering the corresponding manual operations.

The Method Statements and Operating Instructions referred to in Figure 2 are more detailed instructions that translate the principles in the Safety Report into detailed instructions for the operators.

Inevitably some of the methods need to be changed as a result of experience and/or the availability of new information. These changes are controlled by Deviations. Deviations arise at three levels. They are:

Level 1   This level covers:

- Minor deviations from the Safety Case that do not change the safety systems.
- Deviations whose radiological consequences are within the scope of the safety case but outside the current working method statement.

Level 2   This level covers deviations that are either within the Safety Case but not covered by the method statement or involve a change to the Safety Case involving minor changes to safety systems.

Level 3   This level covers all the more major changes and involves formal changes to the Safety case.

The approval route for a Deviation is dependent on the Level of the proposed change. Basically Level 1 lies with the project subject to subsequent report to Amersham. Level 2 involves formal approval by the appropriate member of Amersham staff. Level 3 involves formal consideration and approval under the Amersham safety system.

To date there have been 34 Deviations to the original Safety Case. None have been above Level 2.

CONCLUSIONS

AEA Technology is decommissioning a former Caesium-137 sealed source production facility owned by Amersham.

The chosen method involves the remote execution of the post operational clean out and the dismantling of the facility.

A range of new equipment has been supplied and successfully put to work in support of the decommissioning. The equipment includes:

- A Telerobotic System.

- The extension of the original primary containment to embrace a Maintenance Bay, a Waste Posting Facility and a New Ventilation System.

- The construction of a secondary containment and ventilation system.

To date some 2000kg of waste containing significant quantities of activity of the order of 150 TBq have been removed from the facility.

A number of unforeseen difficulties have arisen during the course of the work. All have been addressed.

ACKNOWLEDGEMENT

The author gratefully acknowledges Amersham's permission to publish this paper

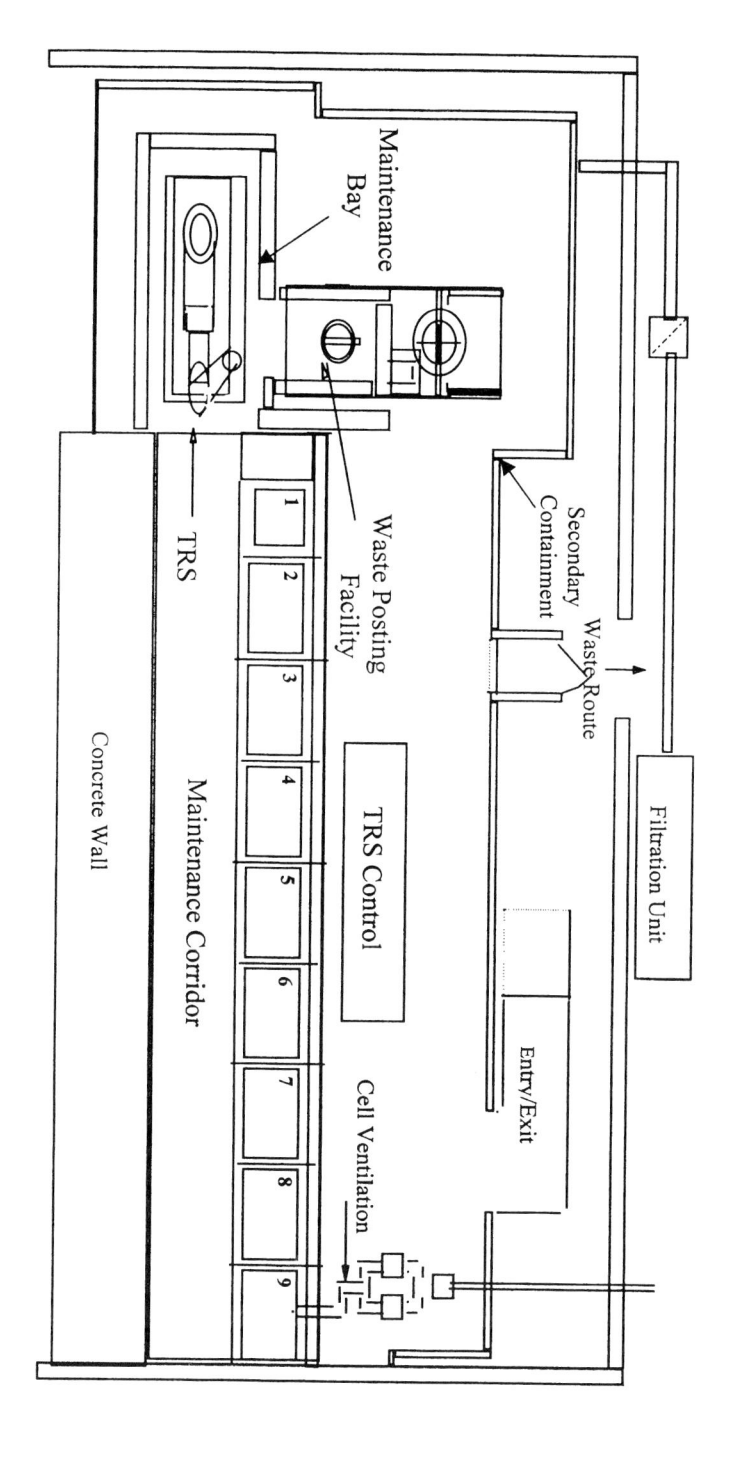

Figure 1 Plan of the facility during the robotic decommissioning phase

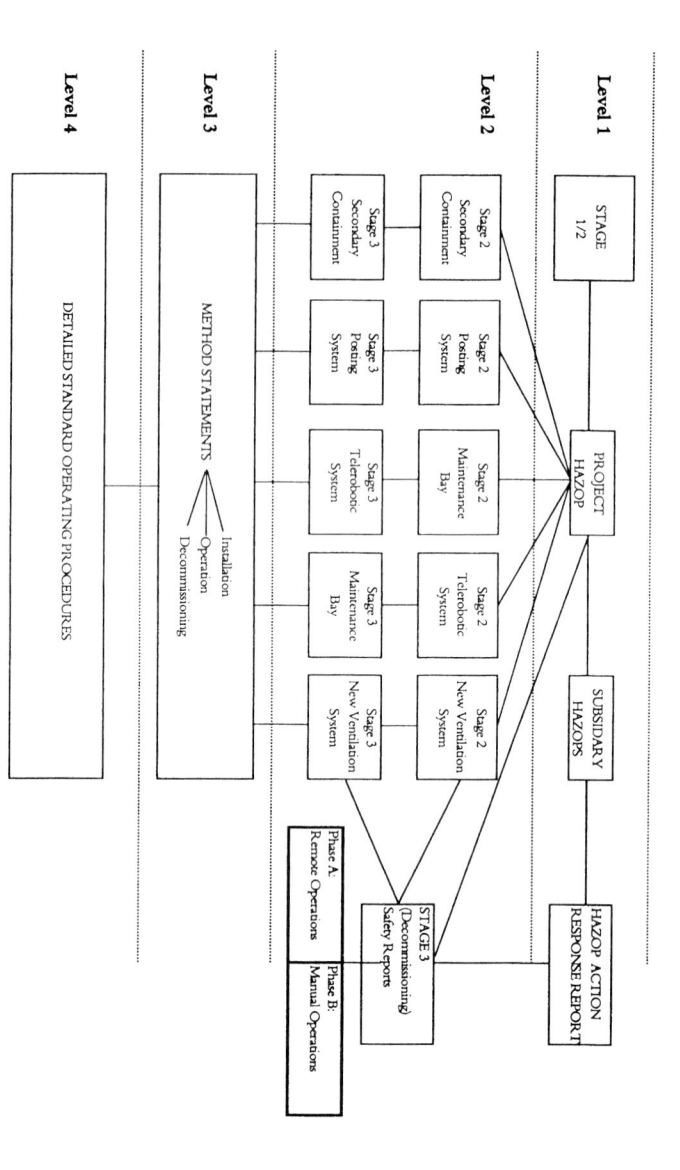

Figure 2 Schematic overview of the safety documentation produced

C596/004/2001

Implementation of NECSA's nuclear liability management programme with special reference to plant decommissioning

P J BREDELL
South African Nuclear Energy Corporation (NECSA), Pretoria, South Africa

ABSTRACT

In 1998 the South African Nuclear Energy Corporation (NECSA) reorganised its entire nuclear decommissioning, waste management and land remediation programme to bring it in line with a nuclear liabilities management approach. The first estimate of NECSA's nuclear liabilities will be submitted to the government during 2001.

In this paper, the emphasis is mainly on the integration of all the nuclear liability management aspects, with special reference to decommissioning. The decommissioning programme is discussed in terms of the current and future dismantling/demolition and decontamination work to be performed at Pelindaba near Pretoria.

1 BACKGROUND

Radioactive waste has been produced at NECSA since 1965. Originally the waste consisted largely of NECSA's own research waste, but also included medical and industrial waste from other radioactive waste generators in the country. The Safari research reactor produced spent fuel and radioactive effluent from the late 1960's onwards. NECSA's nuclear fuel production facilities, generating mostly uranium-based wastes, came into operation during the mid 1980's and continued until the late 1990's when these facilities were taken out of service. NECSA has been storing all its radioactive waste on the Pelindaba site, which is approximately 30 km to the west of Pretoria.

The South African government's decision to sign the Nuclear Non-Proliferation Treaty in 1991 and to enter into a Safeguards Agreement with the IAEA soon afterwards, largely influenced the subsequent course of events leading to the eventual phasing-out of NECSA's nuclear fuel production capability. As a direct result of these political developments, South Africa regained access to international markets for fuelling its nuclear power station at Koeberg near Cape Town. The economic viability of the domestic nuclear fuel production capability was therefore seriously challenged and the need for phasing out the domestic programme became unavoidable.

NECSA commenced with the decommissioning of its nuclear fuel production facilities in the early 1990's. The first facility to be permanently closed down was the Y enrichment plant (high enrichment) in 1990, followed successively by the closure of the so-called semi-commercial enrichment plant or Z Plant (enrichment of less than 5%) in 1995, the fuel fabrication plant in 1996 and the uranium conversion plant in 1998. The withdrawal of NECSA from its traditional nuclear fuel activities left the Government with a large nuclear waste management and decommissioning liability that needed to be addressed as a matter of urgency. In response to this need, the Government made certain appropriations in 1995 for the dismantling and decontamination (D & D) of the above two enrichment plants. A decommissioning team was established and the D & D project was initiated in 1995.

During the period 1996 to 1998, NECSA carried out a preliminary survey of all radioactive waste at Pelindaba with the aim of determining the full extent of its current and past nuclear liabilities. Although inconclusive, the preliminary survey did, however, bring to light the considerable scope of the liabilities assessment task as well as the need for policy decisions to guide the entire process.

In addressing the need for accurately quantifying its liabilities, NECSA established a new division within the corporation called Nuclear Liabilities Management or NLM, which would in future be responsible for all aspects of decommissioning and radioactive waste management, including liability assessment. This department was established in November 1998, and comprises three functional units, namely, a nuclear waste technology unit, a projects and operations unit and a nuclear liabilities assessment unit.

The chief objective of the new department was to accurately assess and systematically discharge NECSA's nuclear liabilities within the funding resources provided by Government. The need to provide best value for taxpayers' money through the application of cost-effective decommissioning and waste management methodologies was the overall guiding principle of the restructured nuclear waste management organisation.

2   THE NUCLEAR LIABILITY MANAGEMENT APPROACH

The approach to nuclear liabilities adopted by NECSA and provisionally endorsed by the Government can be summarised as follows, namely

- Defining the scope of the liability management task
- Defining the mandate of the NLM Division
- Setting the overall management objectives
- Specifying the management system
- Securing the necessary funds for liability management
- Developing and maintaining the capability for implementation
- Responding to stakeholder interests
- Quantifying the risks involved

In order to provide an overview of the development of the NECSA nuclear liability management approach, each of the above aspects is briefly discussed below:

Nuclear liabilities are defined as all the costs which NECSA is expected to have to meet in the future as a consequence of its current and past nuclear activities. The scope of the task, therefore encompasses all activities involved in managing down these liabilities to the point where they are fully discharged (1).

The mandate granted to NECSA's Nuclear Liability Management Division (or NLM) for carrying out its liability management task is derived from the national legal framework, the authorisations granted by Government and the National Nuclear Regulator, relevant international treaties and conventions applying to South Africa, and authorisations generated within NECSA in terms of its delegated powers. The legal regime in South Africa governing nuclear activities is defined principally by virtue of two acts of Parliament: i.e., The Nuclear Energy Act, No.46 of 1999, and the National Nuclear Regulator Act, No. 47 of 1999.

The overall objectives of NLM are defined as follows, namely to

- Provide the necessary *confidence* to the general public that human health and the environment will be adequately protected,
- Manage NECSA's liabilities with respect to the *end-points* defined for liability discharge, i.e., the final destination of the waste and the applicable time frame,
- Discharge the liabilities on a *priority* basis within the Government's funding provisions and
- Annually reassess the *magnitude* of the remaining liabilities.

The management system consists of a liability planning and control system, a liability assessment system, a liability discharge system and a technology support system.

As far as the provision of funding for liability discharge is concerned, a distinction is made between the funds received from Government for discharging historic nuclear liabilities, and the funds provided by commercial waste generators at NECSA toward discharging their own commercial nuclear liabilities. Government has presently provided funds for the three-year period 2000 to 2003, but no undertaking has as yet been given for the discharge of nuclear liabilities in the longer term. Such long-term commitment is expected from Government once NECSA's historic liabilities have been fully quantified and accepted by the relevant authorities. NECSA's commercial business units as well as external customers are required to make financial provisions in a segregated fund for future commercial liability discharge.

NECSA regards stakeholder perceptions to be of paramount importance to the future of the entire nuclear liability management process and therefore goes out of its way to maintain effective stakeholder communication. A wide spectrum of stakeholders needs to be addressed, such as the relevant central government departments, regional and local government councils, the nuclear safety regulator, customers, the general public, etc.

An attempt has been made to quantify the risks involved in the execution of the nuclear liability management programme. The basic risk to the programme is clearly the non-achievement of the main programme milestones. In this regard, the risk of exceeding the estimated cost of liabilities discharge, or exceeding the scheduled discharge period needs to be taken into account in the contingency planning. Two categories of causes of risk can be identified, namely internal and external. Internal causes, over which NLM has some control, include inaccuracies in costing and materials inventory, unforeseen technical difficulties, safety and environmentally related incidents, etc. External causes, which typically fall outside NLM's influence, include contingent aspects such as significant interruptions in Government funding allocations, major departure by Government from existing national policy and strategy stipulations, drastic departures from current licensing standards and procedures by the safety regulator, major shifts in public opinion, etc.

3 LACK OF A NATIONAL POLICY AND STRATEGY: THE NEED FOR MAKING ASSUMPTIONS

In setting overall objectives for nuclear liability management it is necessary to define the points of exit from the liability management system. Generically, there are typically three exit points, namely, the point where materials are released from regulatory control, the point where radioactive materials leave the process system for recycling and reuse elsewhere, and the point where the waste leaves the system when finally disposed of.

The National Nuclear Regulator (NNR) lays down the standards applicable to the free release of materials. Although these standards are still in the process of being formally approved by the NNR Board, nuclear regulatory practice is firmly entrenched in this country and NECSA clearly understands and already applies these standards.

From a liability assessment point of view, the main difficulty, as in most other nuclear countries, is in estimating the cost of final disposal. South Africa is presently operating a low and intermediate level radioactive waste disposal facility at Vaalputs in the Northern Cape Province, for which facility disposal costs are readily available. However, the costs associated with the deep geological facility required for spent fuel and long-lived waste disposal are difficult to estimate, as such a facility is presently still in a very early planning phase. As this country presently does not yet have a formal national position on radioactive waste management in terms of a policy and strategy, NECSA was forced to make certain assumptions in order to be able to estimate the extent of its nuclear liabilities.

The assumptions that NECSA made in the absence of a national policy and strategy for radioactive waste are as follows:

- All nuclear waste would be removed from the Pelindaba site to an appropriate disposal site elsewhere in the country,
- Low and intermediate level waste would be disposed of in near-surface trenches at the National Waste Disposal Facility at Vaalputs in the North Cape Province, as is currently being done for Eskom's Koeberg operational waste,
- Long-lived waste and spent nuclear fuel from the Safari MTR reactor would be disposed in a deep geological disposal facility at a site that still needs to be selected (which could include Vaalputs),
- Very low-level bulk waste would be disposed of at a suitable, but still to be selected, mining tailings dam,
- All decommissioned (and decontaminated) facilities would be reused where feasible, and
- All site contamination outside facilities would be remediated.

NECSA believes that the above assumptions are all in keeping with international norms and practices, and are likely to be endorsed by the national policy and strategy presently being formulated under the auspices of the Department of Minerals and Energy.

4 SPECIFYING THE MANAGEMENT SYSTEM

In defining a nuclear liability management system NLM performed the following activities: Firstly, a survey of all existing nuclear facilities plus associated contaminated materials on the Pelindaba site; secondly, classification and categorisation of all contaminated materials as well as those materials potentially expected to be contaminated as a result of liability discharge activities, thirdly, specification of all unit processing operations; and fourthly, integration of all management activities into an overall management system for liability discharge.

The various activities forming part of the overall liability management system were then prioritised in terms of a set of criteria using incentives (or drivers) and constraints as the main parameters in the evaluation process. Activity priorities were assigned on the basis of the relative strength of the drivers and constraints. Using this procedure, the prioritised activities were then projected into a realistic time frame.

NLM used a liability management planning system based on a three-tier approach: Firstly, a long-term plan developed for executing liability discharge activities at Pelindaba over a period of thirty years, followed by deep geological disposal at a yet to be selected site within a flexible time frame. Secondly, a five-year plan, being a short-term window on the thirty-year plan, mainly taking into account the impact of the funding commitments from Government. Thirdly, detailed project plans developed (and continually updated) in terms of the five-year plan.

The liability assessment system consists of a database containing all relevant information on materials and waste inventory as well as on costing aspects. A programme is used for calculating the cost of liability discharge activities based on the planning system.

In order to keep expenditure within the annual funding levels considered to be affordable to Government, adjustments were made iteratively to the long-term plans until a feasible expenditure profile was obtained. An international review team, under he auspices of Government, has benchmarked in April 2000 NECSA's liability assessment methodology against international norms. The first estimations of NECSA's liabilities have recently been completed and will be submitted to Government for approval in June 2001.

The liability discharge system is organised such that the following main functions can be efficiently performed:

- Decommissioning and demolition
- Waste management
- Site remediation
- Disposal

It should be noted that the National Radioactive Waste Management Disposal Facility at Vaalputs in the Northern Cape Province, where Eskom's low and intermediate level waste is currently being disposed of in near surface trenches, is managed by NECSA on behalf of the

State. It is planned that NECSA's own low and intermediate level waste will also be sent to Vaalputs for disposal. Long-lived waste and research reactor spent fuel will be stored at Pelindaba until a future site earmarked for deep geologic disposal is available.

5 ACCOMPLISHMENTS TO DATE

5.1 Decommissioning

An amount of R150million (£13,5million) has been spent directly on the decommissioning (D&D) of the former nuclear fuel facilities at NECSA. This figure does not include the waste management and other indirect costs involved. The decommissioning project entailed, as already stated, the partial decommissioning of the two former uranium enrichment plants including auxiliary facilities. The feed, tails and product handling facilities of the former so-called semi-commercial enrichment plant (or Z plant) still need to be decommissioned. A total of 72 nuclear facilities, some of which are closed down whilst others are still operating, are scheduled for decommissioning according to the 30-year Pelindaba liability management plan.

Decommissioning activities are subdivided into three phases: The Phase 1 involves plant shut-down; Phase 2, the dismantling and decontamination (D&D) of process equipment inside the nuclear facilities; and Phase 3, the decontamination of the facility structures to the point of deregulation. NLM has to date accumulated valuable experience in terms of Phase 1 and 2 decommissioning. Phase 3 decommissioning is scheduled to commence within the next few years, although the methodology for taking decontaminated facilities out of regulatory control still needs to be fully clarified. The National Nuclear Regulator is currently in the process of approving national safety standards, including the standards for free release of facilities from regulatory control. Facility deregulation is necessary in cases where facilities can be used for non-nuclear purposes. Where reuse is not considered feasible, demolition is indicated and consequently reinstatement of green fields conditions will be pursued.

5.2 Materials Inventory

Figure 1 shows the inventory distribution of nuclear liabilities according to material types as identified during the liability surveys: The data in Figure 1 also include estimates of future waste materials that would be generated during decommissioning. The Safari Reactor spent fuel and redundant sealed sources are not included.

5.3 Total Nuclear Liability

NLM has recently completed a first estimation of the total NECSA nuclear liability. Inasmuch as the responsible authorities are presently considering the implications of the first liability estimate, it would not be prudent at this stage to divulge sensitive information in this regard. However, the main components of the total NECSA nuclear liability, expressed in terms of the activities/waste types, are shown schematically in Figure 2. Alternatively, the overall nuclear liability estimate is broken down in terms of facility types, as shown in Figure 3:

5.4 Risks

There are some areas of technical uncertainty that still need to be resolved, notably the problem of final disposal of spent fuel and long-lived radioactive waste in a deep geological disposal facility. Figure 2 shows that deep geological disposal contributes approximately 43% to the overall nuclear liability of NECSA. The cost estimate of geologic disposal is, however, subject to considerable uncertainty as a geologic repository concept for South Africa is still in a very early stage of development.

The challenges in executing the long-term liability programme are largely of a non-technical nature, notably those involving public, financial and economics issues. It is hoped that once Government has given consideration to NECSA's liability assessment, the financial commitment that is vital to the overall programme would be resolved. On the economics side, NECSA still has to demonstrate that the methodology being used for liabilities discharge is in fact the most efficient application of taxpayers' money. The use of outside contractors as a method for ensuring best value for money has not yet been fully developed by NECSA, mainly due to the lack of adequately experienced nuclear contractors in South Africa.

6 CONCLUSIONS

It is important to note that decommissioning of nuclear facilities makes up only 7% (Figure 2) of the overall NECSA nuclear liability, with waste management being the largest portion. Therefore, decommissioning in the NECSA case should always be considered in the context of the overall liability management programme.

By and large, however, NECSA feels confident that it has positioned itself reasonably well during the past three years to be able to carry out the task of effectively managing NECSA's nuclear liabilities.

7 REFERENCES

(1) OECD Report: *Future financial liabilities of activities*, Paris, 1996.

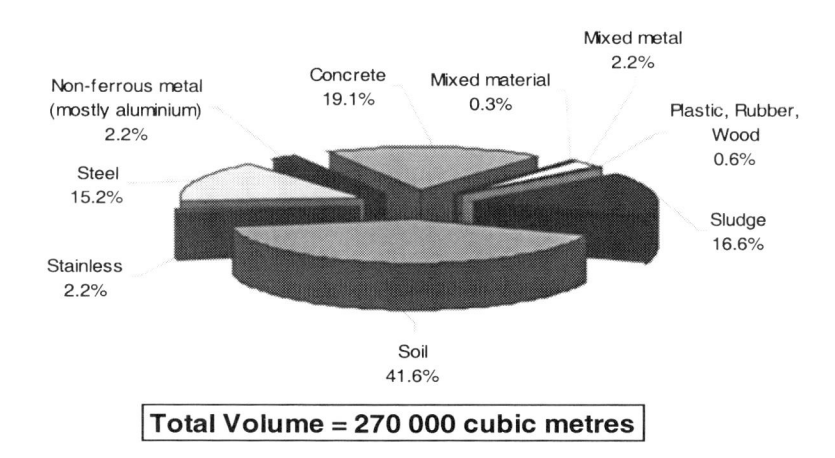

Figure 1: Waste Materials Inventory

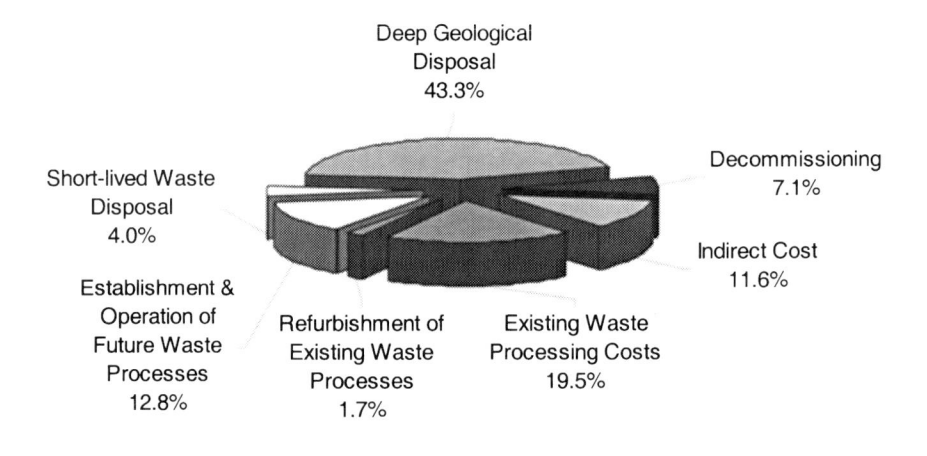

Figure 2: Composition of Total Liability

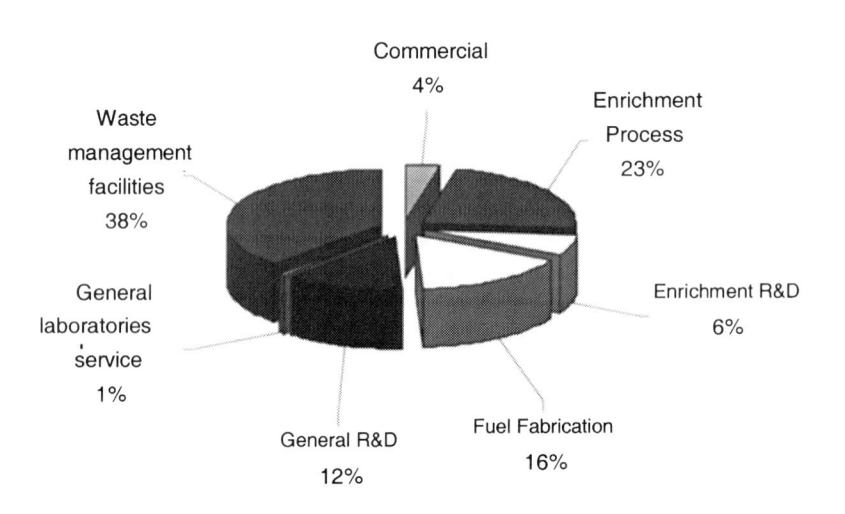

- Deep Geological Disposal 43.3%
- Decommissioning 7.1%
- Indirect Cost 11.6%
- Existing Waste Processing Costs 19.5%
- Refurbishment of Existing Waste Processes 1.7%
- Establishment & Operation of Future Waste Processes 12.8%
- Short-lived Waste Disposal 4.0%

Figure 3: Composition of Total Liability in terms of Facility Types

- Commercial 4%
- Enrichment Process 23%
- Enrichment R&D 6%
- Fuel Fabrication 16%
- General R&D 12%
- General laboratories service 1%
- Waste management facilities 38%

© With Author 2001

C596/030/2001

IAEA views on issues, trends, and development in decommissioning of nuclear facilities and member states' experience

M LARAIA
International Atomic Energy Agency, Vienna, Austria

ABSTRACT

This paper highlights current issues, trends and developments in selecting decommissioning strategies in IAEA's Member States. Radiological conditions, spent fuel and radioactive waste management, funding, economics and the development of suitable technology are important factors for selecting the decommissioning strategy. Although safe enclosure is the selected option for many shut down facilities, delay in dismantling may also have disadvantages such as loss of expertise and cost uncertainties.

Currently, of the many large nuclear installations permanently shut down, only a fraction have been or will be in the near term totally dismantled and decommissioned to unrestricted release state. A trend towards immediate dismantling seems to emerge in some countries, but this is usually due to country-, site- or plant-specific conditions.

1. INTRODUCTION

The conceptual basis for the selection of a decommissioning strategy is to be found in IAEA's Principles of Radioactive Waste Management (1). Principles 4 and 5 refer directly to protection of and burden on future generations (Table I), but are not prescriptive in nature. IAEA's Member States are given the flexibility of evaluating how to implement these principles as reflected in derived safety guides e.g. (2). It can be generally assumed that undue delays in decommissioning of nuclear facilities should be prevented, but the interpretation of "undue" is left to national authorities. The following sections provide information on major factors affecting the selection of a decommissioning strategy, how these factors play a role in Member States' decision-making, and typical examples.

2. FACTORS RELEVANT TO SELECTION OF A DECOMMISSIONING STRATEGY

There are two basic decommissioning options for a nuclear facility for which the decision has been made not to return it to service. These two options are:

- immediate dismantling;
- long-term storage followed by dismantling.

The decision on how to proceed with the decommissioning of a nuclear facility is dependent on a number of factors. These are as follows:

- legislative and regulatory requirements;
- waste arisings and national waste management strategy;
- spent fuel management strategies;
- physical conditions of the plant;
- owner's interest, including planned use of site;
- availability of technology and other resources;
- social considerations;
- decommissioning cost and funding; and
- radiological factors.

Each factor must be examined for the conditions specific to the facility under consideration to arrive at a satisfactory decommissioning plan. Each of the factors is briefly discussed in the subsections that follow.

2.1. Legislative and regulatory requirements

Decommissioning strategies and their timing are regulated in different ways by Member States. In general, three different schemes may be found as described below:

a) fully prescriptive in which national legislation provides for decommissioning planning and timeframe;

b) regulatory arrangements are prescriptive and require preliminary assessment of possible decommissioning options with the final option being selected and justified by cost benefit or multi-attribute analysis;

c) regulatory arrangements are not prescriptive and allow the operator to propose only one option subject to safety justification alone.

Examples of different decommissioning schemes are:

a) Japan requires that facilities go to total dismantling within five to ten years of facility shutdown to allow further use of the site. This is an important consideration in Japan where such land is at a premium. A shorter safe storage period means, however, higher occupational radiation exposure and larger quantities of radioactive wastes, although a Japanese study shows that these factors should not be predominant. It should be noted that many European and US sites are large enough to allow construction of new facilities without the need to demolish those shutdown;

b) Based on technical studies of different nuclear facilities, the US Nuclear Regulatory Commission (NRC) allows possession-only licence only for nuclear power reactors and limits the surveillance period to up to 60 years or as approved. Technical studies showed that for US power reactors there is little benefit in delaying dismantling for longer time periods.

c) In Italy, the operator's decommissioning strategy for shutdown nuclear power plants was long-term safe enclosure. This was essentially based on the lack of a radioactive waste disposal site in the country. Recent developments towards early dismantling are driven by public opinion and strenuous efforts to achieve consensus on siting waste/spent fuel storage or disposal facilities.

It should be noted that other regulatory requirements are essential to safe and cost-effective planning/implementation of decommissioning. Clearance criteria is probably the most important of such requirements. Such criteria are now available in most countries, e.g. Germany, Spain and UK. In some cases they are part of the legislative framework, in others they were established for specific projects.

Large amounts of materials resulting from decommissioning exhibit very low levels of activity or could be readily decontaminated to achieve such levels. Assuming that these materials should be all managed and disposed of as radioactive wastes would result in unnecessary penalties in terms of operational difficulties and significant extra costs. It is generally possible to establish radiological criteria and associated activity levels according to which materials can be released from regulatory control.

Three general classes of ways of removing solid materials/wastes from the facility can be identified as follows:

a) Clearance for unrestricted reuse or disposal

b) Authorized release/reuse within the nuclear industry or in the public domain; or

c) Storage /disposal under radiologically controlled and monitored conditions.

Criteria to be met for these classes vary between countries. Sometimes the criteria are based on nationally applicable regulations, while in other practices, they are based on a case-by-case evaluation. Germany has a full set of clearance criteria, ranging from unrestricted or restricted release to nuclear re-use. International recommendations are under preparation, and interim reports were issued by the IAEA and the European Commission. The availability of suitable disposal facilities can be limited. The strategy for waste management can vary; however, consideration to achieving clearance or authorized release is important to reduce the volumes of radioactive waste for storage/disposal.

2.2. Waste arisings and national waste management strategy

The generation of radioactive wastes in various categories is a direct result of the dismantling of radioactive facilities. The extent of waste arisings in the various categories will be influenced by the timing of dismantling operations. Deferment may reduce the amounts of intermediate level waste (ILW) and increase amounts of low level waste (LLW) while some waste may reduce in activity to categories cleared from regulatory control. This will influence disposal arrangements and costs.

If no suitable disposal facilities for the amounts and categories of waste are available, then the following options exist:

- maintain the facility in safe storage;

- condition the waste and store in appropriate waste stores.

Waste storage arrangements for large amounts of conditioned waste may also be costly or difficult to maintain. These considerations therefore will influence the timing of final dismantling and the period of safe storage.

Safe storage, however, is not considered to be an alternative strategy to the identification and qualification of a disposal site for decommissioning waste. It should be noted that disposal facilities now exist in many countries e.g. France, Spain, UK and the USA. However, other countries e.g. Canada do not have waste disposal sites and therefore have decided for long term safe storage of their shutdown facilities. Experience on disposing of large amounts of decommissioning waste at licensed disposal sites is still limited worldwide.

2.3 Spent fuel management strategies

In some Member States, spent fuel management is not considered part of the decommissioning process, since it is assumed that the removal of fuel from the facility is a prerequisite for the implementation of major dismantling activities. However, experience shows that spent fuel management may strongly affect the selection of a decommissioning strategy. In particular, facilities to store, dispose of or reprocess spent fuel may not be readily available and the fuel must remain in the reactor facility. In several Member States contracts were negotiated with other Member States to transfer spent fuel, but now, for various reasons, this has become difficult or even impossible. The lack of a transfer route for spent fuel may force a licensee into a safe enclosure strategy with spent fuel in the facility or at independent storage facilities. In general, it is desirable to remove spent fuel off-site or to a facility independent of the power plant as soon as possible.

2.4. Physical conditions of the plant

Design, construction and operational aspects of a given plant may be more or less conducive to smooth decommissioning. In addition, the condition of a shutdown nuclear facility influences decisions concerning decommissioning from the viewpoint of integrity and maintainability of the facility and its systems.

2.5. Owner's interest, including planned use of the site

The choice of a decommissioning strategy may also depend on the following considerations:

- the owner may have a shortage of sites for new plant construction and may be forced to re-use a site for a new plant. In that case, immediate dismantling may be chosen;

- if the plant to be decommissioned is co-located with other operating facilities that will continue to be in service, immediate dismantling may be the preferred choice. The necessary security, surveillance and maintenance for the shutdown facility could be provided by the remaining operating facilities, this is the case e.g. at Dresden site in the USA;

- as a factor in his decision to proceed to safe storage, the owner may wish to consider the re-use of some of the plant facilities, for example the cooling water equipment, the infrastructure, and some of the plant process systems, for purposes other than those for which they were originally intended or as part of a new or modified plant;

- if all decommissioning stages are available to him, the owner may wish to optimize his expenditure, depending on the economic situation, in his choice of strategy.

2.6. Availability of technology and other resources

Basic technology for the accomplishment of decommissioning is reasonably well known and tested. However, during the planning stages for dismantling, problems may be identified, for example, poor accessibility, or specific operations to be undertaken during decommissioning. In such cases it may be necessary to develop special tools or means for remote operation or handling.

Long term storage followed by dismantling may take advantage of technological developments during the period of storage. Technological developments in decontamination procedures and techniques, robotics and remote cutting could facilitate future dismantling of the facility.

It should be noted that using decommissioning technology will be easier in countries (France, Germany, UK, USA etc.) that possess a significant nuclear programme. In such countries, a decommissioning "market" developed which should be financially beneficial to the decommissioning budget.

An advantage of immediate dismantling is the retention and utilization of plant expertise on the site during the actual dismantling. This expertise could lessen the potential for accidents and would avoid any dose associated with retraining of personnel. This may be needed particularly in cases where there is a lack of records, where undocumented changes were made during construction or backfitting, and where experimental facilities are to be decommissioned.

As the storage period continues, expertise in the layout, maintenance and operation of the reactor lessens as personnel leave the facility so that at the time of dismantling there may be no one with personal experience of the facility. This expertise will have to be reacquired at the time of dismantling, with a possible corresponding penalty in costs, occupational exposure and other factors.

2.7. Social considerations

The process of deciding between the different decommissioning strategies may take into consideration the possible effects on factors such as:

- environmental factors (e.g. the value of the neighbouring land);
- employment problems; and
- the public's perception of the hazards, whether the installation is maintained in a safe shutdown condition or is dismantled.

Public opinion about the proposed choice is usually taken into account in the procedure whereby the proposals are submitted for the approval of the relevant authorities; the way in which this is done varies from country to country. Social considerations are extremely important in countries having limited opportunities for re-location or re-training of the workforce. The need to re-employ operational staff was a key factor in the selection of decommissioning strategy at Greifswald (Table II). This issue is even more acute in countries resulting from the collapse of the former Soviet Union, where satellite cities were erected to support the operation of nuclear facilities (3).

2.8. Decommissioning cost and funding

The funds required to carry out decommissioning can be split into three categories:

- post-operational costs;
- care and maintenance costs; and
- dismantling and waste management costs.

The first category may include, inter alia, the cost of defueling and the cost of planning for decommissioning. If a facility is held in a state of safe storage, there is an on-going requirement for equipment and resources to maintain the plant in a safe state. The expenditures for these may become increasingly significant with the lapse of time. The third category will include the work of dismantling, waste transport and disposal and may include site restoration.

Whatever choices and decisions are made, it is the responsibility of the owner of the plant to make a financial provision sufficient to cover the costs of all stages of decommissioning up to final dismantling, in accordance with pertinent national legislation and funding requirements. If a long period of safe storage is envisaged, the forecasting of funding requirements may be uncertain because of the variations in costs of regulatory, social and industrial influences.

On the other hand, deferment of dismantling may improve the funding of the task by allowing time to accrue additional funds where these may not previously have existed or by discounted cash flow considerations over a reasonable period of time.

2.9. Radiological exposures

The removal of the reactor fuel or process materials from a facility and, if practicable, from the site would remove the main radiological risk presented by that facility. There will however remain a smaller risk to workers, the public and the environment during decommissioning from the residual radioactivity. This risk will be most significant during the immediate post operational and initial and final decommissioning phases when physical work is being carried out on the facility.

One of the purposes in placing a facility in a prolonged period of safe storage between the initial and final phases of decommissioning is to achieve some radiological advantage for subsequent decommissioning. This will be primarily due to radioactive decay of radionuclides present and include:

the reduction in local dose rates and consequent reduction in operator doses;

the reduction of the radiological consequences of any accidents during dismantling;

the re-categorization of some radioactive wastes.

As far as public exposures are concerned although there will be in general some radiological benefit due to radioactive decay, however these exposures are so low even in immediate dismantling that any benefit will be just marginal.

3. DECOMMISSIONING STRATEGIES WORLDWIDE

The choice between the two prevailing decommissioning options, immediate or deferred decommissioning, depends on a variety of factors, which have been discussed above. Decommissioning costs, waste disposal problems and political aspects are presently considered as major factors governing the decommissioning strategies.

The alternative of leaving a plant in long-term safe storage may cause a specific waste management problem in the future. With future disposal facilities so uncertain, a number of utilities declare to be unprepared to take the risk. The prospect of not having a disposal facility available at any cost may greatly overshadow the economics involved in the long-term build-up of decommissioning funds. It seems that immediate decommissioning will prevail in some countries having limited waste disposal capacities. In fact, recent decisions appear to be driven by the desire to take advantage of existing disposal facilities while the option is still available and before disposal costs escalate to unbearable levels (e.g. in the USA).

The decision to delay the start of dismantling may also depend on other aspects than those mentioned above. To decommission its retired reactors, Electricite de France chose in the past partial dismantling and deferral of final dismantling for 50 years. Although complete dismantling was technically possible, including availability of waste disposal facilities, the utility preferred the delay, which will result in a significant reduction in residual radioactivity, thus reducing radiation doses during the eventual dismantling. Improved techniques were also expected to be available at the dismantling stage, again reducing doses and also costs. A debate is underway in France to evaluate whether a shorter safe enclosure duration is viable. According to recent development, it appears that the French nuclear operator has now selected early dismantling for first-generation reactors (4).

Germany, on the other hand, has chosen direct dismantling over safe enclosure for the closed Greifswald nuclear power station in the former East Germany, where five reactors had been operating, one was nearing operation and two were under construction. Among various reasons for this strategy, the socio-economic aspect of maximizing use of in-house resources played a major role. Other arguments are given in Table II. In mid 1995 the site of the 100 MWe Niederaichbach nuclear power plant in Bavaria was declared fit for unrestricted agricultural use. Following removal of all nuclear systems, the radiation shield and some activated materials, the remainder of the plant was below accepted limits for radioactivity and the state government approved final demolition and clearance of the site.

In Japan, where suitable nuclear sites are scarce, the official policy is that commercial power reactors should be dismantled and removed as soon as possible after shut down (usually within some 5-10 years) and the site should continue to be used for nuclear power plants. As a first case, the Japanese BWR at Tokai (JPDR) has been dismantled in 1996 and the site cleared for another nuclear use.

Various factors influenced the decision about decommissioning of some shutdown US nuclear power plants. While some plants have been or are being dismantled without putting the facility in a safe enclosure state (e.g. Trojan, Fort St.Vrain), the long safe enclosure periods for Dresden-1, San Onofre-1 or Indian Point-1 have origin in the utilities' considerations not to start dismantling unless other units located on site are also shut down.

Independent from factors which are likely to prevail in the individual cases, it can be seen that the strategies eventually selected vary from country to country and even within one country. This is apparent from Table III showing differences between selected strategies for shutdown reactor units in the United States.

Table I Radioactive waste management principles relevant to the selection of a decommissioning strategy (1)

PRINCIPLE 4:	PROTECTION OF FUTURE GENERATIONS
	Radioactive waste shall be managed in a way that the predicted impacts on the health of future generations do not exceed relevant levels that are acceptable today.
PRINCIPLE 5:	BURDEN OF FUTURE GENERATIONS
	Radioactive waste shall be managed in a way that will not impose undue burden on future generations.

Table II Specific arguments for Greifswald to go to immediate dismantling

Need to re-employ exceedingly large operational staff (thousands of workers) in an economically depressed region.
Prompt decision on decommissioning strategy allowing re-employment of key staff.
Availability of waste disposal facility at the beginning of the decommissioning project.
Funds made available by State.
Full set of clearance criteria harmonized to international standards.
Difficulty of installing safe storage for WWERs (no secondary containment).

Table III Strategies selected in the US for shutdown nuclear power plants

Reactor unit	Type	Shutdown	Status
Indian Point 1	PWR	1974	SAFSTOR
Dresden 1	BWR	1978	SAFSTOR
Fermi 1	FBR	1972	SAFSTOR
GE VBWR	BWR	1963	SAFSTOR
Yankee Rowe	PWR	1991	DECON
CVTR	PTHW	1967	SAFSTOR
Big Rock Point	BWR	1997	DECON
Pathfinder	BWR	1967	SAFSTOR + DECON (Licence terminated)
Humboldt Bay 3	BWR	1976	SAFSTOR
Peach Bottom	HTGR	1974	SAFSTOR
San Onofre 1	PWR	1992	SAFSTOR
Fort St.Vrain	HTGR	1989	DECON (Licence terminated)
Rancho Seco	PWR	1989	SAFSTOR
TMI 2	PWR	1979	SAFSTOR
Shoreham	BWR	1989	DECON (Licence terminated)
Trojan	PWR	1992	DECON
LaCrosse	BWR	1987	SAFSTOR

Legend: SAFSTOR = Safe Storage; DECON = Immediate Dismantling

REFERENCES

(1) INTERNATIONAL ATOMIC ENERGY AGENCY, The Principles of Radioactive Waste Management, Safety Series No. 11-F, IAEA, Vienna, 1995.

(2) INTERNATIONAL ATOMIC ENERGY AGENCY, Decommissioning of Nuclear Power Plants and Research Reactors, Safety Guide No. WS-G-2.1, IAEA, Vienna, 1999.

(3) INTERNATIONAL ATOMIC ENERGY AGENCY, The Decommissioning of WWER Type Nuclear Power Plants, IAEA-TECDOC-1133, IAEA, Vienna, 2000.

(4) NUCLEONICS WEEK, EDF Confirms Early Dismantling of First-Generation Reactors, Vol. 42, No. 2, 11 January 2001.

C596/041/2001

Progress in decommissioning of a major alpha processing facility at AWE Aldermaston

J STARKEY
AWE, Reading, UK

1   INTRODUCTION

AWE has been at the heart of the UK Nuclear deterrent since it was established in the early 1950's. A number of buildings from that early era are either undergoing decommissioning or are in care and surveillance awaiting decommissioning. The building that is the subject of this paper was a manufacturing facility located on the Aldermaston Site. The building represents the largest decommissioning project to be undertaken at AWE, both in terms of size of Facility and complexity of the plant. The first stage of the decommissioning is underway with three gloveboxes removed and physical preparations underway for the removal of a further five. This first stage represents only a small percentage of the total decommissioning project, which is scheduled to be completed over a 20 year period.

2   DESCRIPTION OF THE FACILITY

The original building is a two-storey building with a ground floor area of approximately 3500m^2 and was commissioned in 1952. It consists of a central access corridor with three research laboratories containing various glovebox suites, located down the one side of the building, with a line of gloveboxes used for production purposes known as the Box Bay Line (BBL) to the other side of the corridor. The BBL is unusual in that it is located in a purpose built Lower Pressurised Suit Area (LPSA), the purpose of which was to facilitate decontamination and maintenance of the plant and equipment located in the gloveboxes. All of the building gloveboxes were supplied with an inert gas atmosphere by an Argon plant located adjacent to the LPSA. The first floor of the building contained the Extract and filtration plant for the gloveboxes, LPSA and general spaces. This plant was located in the Upper Pressurised Suit Area (UPSA).

The building has undergone a number of modifications and additions since construction, the most significant of which was the addition of an extension in the early 1960's. This was approximately the same size as the original building and contained a new inert gas plant (Nitrogen instead of the original Argon) and a large suite of gloveboxes. These "extension" gloveboxes were designed to be moved to an area which contained an extension of the LPSA, this allowed access to the glovebox internals for maintenance purposes without the requirement to place the whole box in the LPSA. The first floor of the extension contained an overhead transfer tunnel system (itself an extensive glovebox) which facilitated movement of materials between each of the ground floor gloveboxes and a large network of services for the boxes. The ventilation for the extension was provided by two separate systems, a double HEPA filtered system for the glovebox area and an unfiltered system for the Nitrogen plant and tunnel/services area.

During construction of the extension, one of the original research Labs was divided into two smaller Labs and two of the (then) modern extension type gloveboxes placed into half of the Lab.

Following completion of the extension, the redundant Argon gas plant was decommissioned and replaced with an electro-refining glovebox suite, which was served by a separate ventilation system from the rest of the building.

Although production work continued up until 1997, certain parts of the building have been redundant for a number of years. The BBL and the research laboratories ceased operations in the 1970's and have remained under care and surveillance since that time, although two of the laboratories underwent varying amounts of decommissioning in the early 1980's, and shortly after this a new glovebox suite was introduced into half of one Lab.

3   STATUS OF THE PLANT

As is typical for buildings of the 1950's and 60's, a large amount of changes to plant configuration occurred over the years. Much of this work was well conceived with good records/drawings kept to show the extent of the modifications. Unfortunately there were also a number of modifications carried out on the plant for which few records were ever produced. As AWE moved towards reaching Licensed Site Status in 1997, considerable work was undertaken in amending Facility drawings to try and reconcile and better understand the plant configuration. Partly as a result of this work and partly as a consequence of the transition towards commencement of decommissioning, it was recognised that the building required some significant remediation works to be undertaken in order to bring the building up to a standard where it could continue to remain a safe environment for a 20 year decommissioning programme. This work focussed on the building infrastructure and consisted of the following major elements:-

- Ventilation upgrade
  - As previously described, the building has a number of different ventilation systems. In principle the upgrade scheme will fit a "tail end" extract and filtration plant to the original

building system; shut off the extension unfiltered extract system; re-route and enhance the extension filtered extract system. A new dedicated control system will completely replace the existing, difficult to maintain, disparate control systems. This project is underway and is scheduled to be completed towards the end of 2002.
- Containment boundary
  - Investigations revealed a number of pathways for airflow under fault conditions to go from high to lower hazard areas. As a result the building and LPSA/UPSA boundaries were thoroughly examined and all pathways sealed up where possible. Where it was not possible to seal penetrations (due to ventilation requirements) they were protected by self-closing dampers.
  - The entrance to the LPSA and UPSA relied on dynamic containment provided by the ventilation system. This has been supplemented with a pair of enclosures fitted with self-closing dampers.
  - The physical integrity of the BBL containment was in question and so a robust mechanical protection system was installed.
  - Glovebox containment was improved by the fitting of "top hats" over each of the approximately 2000 gloveports in the building.
- The fire loading located in the LPSA was reduced following a number of pressurised suit operations, and the removal of 50 drums of ILW.
- Depression monitoring was fitted around the facility to provide positive indication that the pressure gradients in the facility provide airflow from low to high hazard areas.
- The roof, which had been deteriorating for some time was completely re-covered. This work also included sealing and protecting skylights, which provided a potential pathway from the UPSA to the environment.
- Approximately 350 electrical circuits were identified as being redundant and isolated. This both removed a significant maintenance burden, but also significantly reduced fire risks.
- Improvements were made to the fire detection and alarm system.
- Post Operational Clean Out (POCO) was performed, which reduced material hold up, removed sensitive tooling, and minimised fire loading in the gloveboxes.
- A two-stage re-wire project was instigated.
  - Stage 1 consisted of installing a brand new 3000 amp electrical supply to the building from a sub-station to a new switch room located in the building. This supply was sized to cater both for the new ventilation upgrade and future decommissioning loads.
  - Stage 2 includes the replacement of all remaining Vulcanised India Rubber (VIR) circuits, as well as certain improvements such as facility lighting. This part of the project is scheduled for completion in 2002.

4   DECOMMISSIONING PHILOSOPHY

Due to the size of the facility and the overall complexity, as well as a number of (not always complimentary) external influences, an overall decommissioning plan was prepared for the facility. This plan was developed to cover the entire decommissioning process through to brownfield site.

The objective was (and is) to carry out the decommissioning programme safely and with minimal environmental impact, using proven methods and technologies that are continuously developed in line with best practice. Subject to this, the objective is to achieve the decommissioning programme to time and cost and to minimise project risk.

4.1 Technical Strategy

As the project was so large a decommissioning plan was prepared to reduce the project into manageable stages. The facility was broken down into a number of logical areas / functions. Each of these were assessed against the following criteria to assess their relative priority against each other:-

- material hold-up and criticality considerations
- containment integrity
- adequacy of existing process ventilation
- value of area in providing storage or working space
- waste minimisation/optimisation
- whether or not the area is required to provide continuity of a service function
- low hazard practical experience to be gained for later higher hazard operations
- practicality - difficulty of decommissioning
- what conventional (non nuclear) hazards will be reduced by decommissioning

The output of this assessment was a prioritised list of 21 areas, some of which were logically related, others of which were not. Following a substantial scrutiny process the 21 areas were grouped into project stages. The grouping of the areas does not slavishly adhere to the priority order previously identified, but also considers geographical location and ease of control. The priority order of the stages was evaluated against the following factors:

- The need to systematically and progressively reduce the hazard
- The degree of risk of a release of RA material to the Facility or the environment
- The time at risk
- The need to maintain a service or capability to support decommissioning tasks in other areas
- Practicability - degree of difficulty in performing a task sooner compared with later
- The need to keep exposure of personnel to RA Hazard, additional or secondary waste generation, time taken, ALARP
- Geographical location and ease of control of the tasks.

All of the above were balanced against each other and the need to control the degree of project risk. The grouping of areas resulted in ten stages of decommissioning including demolition.

4.2 Safety Case Strategy

As the facility completed its production commitments and moved into the POCO stage of decommissioning, it became clear that the extant safety case was becoming less relevant to the future operations of the Facility and by coincidence was approaching an age where the AWE systems demanded a fundamental review of the document. At the same time it was recognised that the decommissioning operations would need a fully developed safety justification prior to implementation. To achieve this a safety case strategy was developed to support the programme

The strategy was to develop the safety case in successive Stages over the duration of the programme to take account of technical developments and experience gained from work completed. The emphasis being on continually feeding back the lessons learned from each Stage into the planning for the next as well as gaining from the experience gained on other decommissioning projects. This accords with the "just in time" principle favoured by the NII and maximises the potential to meet the objectives described. This process is shown in Figure 1.

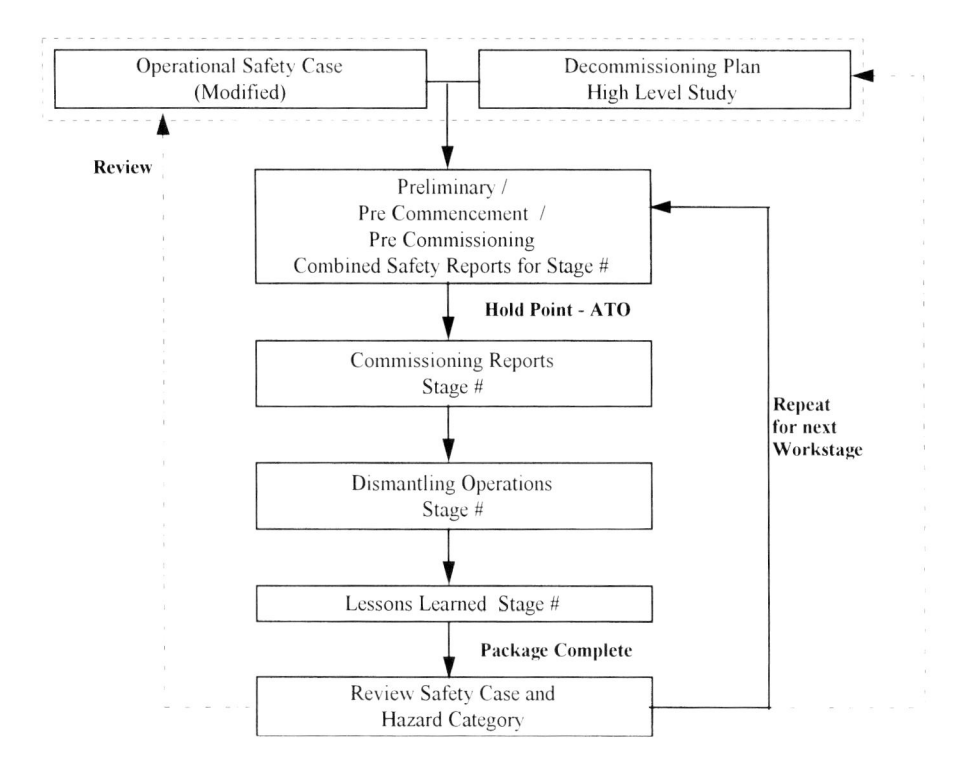

Figure 1 Safety Case Process

5 STAGE 1 DECOMMISSIONING

The first of the decommissioning stages to be undertaken in the facility consisted of three laboratories from the old part of the facility. The principles behind why they were selected are explained in section 4. At the time of preparing this paper the first of four planned Modular Containment Systems (MCS) has been erected, two gloveboxes completely removed, the MCS decontaminated and removed, with work well advanced on the preparation of a further two MCS' to deal with a further six boxes and seven fume cupboards.

5.1 Description of the plant to be decommissioned

The area of the building undergoing decommissioning is described below and shown in Figure 2

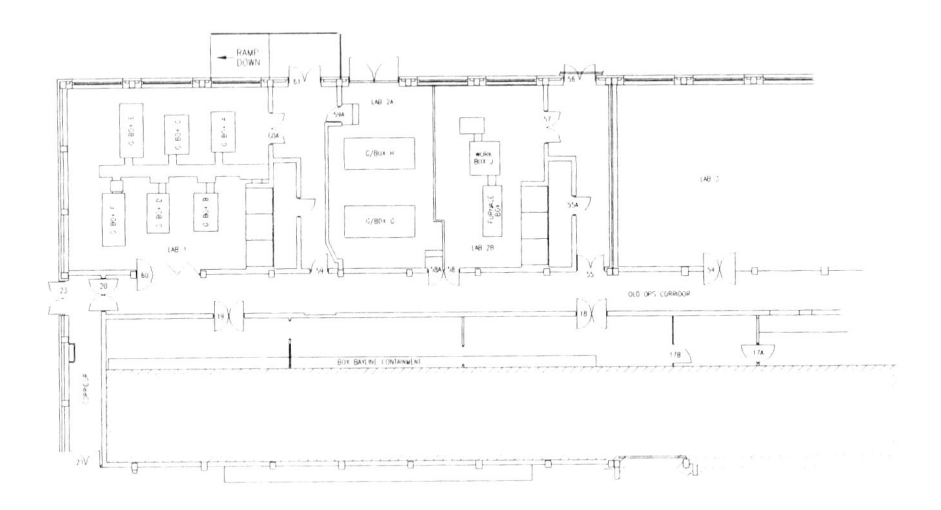

Figure 2 Area of the Facility Currently Undergoing Decommissioning

Lab 1
The five active gloveboxes in Lab 1 (Boxes A,B,C,D and F) are of identical construction, although the sizes vary. The box frames are of welded steel angle which forms both the box structure and its supports and onto which the Perspex panels are bolted.
The gloveboxes are connected by a tunnel system, which is approximately 610mm square. A short tunnel stub connects each box to the main tunnel that runs the length of the Lab. Circular

sliding doors enables each box to be closed off from the tunnel. Permanent isolation of boxes from the tunnel can be achieved by bolting blanking plates onto a flange at the tunnel end of the box stub. Above the tunnel and slung from the ceiling is an overhead gantry which supports the box nitrogen and HPE pipework and valves.

The three fume cupboards are positioned in a recess at one end of the Lab. These are manufactured from steel plate and are simple units with no air flow compensation for the sash opening. The sashes are currently set almost closed. The extract duct connection is from the rear of the cupboards.

Supply ventilation air to the Lab is via three inlet grilles mounted in the ceiling at the opposite end from the fumecupboards. Extract ventilation is entirely via the fume cupboard extract flows.

Lab 2A

The two gloveboxes are of identical construction and have external dimensions of 3.5m x 1.5m x 3.4m high. The box frames are fabricated from 'Zintec' (mild steel, electrolytically deposited zinc coating) sheet and plate of 5mm thickness. The frame provides structural support for all in-box items, including the machine tools and overhead hoist. For connection of the 13mm thick Perspex panels, the Zintec plate is formed into a U section to provide the surface onto which the panels are attached. The base of the box is a single, ribbed, iron casting some 205mm thick. On top of this casting sits a 'Colclad' plate (stainless steel clad (3.5mm), mild steel plate (13mm)) onto which the box frame is welded and which itself is bolted onto the base casting. The base casting is bolted onto a cast-in grillage. The total weight of the box and contents is in the order of 10te.

The principal contents of each box are a large hemiturn lathe, associated vacuum and coolant systems and an electrically operated overhead hoist. In addition the boxes contain a variety of hand tools and other accumulated wastes.

Supply ventilation air to the Lab is via two inlet grilles mounted in the ceiling.

Lab 2B

Box J is in fact two boxes connected by a short tunnel. The work box is connected to an entry/exit fume cupboard also via a short tunnel. The main box structure is manufactured from a stainless/lead/stainless sandwich. The Perspex panels are sealed into holes cut in the structure using rubber sealing strips. The fume cupboard attached to Box J is of a conventional steel plate design with gloves set in a Perspex panel above the sash. The main box contains a HF induction furnace and associated equipment, the work box contains a number of hand tools and other furnace equipment.

The three original fume cupboards are positioned in a recess at one end of the Lab. These are manufactured from steel plate and are simple units with no air flow compensation for the sash opening. The fronts of these fume cupboards were sealed with aluminium plate in 1990 to enable adequate velocities to be achieved at the sash opening of the Box J fume cupboard, although the units are still connected to the extract system and are consequently held at a depression.

5.2   Decommissioning Methodology

The sequence of events adopted for the decommissioning of the gloveboxes in the three laboratories was to divide the boxes into four areas, around which a MCS could be erected, this work consisted of a number of elements:-

- Installation/Commissioning of Pre-works
- Provision of MCS
- Re-circulatory filtered ventilation for MCS containment
- Modifications to building ventilation system
- Provision of breathing air panel
- Temporary electrical supplies for decommissioning works and equipment
- Emergency lighting for MCS's
- Temporary modifications to building structure
- Radiological monitoring equipment

5.3 Containment and Ventilation

The principal containment for the above dismantling activities is a MCS, which covers the area where the gloveboxes and fume cupboards to be dismantled are located. A typical MCS configuration, which was used for Boxes C and A, is shown in Figure 3. The MCS re-circulatory ventilation is provided by a Mini Mobile Filtration Unit (MMFU), which includes a terminal HEPA filter at the point of extract and a single stage of HEPA filtration on the MMFU. The return from the MMFU to the Labs is via a sock diffuser located at high level within the room. Access to the MCS is via an entrance tunnel providing dynamic containment via the ventilation system. A door, fitted with self-closing dampers provides containment when operations are not being carried out.

5.4 Size Reduction Techniques

The chosen option was to size reduce the gloveboxes and fume cupboards manually using a combination of cold cutting techniques, such as manual or power sawing, as well as plasma arc cutting. The most appropriate technique being selected for any particular task. For all activities involving the breaking of glovebox/tunnel containment, the operatives wear Pressurised Air Suits. The day to day size reduction activities are undertaken in accordance with a safe system of work.

5.5 Dismantling Operations

The dismantling operations follow a logical sequence of events:-
- All external services and pipework disconnected and removed.
- A strippable tie-down coating is applied to the internal surfaces of the glovebox and contents, via a number of the gloveports (following removal of the containment covers), with the exception of boxes G and H where oil present in the boxes will prevent the tie down solidifying.
- Where tunnels are connected to a glovebox, the door connecting the tunnel section to the glovebox is to be closed and sealed. The gloveboxes are then cut away from the transfer tunnel around the sealed door and size reduced. The transfer tunnel will be size reduced as one item inside Lab 1 MCS 3.
- With the exception of Box J, a Perspex panel from the glovebox will be manually removed to gain access. In the case of Box J, access will be gained by removing the windows/seals.

- Any loose equipment will be removed from the glovebox.
- Any contamination hot spots, identified from earlier monitoring, will be removed or shielded where practicable.
- The remaining Perspex panels will be removed and size reduced.
- Where large items of fixed equipment are present (e.g. lathes in Boxes G and H) any oil will be drained, removed and stored elsewhere in the Facility until an appropriate disposal route is found.
- The equipment will be disassembled/size reduced into manageable pieces for disposal in 200 litre waste drums; large pieces may be moved to another area of the containment for further size reduction.
- The gloveboxes will be progressively size reduced into manageable pieces, starting at the roof, using appropriate techniques. Each item will be subject to monitoring to provide an estimate of fissile material content prior to placement in the waste drum. For large pieces, where appropriate an attempt will be made to decontaminate the surface to allow it to go as LLW.
- For Boxes G/H, the dismantling will stop at the base plate level. The securing bolts will be drilled out and the Colclad plate will be removed. Tie down will be applied to the glovebox base plate on removal of the Colclad plate. The glovebox base plate will then be wrapped in PVC sheeting and stored locally within the MCS until the Colclad plate is size reduced. If monitoring indicates the base plate is ILW (following any attempt to decontaminate), it will be size-reduced in-situ. If it is LLW, it will be removed intact from the containment area on an 'air skate' (or similar) arrangement, and placed in a Half Height ISO-container for subsequent disposal.
- In the case of fume cupboards, internal surfaces will be coated with a strippable tie-down, and starting from the blanked-off connection to the extract ductwork, the structure will be progressively size reduced into manageable pieces.

6   DECOMMISSIONING PROGRESS

The Area shown in Figure 2 was broken down into four MCS's for the decommissioning work. As previously described the intention was to keep the approach simple, making use of existing tooling and techniques wherever possible. The first of these areas to be undertaken (MCS1) successfully dismantled boxes C and A within Lab 1. The layout of MCS 1 and the position it occupied within the Lab is shown in Figure 3.

6.1   MCS 1 Installation

One of the significant problems with the decommissioning works for MCS1 was the lack of Space. The area which was planned to occupy the MCS entrance was blocked by another glovebox, box E. Accordingly part of the preparatory works for the MCS1 required the size reduction of Box E. This box had never been fully commissioned and therefore posed no radiological problems. The opportunity was taken to utilise this box as a training aide and an enclosure was placed round it to simulate an MCS. The dismantling of the box was performed as a representation of future decommissioning operations, utilising pressurised suit operatives.

Although the workers who were involved in dismantling box E were experienced in decommissioning operations, the operation did give some valuable experience, which was to be utilised on the (contaminated) boxes A and C.

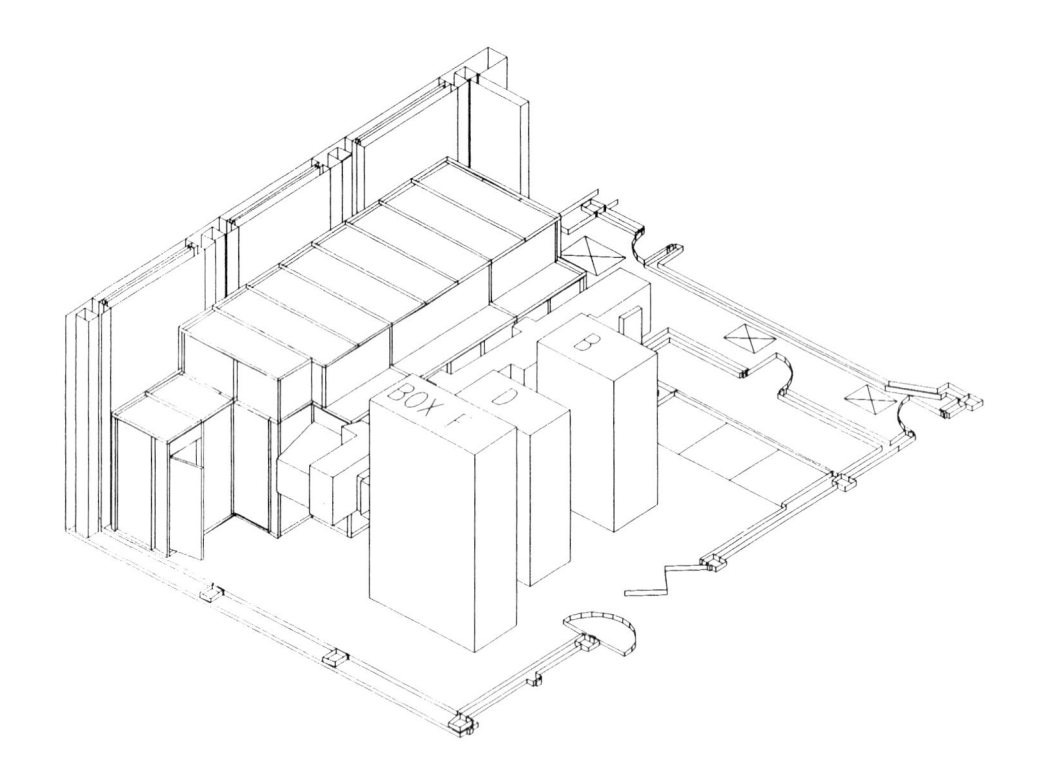

Figure 3 Layout of MCS 1

Even when box E had been removed the space available for MCS1 was extremely limited. The gap between the end of each box and the outside wall of the Lab was limited to approximately 1m. To form the wall of the MCS in this position required modified panels for the MCS, with the connection flanges being located inside the containment area. Obviously this caused the creation of a potential contamination trap within the MCS. Another area of significant space constraint was on the side of the MCS that was adjacent to the glovebox suite transfer tunnel. Each of the boxes was connected via a stub section and door mechanism to the tunnel system, and the MCS panels were required to go round this. Additionally directly above the tunnel, there was a gantry, containing glovebox services required for boxes B, D & F which were to remain and be dealt with as part of a later MCS.

The MCS was fitted with a standard 200 litre drum posting port for transfer of waste items.

As well as completion of the MCS construction a number of other systems were required to support the dismantling operations, these included:-

- MCS ventilation
  - This consisted of a MMFU which contained a single stage HEPA filter. This was connected to a twin bank filter housing located in the one end of the MCS, again containing a single stage HEPA filter. This ventilation system was configured to discharge into the Lab via a high level sock diffuser.
- Air Monitoring
  - Consideration was given to the installation of a Constant Air Monitoring (CAM) system in the ventilation ducting, but this was considered impracticable, so two portable CAM units were located within the Lab to monitor the breathing zones.
- Electrical Supplies
  - A number of 63A outlets were provided, into which mobile distribution towers were plugged in. These provided for services such as ventilation system, lighting, CAMs, communication equipment, as well as the tooling required for the dismantling work.
- Breathing Air
  - The existing breathing air system for the Facility was considered inadequate for decommissioning operations using pressurised suits, therefore a new system serving the Lab 1/2 area was installed. The system was modular in approach and will be extended into the rest of the building as decommissioning progresses.
- Evacuation building
  - The building is fitted with a criticality alarm, which signals the requirement for an immediate evacuation of the building. The pressurised suit workers would be unable and it would also be undesirable from a contamination control point of view to evacuate to the normal evacuation building. It was therefore decided to convert an existing store, located outside the evacuation zone into a Pressurised Suit (Frog) Evacuation Building (FEAB)

6.2 MCS 1 Commissioning

Prior to gaining Authority to Proceed from the AWE Nuclear safety Committee and the NII, a Commissioning Safety Report was required. This report focussed on three aspects of the commissioning process namely – Plant; Procedures; People.

6.2.1 Plant

Plant commissioning was relatively uneventful. During the planning stages of the work a commissioning schedule was defined for the MCS. This schedule chiefly related to functionality of the equipment, although in some areas there were specific requirements such as filter efficiency, Airflow through the MCS, Breathing air quality and availability/quantity, lighting levels etc.

As part of this process some of the parameters were not immediately achievable, but with minimal work compliance with the original Commissioning Schedule was achieved.

6.2.2 Procedures

In a highly regulated industry such as this, there is a large reliance placed upon a procedural approach where engineering controls are unavailable. Typically in decommissioning of alpha contaminated plant, although we have engineering controls to protect the environment and the general building population, the worker undertaking the dismantling operations is protected by a relatively fragile pressurised suit. To mitigate the risks, procedural controls were put in place to ensure the safety of the workforce and these procedures were validated as part of the commissioning process.

6.2.3 People

Probably the most difficult aspect of commissioning is that of people. Unfortunately they are all different in terms of ability, experience and understanding of the requirements of the decommissioning task. Some elements of the people commissioning process were easy – passing tests on building knowledge, RA safety/ industrial hazards etc. Much more subjective is the ability to respond to issues, both planned and unplanned as they arise during the works. To achieve this a large number of exercises were performed to "test" the people under varying scenarios. Each of these were treated as a learning opportunity and valuable lessons were fed into subsequent exercises. Only when complete confidence was gained in the ability of the team to react in an appropriate way, were they deemed competent to undertake the decommissioning work.

6.3 Decommissioning Operations

The first of the decommissioning operations commenced at the end of 2000, and although only planned to last three months, in total took around five. Examination of the data behind this demonstrated that the total number of operations required were in accordance with the original plan, although there were a number of delays incurred due to:-

- Failure of the breathing air system.
  - Although there was a new system provided to support the decommissioning operations, it relied in part on some old elements, which when put to the test proved unreliable
- Resources
  - As the move into decommissioning operations occurred, it coincided with a change in resource deployment across the local area which caused a shortage at certain times of appropriately "commissioned" people

- Working Time
  - Due to the nature of the decommissioning operations, each "shift" took approximately four hours. This meant that there was not enough available time to undertake two operations in a day. This was resolved towards the end of MCS 1 by the adoption of a longer working day over a shorter week.

Overall the decommissioning of the first MCS should be viewed as a success. The team involved gained more experience that will obviously be of use for the future operations.

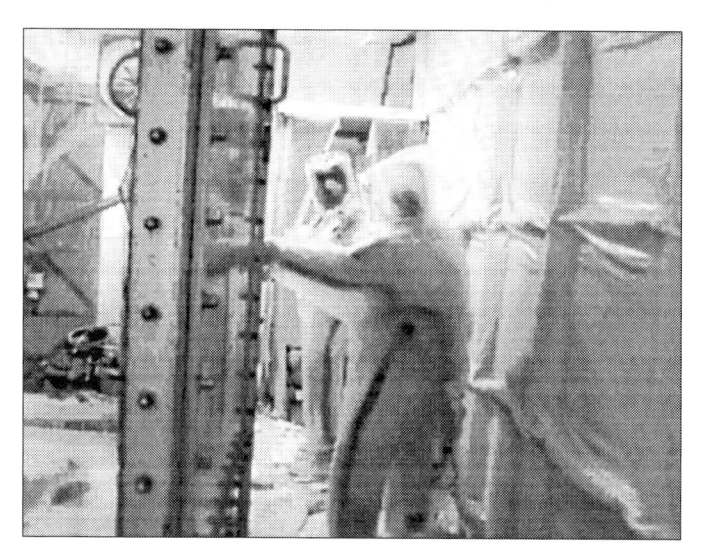

Figure 4 Dismantling of Box C

6.4 The Future

At the time of writing preparations are well advanced for MCS2 and 3, which will be undertaken in a very similar manner to MCS1. In approximately 12 months MCS 4 will be commissioned which will signal a significant change to decommissioning operations at Aldermaston with the adoption of Plasma Arc cutting. Design work for the installation of a three element ventilation system is complete for this work and clean commissioning trials are scheduled to be carried out prior to installation in the building.

© Crown Copyright 2001

C596/014/2001

Soil sorting gate equipment for site remediation

B G CHRIST, K FROSCHAUER, and **G G SIMON**
NUKEM Nuklear GmbH, Alzenau, Germany

ABSTRACT

Besides the dismantling of components and equipment and the decontamination and demolition of building structures, the remediation of outside areas of nuclear facilities becomes of increasing interest. For the treatment of suspected contaminated ground material, from its former fuel element fabrication factory, NUKEM has recently developed a soil sorting system. The system, being mobile, may be applied at various nuclear sites for site remediation purposes.

INTRODUCTION

Nuclear facilities by the end of their life time have required the decommissioning of contaminated installations, the decontamination of surfaces and the demolition of building structures.
The final activity for a significant number of nuclear decommissioning projects before the return of the site to the public domain is site remediation. Soil sorting turns out to be the most economic way to clean-up the outside areas.
For more than 20 years NUKEM in its facilities at Hanau, Germany, has processed Uranium and Thorium materials and produced different types of fuel elements for research and material test reactors (MTR).

In 1987 the company withdrew from these fuel element production activities. The redundant contaminated equipment and buildings had to be decommissioned. Within several licensing stages NUKEM sold or handed over useful process equipment to other nuclear users. All other remaining installations were dismantled. Building structure surfaces had to be decontaminated. After the final measurement of building contamination, mainly by applying in-situ γ-spectrometry, and the approval for free release, or landfill disposal of the material, NUKEM is currently demolishing the buildings.
The soil in the surrounding areas and under the buildings is to a greater of lesser extent contaminated. This was ascertained through an intensive radiological survey. Before being able to sell the site to a conventional user NUKEM has to remove contaminated foundations and soil from the site to leave an area free of radioactive contamination.

Figure 1: MTR-fuel element fabrication facility, NUKEM, Hanau

OBJECTIVE

NUKEM is to decontaminate and to pull down the buildings of their former MTR fuel element factory and finally to clean up the site ground. Since contamination has penetrated down to two meters into the soil and below the building's foundations, site remediation will be completed with the excavation of foundations and soil. Special attention will be given to potentially contaminated underground drain-pipes.

Figure 2: Demolition of NUKEM's MTR factory buildings

For the purpose of minimisation of volume of radioactive waste – and therefore decreasing the costs for packaging, transportation and storage - NUKEM has developed and is to apply a special soil sorting and activity measuring system, the „Soil Sorting Gate Equipment" (SSGE).

TECHNICAL APPROACH

For the purpose of sorting and radiological measurement of contaminated soil and foundation material, NUKEM has developed the continuously operating sorting and activity measuring system. The process is based on γ-spectrometric measurement of the excavated material whilst it is being transported via a belt conveyor system. Material is separated and classified into: radioactive waste (RW), land fill material (LFM), and material for free release (FRM) depending on mass specific activity of material.

ISOTOPE AND LIMIT VALUES

The radiological relevant nuclides for the NUKEM fuel element fabrication site are Uranium (U-234) and Thorium (Th-232). Necessary limit values for free release are: U-234 3.6 mBq/g and Th-232 37.9 mBq/g (20 ton batch).

REQUIREMENTS FOR MEASURING PROCESS

To ensure an effective measuring process the following criteria have to be met:
- a throughput of up to 50 metric tons per hour is required, to process the given soil amount within acceptable time
- to guarantee acceptance criteria of the landfill repository, material shall not exceed a particle size of 50 mm.
- the weight of material must be known.
- the thickness of material layer on conveyor belt will be adjusted to 100 mm.
- material and activity are homogeneously distributed on conveyor belt (ensured by mixing during excavation, loading, crushing).
- a steady flow of material is presumed (acheived through continuous loading).

MEASURING PROCESS

γ-radiological measurements are suitable for the detection and the assessment of relevant nuclides U-234 and Th-232. During measurement the radioactive nuclide U-235 will be detected as representative for U-234 and Ac-228, with Pb-212, as representative for Th-232. Measuring sensitivity is 3.6 mBq/g for U-235.

DESCRIPTION OF SOIL SORTING GATE EQUIPMENT

A sketch of the equipment is shown in figure 3. All transportation units selected are proven and reliable standard products from civil transport/construction machinery. The following units will be used for crushing, weighing, charging and distribution of material to be measured:
- crushing device; semi-mobile, incl. supply chute, jaw crusher, water spray
- band conveyor from crusher to screening machine
- rolling screening machine; grain size 0 - 50 mm, runoff for oversize particles, arranged above the charging box

- weighing band; conveyor band arranged on electric weighing cells, connected to the metering processor
- scanning band; trough belt conveyor, lead shielding in the area of the scanner, loading capacity of one metric ton after scanner
- intermediate band; conveyor similar to type of scanner band
- reversible band; conveyor band, 90° rotatable, high speed engine, placed above charging boxes for radwaste/undecided other material
- radwaste charging box; with underneath placed drum filling device
- landfill/free release material charging box
- landfill heap band; conveyor similar to scanner band
- free release material heap band; conveyor similar to scanner band.

Figure 3: Sketch of Soil Sorting Gate Equipment principle

MEASURING SYSTEM

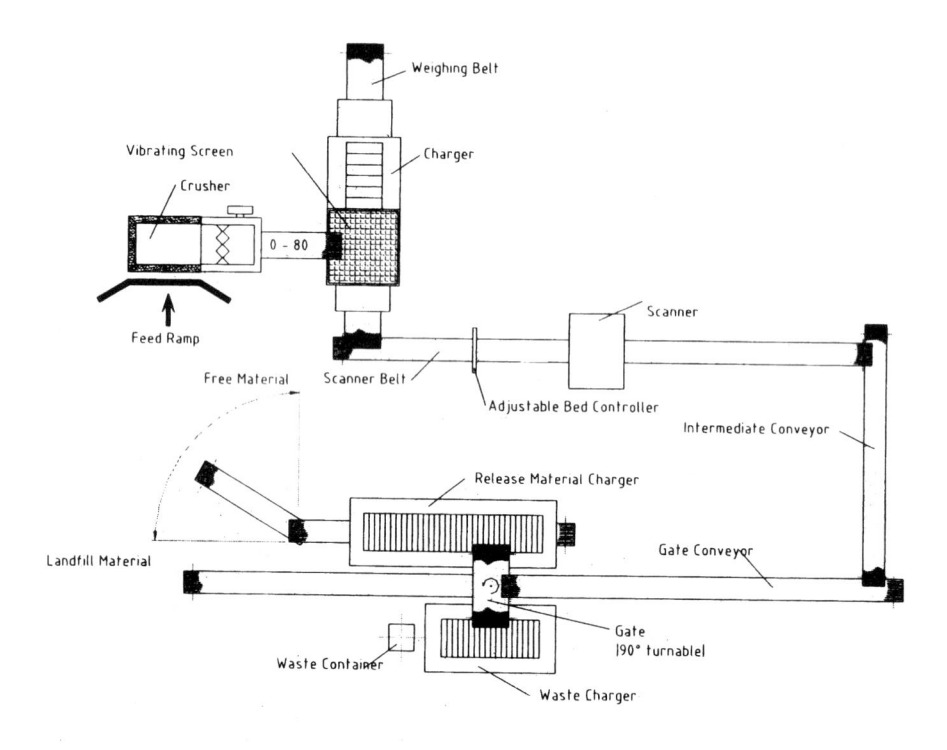

The soil sorting gate equipment's (SSGE) measuring system will consists of:
- measuring housing with scanner, four semi-planar HPGe-detectors mounted on a bridge above the conveyor belt, lead shielding

- central coolant supply system
- cabinet for electronics and data collection
- operator housing

Pre-amplified signals from detectors will be led to the main amplifier in the electric cabinet. Signal processing and gating will occur separately for each channel. Results will be fed to the central processor (PC). This central processor not only will serve to evaluate the detector signals (conversion of cps into Bq/g) but also to control the conveyors and records the relevant data.

DESCRIPTION OF SORTING AND MEASURING PROCESS

The principal arrangement of the soil sorting gate equipment (SSGE) is given in figure 3. The sorting amd measuring process is described as: Demolished and excavated material with maximum edge length of 800 mm is loaded by a wheeted dozer into the jaw crusher's chute and crushed to grain size < 80 mm. Grains are fed via a band conveyor to the screener, positioned above the charging box, which will also segregate loamy and sticky material. Oversized grain (> 50 mm) is separated and fed back to the crusher.

Following the start signal from the metering processor, and by means of the charging box's discharge and metering device, the material is poured on the weighing conveyor, which is located on weighing cells and connected to the metering processor. The weighing conveyor with a steady band velocity discharges all the weighed material onto the scanner belt and is then ready for the next weighing batch.

Weighed material is distributed to form a steady material flow of 800 mm width and 100 mm height. By means of the scanner belt the material continuously passes the detectors of the scanner arrangement positioned above the band. The emitted γ-rays are analysed, interpreted and compiled to classify the activity content. Measuring starts when ultrasonic sensors detect material arrival and is interrupted when all the material of the batch has passed the scanner. Material flow from weighing conveyor to scanner conveyor is adjusted in such a way that sufficient large gaps between one metric ton batches on the scanner band are produced.

From the scanner belt the material is transported to an intermediate conveyor and to the gate conveyor. In accordance with the determined mass specific activity the processing unit directs the ongoing material flow. The one metric ton batches, after measuring, are classified into the categories, radwaste (RW), landfill material (LFM), and not yet decided landfill/free release material (LF/FRM). Radwaste material and LF/FRM, respectively, is discharged into the corresponding charging boxes. LFM is sent via a dump conveyor to a heap area. RA material is directly filled into drums by means of the charging box's discharge device.
Not yet determined LF/FRM is accumulated in batches of up to 20 metric tons to enable unmistakable and significant classification into LFM or FRM. Via the dump conveyor and the FRM heap conveyor these materials are conveyed to either the LFM heap or to the FRM heap area as appropriate.

The complete SSGE is controlled by a SPS and automatically operated. All levels in the charging boxes are permanently surveyed by ultrasonic detectors; min/max levels steer the material flow. Activity measurements are performed per one metric ton batch, data is collected and processed for final documentation.

The system described provides the ability to separate and measure foundation material and soil into material for free release, land fill material and radioactive waste. It has to be emphasised that, depending on site requirements, the license situation, and waste and storage strategy, a modified system may be used for other site remediations (e.g. with reduced equipment).

CONCLUSION

The start up of the NUKEM site clean-up is scheduled for summer 2001. The machine design and measurement philosophy of the SSGE, developed for sorting and measuring of some 21,000 metric tons of excavated ground material and concrete from NUKEM's decommissioned fuel element factory, has already passed the official licensing procedure. Due to the recently detected historical hazard chemical inventory of the soil, the storage philosophy of treated material is currently being reinvestigated. There is a potential to store radwaste at a nuclear repository and all other material in an industrial underground dump. This would effect NUKEM's measuring program and reduce the number of band conveyors.

With the SSGE there is now available a method to safe and economically remediate contaminated ground of nuclear sites. The SSEG, being a mobile system, can be applied at various sites and adopted to the specific customers needs.

C596/042/2001

Decommissioning of the nuclear fuel storage ponds at Berkeley Power Station

D M WILLIAMS
Rolls-Royce Nuclear Services Limited, Gateshead, UK

SYNOPSIS

As part of the preparatory work for the Care & Maintenance 1 of the Berkeley Power Station decommissioning programme, the works associated with the Cooling Ponds building and associated irradiated fuel Transfer Tunnels was segregated into three phases by BNFL. Between April 1997 and May 2000, Rolls Royce Nuclear Engineering Services (R-RNESL) carried out the Phase 2, the largest of the three phases of the project. This involved the design, manufacture, installation and operation of equipment to remove the contaminated material from the building and associated transfer tunnels. The ponds building and transfer tunnel ventilation systems were also completely removed.

This paper outlines how R-RNESL and its subcontractors have approached the complete removal of contamination and remaining plant and equipment from the Cooling Ponds building and Transfer Tunnels in order to meet the Radiological End Point (REP).

1.0 DESCRIPTION OF COOLING PONDS BUILDING

1.1 Background

The cooling ponds were formally used for the storage of irradiated fuel elements prior to their dispatch for processing.

1.2 Building layout

The Cooling Ponds were situated at ground level between both reactors in a purpose built building which was "Crucifix" shaped in plan and at ground level was approximately 77 meters long by 42 meters wide. The building had a reinforced ground slab supported on beams and piles. Above ground, the building was of steel portal frames with precast concrete roof units providing a weatherproof enclosure. The external walls were mainly of solid brickwork. The below ground structures were reinforced concrete. Figure 1 shows an external view of the cooling ponds building.

There were two main ponds, Pond 1 and Pond 2, which were situated within the North end of the building. The depth of each main pond from surrounding wall to pond bottom was approximately 9 metres. Within the centre of the building, a smaller pond existed, this Burst Slug Pond (BSP) was provided for the receipt and isolation of suspect fuel elements. Adjacent to the BSP, located at the west side of the building, was the Sludge Tank Chamber and Dry Canning Machine Plinth.

At the South end, the lay out of the building provided for road transport flask receipt and dispatch and collection of operational wastes via a through way. An extension on the West side of the building accommodated an active effluent tank (Active Drains Tank D), utilised for the collection of pumped arisings from both reactors and pond building drainage system.

A fifty ton gantry crane was provided to lift the fuel transit flasks via access shafts into the main bay of the pond building and to lift the road transport flasks from the ponds and transfer them through the building. Figure 2 shows the layout inside the ponds building prior to decommissioning.

2.0 DECOMMISSIONING WORKS

2.1 Cooling pond decommissioning phases

Phase 1 of the Ponds decommissioning programme was to remove redundant plant and equipment, to remove loose and near surface contamination within the pond tanks to a depth of 2mm and to clean the building superstructure and all other remaining surfaces to a level approaching contamination C2 conditions.

Phase 2 of the Ponds decommissioning programme was the design, manufacture, installation and operation of equipment to remove the radioactively contaminated material from Cooling Ponds 1 and 2, the Burst Slug Pool, Irradiated Fuel Transfer Tunnels, sumps, building walkways, etc. The Ponds building and Transfer Tunnel ventilation system was completely removed as was any other area within the building and transfer tunnels where contamination had occurred.

In addition the above decommissioning activities the modification of the Active Drains system was required. This was to involve the removal of Active Drains Tank "D" and re-routing of pipework from Tanks A, B and C.

Phase 3 of the Ponds decommissioning programme is the controlled dismantling of the Building and remediation of the surrounding area. Jordan Engineering Ltd undertook Phase 1 and the Phase 3 works, was completed in April 2001 by Mowlem Ltd.

3.0 CONCRETE PLANING

3.1 Background

Following on from phase 1 operations, a sampling regime was carried out within the ponds building to assess the extent of contamination ingress within the main areas i.e. ponds, burst slug pool, etc. The specified depths for concrete removal was as follows:-

Pond 1 and associated Wet Sump	-	60mm
Pond 2 and associated Wet Sump	-	40mm
Burst Slug Pool	-	30mm
Walkways	-	20mm
Transfer Tunnel Floors	-	10mm
TT Walls, Ceiling and Building Superstructure	-	Remove Paint
Other areas i.e. Washdown Bay, Sludge Tanks, etc	-	Decontam to REP

It was considered that removal of concrete to the above depths would be sufficient to enable clearance monitoring of the ponds building. The specified depths were best estimate and it was thought probable, there may be certain areas where contamination may have penetrated beyond the designated depth, i.e., cracks in concrete surfaces, civil construction joints and areas where surface coating had deteriorated.

3.2 Pond walls & floors

In order to minimise waste volumes and negate the production and subsequent processing of wet waste arisings, our main civil works subcontractor, devised a remotely operated dry concrete planing technique. Apart from the above advantages of using a dry process, the planing process produced an even concrete surface, which aided the final radiological monitoring operations.

Initial trials of the planing equipment were carried out within the Burst Slug Pool, which resulted in modifications to the equipment, in preparation for concrete removal within the cooling ponds.

The modified planing equipment was able to remove concrete at the specified depths from the wall surfaces within the pond areas. The planing head was supported by a frame (via a hoist which controlled the vertical movement) sited on two rails, one located at the pond floor and the other located at the walkway level. The planing head travelled vertically throughout the height of the wall while the frame travelled horizontally along the length of the wall.

Due to the deployment framework, the planing head could not access the corners of the pond walls. The concrete in these areas had to be removed manually with jackhammers.

The floors of the ponds were planed at the specified depths by attaching the planing head (used for the walls) to a skid steer unit which was driven over the pond floors.

4.0 DEMOLITION OPERATIONS

4.1 Structures requiring removal

As the concrete removal works progressed within the ponds areas and a greater appreciation of the contamination ingress in certain structures was achieved, it became apparent that removing the surface concrete from walls associated with the Pumphouse and wet sumps had not been sufficient to meet the REP. Additional concrete removal to achieve the REP for these structures, required the removal of large areas of reinforcement bar which would have resulted in structural instability. The following structures needed to be removed completely:-

- Pond 2 Destrutter Wall
- Wet Sump Walls
- Removal of Pumphouse Walls down to 5 meters from the pond floor.
- Destrutter Bay Plinth areas within Ponds 1 & 2
- West & East Parapet walls in Ponds 1 & 2 respectively

4.2 Demolition equipment

The first major structure removed was the pond 2 destrutter wall. Due to the access restrictions it was considered that conventional demolition of this structure was out of the question and procedures were put in place to dismantle the wall in a controlled manner. The equipment utilised would be a mixture of hydraulic crunching and drill and burst techniques.

Scaffolding was erected around and over the wall, which in turn acted as the framework for the containment structure. A lifting beam located on the scaffold directly over the wall was utilised to support the hydraulic cruncher. The cruncher sequentially demolished the wall followed by rebar cutting down to the pond floor level.

The hydraulic crunching method was also used in conjunction with drilling and bursting for the controlled demolition of the pumphouse and wet sump walls. For this operation a large containment structure was erected around the pumphouse which was modified to cater for the decreasing size of the pumphouse and various working faces.

Figure 3 shows operatives positioning the cruncher during pumphouse demolition works. Figure 4 (taken from a similar location as figure 2) shows the extent of concrete removal associated with the pumphouse.

Mass concrete removal such as the plinth areas and sumps located on the floor of the ponds was carried out using a hydraulic breaker mounted on a skid steer. Figure 5 shows ponds 1 and 2 after concrete removal with remaining pumphouse structure.

4.3 Hotspot removal operations

There were various localised contaminated areas within the building. These areas were associated mainly with construction joints in the ponds and building floor slabs and concrete removal operations were mainly carried out using manual methods i.e.

jackhammers. This equipment was ideal for chasing contamination into construction joints and operator control and awareness negated damage to the water bar which would allow water ingress into the ponds, (the pond floor in certain areas and conditions was considered up to 6 meters below the water table).

4.4 Containment structures

The strategy for the Phase 2 decommissioning works was to maintain the building as a C2 designated area so as not to restrict operations within the building. Using a dry concrete removal process it was necessary to contain these areas to enclose any generated airborne activity that may have caused the whole of the building to be designated as C3.

The main containment structures with dedicated ventilation were placed over each pond, the Pumphouse, Pond 2 destrutter wall and the Burst Slug Pool. With the exception of the BSP, the large containment structures (constructed with scaffolding and associated working platforms) were continuously modified to assist in concrete removal and optimise airborne control measures. The containment areas were designated C3 and the decommissioning operatives were dressed appropriately. Each containment had its own change barrier where operatives would change to the appropriate dress depending on entry or egress.

Two large containment structures were erected to contain the screed and concrete removal on the building walkways south of the cooling ponds. Smaller localised containment structures were erected to enclose localised concrete removal operations throughout the ponds building.

4.5 Waste handling

Concrete waste arisings were loaded into 200 litre drums at the point of origin and transferred to the ISO loading area where they were weighed and monitored prior to the contents being tipped into the ISO container in order to achieve maximum waste packaging efficiency. The waste handling operations were carried out by a dedicated team who worked in conjunction with the concrete removal team in order to optimise the most efficient and safe handling and transfer of the waste from the contained areas to the waste packaging area.

5.0 REMOVAL OF REDUNDANT PLANT & EQUIPMENT

Throughout the Phase 2 works, various redundant plant and equipment was removed from the building. An inventory of the main items decommissioned is as follows :-

- active drains pipework and ancillary equipment
- effluent sumps and associated steel and brick liners
- various encast steelwork, i.e. crane rails, flask supporting steelwork, etc
- electrical panels, cables and ancillary equipment and services
- Ponds building & Transfer Tunnel ventilation and filtration plant
- Ponds building 50 Ton crane

6.0 DECONTAMINATION OPERATIONS

6.1 Grit blasting

The dimensions and layout of the transfer tunnels were ideal for the use of grit blasting as it was easy to contain the grit blasting area, (10m lengths of tunnel were segregated with dedicated ventilation and services). This segregation facilitated the retrieval of the grit from the floor for recycling. Figures 6 and 7 show the transfer tunnels prior to and after grit blast operations.

From the experience gained in the tunnels, the option for grit blasting within the main building was assessed. It was quickly realised that the layout of the main building superstructure, when considering the amount and layout of steelwork and other obstructions around the building, the paint removal from the walls using grit blasting would prove difficult. The layout of the building with particular respect to the floor areas did not facilitate easy retrieval of the grit material for recycling.

The successful campaign within the tunnels, led to successful grit blasting trials and eventual decontamination works within certain inaccessible active drains pipework within the building. For example, some drains pipework was buried up to 8m under the building base slab and mass concrete.

A containment was set up within the building for grit blasting operations and was used to decontaminate any plant and equipment used for the works. The greatest achievement was the decontamination of the active ventilation ductwork which reduced the waste disposal volumes considerably. Out of all the active ventilation ductwork, only the welded seams, (<1% of overall ductwork volume), which were deemed difficult to prove as free release material, were disposed of as LLW.

6.2 Scabbling operations

It was demonstrated that the contamination associated with painted surfaces within the main building i.e. walls and steelwork was mainly attributable to 2 metres above the walkway level and this paint required removal. Air operated scabblers and rotopeens were used for this operation as these methods were considered appropriate to minimise generation of dust levels (airborne contamination) within the building.

Rolls-Royce together with Trelawney, developed a 10" rotopeen, which were supported from the building superstructure beam via a tool balancer, was able to scabble large wall areas in a short timescale.

7.0 RADIOLOGICAL END POINT MONITORING

7.1 Radiological monitoring

After the specified depths of concrete had been removed, extensive health physics monitoring activities were carried out on the ponds wall and floor surfaces in order to ascertain if the radiological end point had been met. This operation was mainly carried out manually, as 100% surface monitoring was required. Prior to monitoring activities, each wall and floor surface was "marked up" into m^2 segments to assist clearance-monitoring objectives.

7.2 Routine monitoring

Routine monitoring operations were carried out on a daily basis to assess:-

- airborne levels within the main building airborne contamination levels within contained areas
- identify high dose rate areas resulting from concrete removal operations
- monitor personnel doses during decommissioning operations
- change barriers
- bio assay

8.0 CONCLUSION

Various methods of working, together with various types of plant and equipment was utilised and modified to suit the specific work activities in hand in order to achieve the removal of remaining contamination within the ponds building and associated transfer tunnels.

The Phase 2 decommissioning works proved a great success, which paved the way for a successful demolition stage. This was due to the teamworking spirit between BNFL, Rolls-Royce and our subcontractors and in particular the spirit, competence and commitment of the site team completing the works. As in any decommissioning works there are always 'surprises' which need to be resolved, usually, on an urgent basis. This is only achievable if the teamworking attitude is adopted.

ACKNOWLEDGEMENTS

BNFL Magnox Generation
Rolls-Royce Nuclear Engineering Services Ltd
Longholme Construction

Figure 1. Showing external view of ponds building.

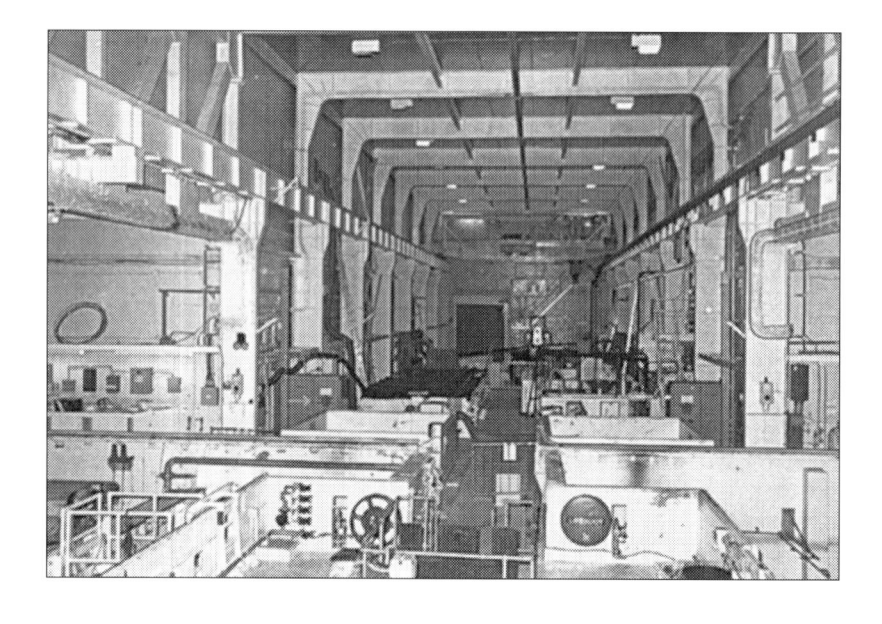

Figure 2. View inside building looking south with pumphouse in foreground.

Figure 3. Operatives positioning cruncher on pumhouse wall.

Figure 4. Showing extent of pumhouse structure removed.
(photograph taken in similar position as Fig 2.)

Figure 5. View of cooling ponds and remaining pumphouse structure after concrete removal operations

Figures 6 & 7. Transfer tunnels prior to and after decommissioning.

Technical Visit

C596/099/2001

JET decommissioning project

K A WILSON
Fusion Decommissioning Group, UKAEA, Culham, UK

SYNOPSIS

The Joint European Torus (JET) at Culham in Oxfordshire is an experimental facility investigating the conditions necessary for the initiation and stable operation of a nuclear fusion reactor. It is the largest project of the European fusion programme.

JET is currently operated by the UKAEA for the European Fusion Development Agreement (EFDA) under the JET operation contract (JOC).

The UKAEA is responsible for the safe decommissioning of the JET facility once the experimental programme is finished. Local Authority Planning Consent then requires removal of the associated equipment and buildings by the end of December 2018.

This paper describes the stages involved in decommissioning the JET facility and the areas of uncertainty when planning the decommissioning activities. It also describes the work that is being undertaken by the UKAEA Fusion Decommissioning Group in preparation for the eventual decommissioning of the facility.

1 INTRODUCTION

The Joint European Torus (JET), situated in Oxfordshire at the UKAEA's Culham Science Centre (Figure 1), is an experimental facility investigating the conditions necessary for the initiation and stable operation of a nuclear fusion reactor. It is the world's largest fusion experiment and currently the only fusion experiment in the world able to operate with a deuterium-tritium fuel mixture.

The JET Joint Undertaking was established in June 1978 to construct and operate JET. The facility came into operation in 1983, five years after construction started. In November 1991 JET became the first experiment to produce controlled fusion power. Further experiments carried out since then have provided useful information about the parameters needed for fusion power stations. In 1997 JET operations included successful experiments using the mixed deuterium-tritium fuel planned for future commercial power stations.

In December 1999 the JET Joint Undertaking came to an end. The UKAEA has taken over responsibility for the safety and operation of the JET facility on behalf of its European partners. The experimental programme is being co-ordinated by the European Fusion Development Agreement (EFDA) Close Support Unit (CSU), led by an EFDA JET Associate Leader based at Culham.

Culham was part of the Oxford green belt when JET was built and a condition of the planning consent at the time of construction was that, at the end of the experimental programme, the UKAEA must clear and restore the site to its original condition. The clearance date for most buildings is at present the end of December 2018.

2 THE JET FACILITY

2.1 JET machine

The JET device (Figure 2) is of the tokamak design and comprises a central toroidal vacuum vessel surrounded by substantial structures which provide the necessary magnetic fields which control the plasma (Magnetic Configuration), the means to add additional heat into the plasma (Additional Heating Equipment), a means to measure machine performance (Diagnostics) and machine services (Vacuum and Cooling Systems). The total mass of the JET machine and associated equipment is approximately 4,900 tonnes. The machine stands approximately 12 metres high and occupies a floor area of 400m^2.

The main component of the magnetic field, the so-called toroidal field, is provided by 32 D-shaped coils with copper windings surrounding the vacuum vessel. This field combined with that produced by the current flowing in the plasma, the poloidal field, form the basic magnetic fields for the tokamak magnetic field confinement system. The massive forces created when the toroidal coils are energised are resisted by a tightly fitted mechanical shell. Additional coils positioned around the outside of the mechanical shell are used to shape and position the plasma (poloidal field coils).

Beryllium is deposited onto the first wall components within the vacuum vessel using beryllium evaporators. Beryllium is used as a first wall material because of its excellent plasma facing material properties and low atomic number.

2.2 Shielding and containment

The JET machine is contained within a concrete biological shield (Torus Hall) which is within part of the main operations building, J1, on the JET site (Figure 3).

The Torus Hall is 35m by 35m by 23.5m high (internal dimensions) with the JET machine located in the centre. The walls, ceiling and floor are constructed from reinforced concrete with an internal boronated skin for neutron shielding.

The walls are 2.5m thick ordinary concrete internally clad with 300mm boronated concrete blocks. The roof is constructed from 2.25m thick concrete internally clad with 40mm boronated concrete screed and comprises pressurised concrete beams spanning 37.5m to the perimeter walls. The floor is 800mm thick reinforced concrete with 200mm boronated concrete screed topping and a central circular hole to accommodate the services from the basement.

2.3 Active Gas Handling System (AGHS)

During operations, hydrogen isotopes and other gases arising from plasma operations are handled, recycled and stored within a purpose built plant known as the Active Gas Handling System (AGHS) contained within building J25, situated to the west of building J1.

The AGHS consists of a number of sub-systems, which have the overall function of receiving, purifying and isotropically separating Torus exhaust gases. The purified deuterium and tritium gas streams are stored on uranium beds and re-used in the Torus. The detritiated impurity gases are discharged to the environment (Figure 4).

The mass of tritium on the JET site is quite small, a total inventory of 20g, but this has to be supplied to the experiment and recovered with precise control.

2.4 Operation

The JET machine has basically two modes of operation: one in which deuterium is used to fuel the fusion reaction and a second in which a mixture of deuterium and tritium is used. These two modes of operation are called D-D and D-T respectively. The two main differences between D-D and D-T operation are that the maximum neutron energy group for the former is 2.5MeV compared with 14.1MeV for the latter and that the D-T reaction rate is two orders of magnitude higher. The consequence of these differences is that D-D operation, although in overall operation terms more numerous than D-T operation, does not result in significant activation of the machine structure and support equipment. D-T operation however, because of the higher energy of the resulting neutrons and greater numbers, does result in the activation of the machine and ancillary equipment. Also, because of the use of tritium during D-T operations, components comprising the vacuum envelope and the AGHS will become heavily tritium contaminated. Other areas exposed to tritium will also become contaminated.

3   DISMANTLING PROCESS

The decommissioning of the JET facility represents many new challenges compared with decommissioning conventional fission reactors.

At the end of the experimental programme the dismantling will be undertaken in 3 stages. Stage 0 is currently ongoing and is involved with the preparation for decommissioning.

3.1   Stage 1 – phase 1

Stage 1 – Phase 1 decommissioning commences on the termination of the JET Experimental Programme. It is mainly concerned with the decontamination and clean up of the JET machine and facilities. One of the first activities will be the removal of bulk tritium from the site. In parallel to this, de-tritiation of the machine components contained within the vacuum envelope will be carried out by a programme of bake-out operations. This may be followed by the removal of all flight tubes, which penetrate the Torus Hall walls, and sealing of the penetrations to complete the outer containment boundary. High voltage supplies or services no longer required will be disconnected and removed. Information on the levels of tritium contamination and activation of materials will be obtained during this stage by a programme of survey, sampling and calculation work.

During this period there will be continued operation of:

- All systems necessary to support Stage 1 decommissioning activities;
- The AGHS (particularly the Exhaust Detritiation Plant);
- All necessary control and data acquisition systems (Health Physics monitoring) and remote handling systems;
- All ventilation plant necessary for radiological safety.

Once the operational plant has been defined then the redundant systems, equipment and buildings would be progressively decommissioned and disposed of.

The design for the decommissioning waste management facility will also be undertaken in this phase.

The duration of this phase is expected to be 3 years.

3.2   Stage 1 – Phase 2

This phase of the project will complete the removal of equipment and buildings on the JET site to meet the early removal planning consents. The removal process will be determined by the requirements during the detritiation stage. All of the Post Operation Clean Out (POCO) processes will be completed. This phase will also address the equipment and facilities required to carry out JET machine and systems decommissioning, dismantling, size reduction, packaging and disposal.

The duration of this phase is expected to be 2 years.

3.3 Stage 2

Because of the finite, relatively short, overall duration of the decommissioning programme it is not considered that there will be a dedicated period of care and maintenance between the end of Stage 1 – Phase 2 and the beginning of Stage 2. Stage 2 will comprise:

- The development, procurement and commissioning of remote handling equipment;
- The completion of the design, construction and commissioning of a decommissioning waste packaging facility (DWPF);
- Obtain approval for and then undertake the design and construction of an ILW interim storage facility;
- The establishment and operation of a radioactive waste transport infrastructure;
- The removal, packaging and disposal of in-vessel equipment, diagnostics and support systems, main machine assembly, basement plant, and services;
- The removal, packaging and disposal of non-machine radioactive material;
- The shutdown and decommissioning of the AGHS;
- The decommissioning of the waste management facilities once they are no longer needed;
- The removal of redundant support buildings.

Regular surveillance of the radiological and physical state of the JET site and the functional state of the operating systems will be undertaken during this period.

At the end of Stage 2 most of the equipment will have been removed and the main operations buildings will be ready for decommissioning.

The duration of this stage is expected to be 7 years.

3.4 Stage 3

The final decommissioning Stage will consist of:

- The decontamination of contaminated structures;
- The demolition of J1 and J25 and the clearance of remaining support buildings;
- A detailed radiological survey of the site;
- Appropriate landscaping in accordance with planning consent requirements.

The Torus Hall floor and basement columns are heavily reinforced and are expected to present difficulties during demolition. However, subject to appropriate approval, it may be possible to leave the 3m thick basement slab in place together with some other building elements provided that this does not compromise the final restoration of the site. The aim is to remove all structures above ground level.

The duration of this stage is expected to be 4 years.

4 AREAS OF UNCERTAINTY

The JET facility is still operational and therefore there are several areas of uncertainty which impact on the proposed decommissioning plan. These include:

- Machine performance;
- Changes to the experiment;
- Experimental end date and subsequent further use of the JET facility;
- Radiological condition at the start of Stage 1 – Phase 1;
- Waste management.

4.1 Machine performance
The JET experiment has been running successfully for about 20 years. If the Torus device was to have a serious breakdown and it was decided not to continue with the experimental programme, Stage 1 – Phase 1 decommissioning would commence earlier than expected.

4.2 Changes to the experiment
In the past 20 years there have been numerous changes to the JET facility, for example new diagnostics have been added and replaced, a divertor was fitted and changes have been made to it. There will probably be many more between now and the end of the experimental programme.

There is currently 4,900 tonnes of equipment within the J1T Torus Hall that is subject to neutron activation, some of which is also subject to tritium contamination. There is also tritium contaminated pipework and equipment in the AGHS and the J30 Waste Handling Facility.

Any new equipment installed in radiological areas will present an additional decommissioning liability when it comes time to decommission the facility.

4.3 Experimental end date and subsequent further use of the JET facility
The JET Experimental Programme is due to end in December 2002. There is currently a plan to extend the experimental programme, for a further 4 years, until the end of December 2006. There have previously been three extensions to the experimental end date and after each one the planning consent for site clearance was extended. The current site clearance date is the end of December 2018.

4.4 Radiological condition at the start of Stage 1 – Phase 1
The extent of the induced activity is directly proportional to the number of 14.1MeV neutrons produced. The JET machine was designed for a maximum credible D-T neutron production of 1×10^{24}. If this were achieved, estimates suggest that personnel access into the Torus Hall would not be possible and dismantling activities would need to be undertaken remotely. Also, at this level of neutron production, significant quantities of the Torus Hall bioshield would be activated and need to be disposed of as LLW.

Materials will fall broadly into three categories as follows:

- That which are radioactive by virtue of neutron irradiation alone;
- That which are radioactive by virtue of tritium contamination alone; and
- That which are radioactive by virtue of a combination of neutron irradiation and tritium contamination.

The amount of waste in category 1 is anticipated to be small and, as a result of decay, will fall below the ILW threshold relatively soon (4 years) after shutdown (based on a neutron production of 5.21×10^{21}).

Category 2 wastes are expected to arise from system components not subject to neutron irradiation and are dominated by components from the AGHS. Although the AGHS will be detritiated as part of the decommissioning work a significant quantity of waste is anticipated to remain above the ILW threshold for many decades.

Category 3 wastes include components contained within the vacuum vessel (Pumped Divertor, RF Antenna, Limiters etc.), the vacuum vessel itself and machine components which come into direct contact with tritium i.e. parts of the Neutral Beam Injection Systems and the Pellet Injection System. The presence of tritium contamination pushes the specific activity of this material above the ILW threshold for many years after shutdown. Most of the ILW generated is expected to fall into categories 2 and 3.

The 1997 DTE1 experimental campaign produced approximately 2.4×10^{20} 14.1MeV neutrons. The neutron budget agreed under the JET Implementing Agreement (J1A – Article 11) is 2.01×10^{21} 14.1MeV neutrons during the total life of JET, hence the remaining D-T programme has available a neutron budget of approximately 1.77×10^{21} 14.1MeV neutrons.

It is expected that there will be more D-T operations before the end of the experiment.

The level of radioactivity which will be present when decommissioning commences is therefore uncertain as it is dependant on future experiments.

4.5 Waste management

The present decommissioning programme requires clearance of the site by the end of 2018. As a consequence of the present situation in the UK regarding ILW disposal, the decision has been taken for the interim storage of this waste off the Culham site until there is a national repository to accept it.

All LLW arising from the decommissioning of JET is assumed to be consigned to Drigg. The majority of the waste from the site will be free release.

An alternative solution to the interim storage of ILW would be to delay the production of this waste on a time-scale, which would allow direct consignment to the deep repository. This would necessitate an extension of the current planning consent clearance dates for some of the buildings on the site.

5 PREPARING FOR DECOMMISSIONING – STAGE 0

Work on the planning and preparation for decommissioning the JET facility is currently being undertaken by the Fusion Decommissioning Group in UKAEA Culham Division.

The work undertaken is divided into a number of areas:

- The monitoring of JET experimental operations and plant modifications for changes that may affect decommissioning. This is to provide advice regarding changes to the liability;
- The formulation and development of plans for decommissioning and the management of all of the wastes that will be produced;
- Undertaking development work to prepare for decommissioning – this includes collaboration with other organisations (e.g. on tritiated materials management);
- Ensuring that experiences from other decommissioning work (in particular from other parts of UKAEA as well as from other organisations, both in the UK and overseas) is applied to decommissioning JET;
- Collection and storage of information relevant to decommissioning and site restoration;
- Ensuring that when any changes are made that decommissioning needs are taken into consideration.

5.1 Information collection

Information collected in support of decommissioning the JET facility is entered into the JET Decommissioning Database (JDD). This is the "information backbone" of the decommissioning project and helps the Fusion Decommissioning Group carry out the tasks of recording, storing and reporting information relevant to the decommissioning of the JET facility.

The database is split into a number of different categories:

- Building survey register;
- Drawing register;
- Material inventory;
- Infrastructure costs;
- Staff management;
- Document management.

The JDD is an ORACLE database with an ADMiT interface, a user-friendly system allowing easy interrogation and extraction of data.

In addition to the JDD there is also a Quality Assurance Archive containing historical information about the JET project, such as assembly procedures, welding records and other early information about the facility, that will be useful for the planning and eventual decommissioning of JET.

5.2 Decommissioning options studies

Following on from the production in 1990 of a Preliminary Decommissioning Plan (PDP) for JET a number of studies have been undertaken to provide more information on the JET facility and methods of decommissioning it.

These include:

- J1 building demolition study;
- JET machine dismantling study;
- AGHS decommissioning study;
- PRICE cost estimate for decommissioning the facility.

5.3 Divertor coil cutting trials

During previous decommissioning studies the four divertor coils in the bottom of the vacuum vessel were identified as being difficult to remove (Figure 5).

The divertor coils were added to the reactor to control the impurities generated within the plasma. The coils are a continuous winding with no joints and will therefore have to be cut from the vacuum vessel and removed in sections.

Each coil consists of an assembly of insulated copper conductors encased in an inconel 600 casing with the interspaces filled with epoxy resin. Each conductor is insulated from its neighbours and the inconel skin by a composite insulating material made of a boron free glass fibre material known as E-glass and also Kapton ground insulation. Because of the operating temperatures within the vacuum vessel each copper conductor has a 16mm diameter hole down its centre for the passage of coolant.

The application of basic cutting methods to any reactor component is made more difficult if the section to be cut comprises materials with widely differing mechanical, thermal and electrical properties since these properties form the main criteria by which cutting techniques are selected.

A series of cutting trials were undertaken, using a section of divertor coil identical to the one in the machine, to identify the best method for cutting the coils so that they could be removed from the main machine during decommissioning.

Of the nine methods tried out on the coil mock-ups only four were successful at cutting them, these were bandsawing, disc grinding, reciprocating sawing and milling using a side face mill. When considering the feasibility of remotely deploying the four successful methods within the vacuum vessel it was concluded that disc grinding would be most easily adapted for use. It was also found that disc grinding was the best method with regard to blade wear.

After these trials were undertaken, technological developments in both diamond wire cutting and tungsten carbide toothed blade sawing have meant that these techniques were further investigated with respect to the cutting of the divertor coils.

These extra trials concluded that the tungsten carbide toothed saw should be adopted as the preferred method for cutting the JET divertor coils on the grounds of cut speed, durability and minimal secondary waste production in comparison with the abrasive disc cutter previously tested.

Diamond wire sawing was disregarded on the grounds that it would not reliably cut the JET divertor coils due to the durability of the wire and the diamond wire material dulling the cutting face.

5.4 Radioactive inventory assessment

A sound understanding of the radioactive inventory is essential in the formulation of plans for both operation and the ultimate decommissioning of the JET facility. To this end, the radioactive inventory of the equipment within the J1T Torus Hall has been calculated using neutronics data in conjunction with the FISPACT inventory code and data libraries developed specifically for fusion reactions.

FISPACT is an inventory code that has been developed for neutron-induced activation calculations for materials in fusion devices.

A programme of work was undertaken to validate the neutronics data in 1997. 50 Activation Foil Pack (AFP) neutron monitors, containing specially selected materials, were installed in the J1T Torus Hall before the Deuterium-Tritium Experiments (DTE1) in February 1997. They were activated by neutron bombardment during the experimental campaign and subsequently retrieved from the Torus Hall after DTE1 and analysed to determine the activation of the materials.

The data was then used to determine the flux, fluence and spectra at the 50 AFP locations. This information was used in conjunction with the material inventory database to determine the nuclear inventory within the Torus Hall for a given neutron budget.

It is estimated that there will be 350m^3 of conditioned ILW and 2800m^3 of conditioned LLW from decommissioning JET, when subjected to 5.2×10^{21} 14.1MeV neutrons.

5.5 Analysis of the Torus Hall biological shield for tritium contamination

This analysis is being undertaken to improve our understanding of the presence of tritium in the Torus Hall biological shield following the DTE1 experimental campaign. This is a significant structure and will represent a large decommissioning cost if significantly contaminated with tritium. The Torus Hall is made up of 33,200m^3 of boronated and plain concrete.

A number of cores are being taken from the walls, ceiling and floor of the Torus Hall during the 2001 shutdown. These cores are being sectioned and analysed for their tritium content to determine the tritium levels at the front face and the degree of diffusion into the concrete.

5.6 Information management and geographical evaluation system: Integration of data management with ArcView GIS

The decommissioning of the JET facility will involve:

- The demolition of buildings and facilities;

- The assessment of land quality and management and clean-up of contaminated land.

Many sources of information contribute to the overall assessment of land quality or potentially contaminated land. A data management system known as the Information Management and Geographical Evaluation System (IMAGES) is being used to collate site information so that it can be input into ArcView GIS (Geographical Information System).

A Geographical Information System (GIS) enables spatially referenced data to be viewed, interpreted and modelled.

The information used includes records of:

- Buildings – particularly past and current which may have caused, or could potentially cause contamination e.g. records of spills, leaks, building fabric;
- Any waste disposals;
- Drainage systems;
- Groundwater monitoring;
- Routine land surveying;
- Non-routine investigations;
- Photographs;
- Drawings.

Work has also been undertaken to collect information on the site before the UKAEA took it over, when it was a Royal Navy Fleet Air Arm Station.

This system is also currently in use at Harwell, Dounreay and Winfrith to assist in their site de-licensing work.

5.7 Future

Future work will involve:

- Further research into techniques for decommissioning/dismantling fusion facilities;
- Re-visiting the decommissioning options studies which have previously been undertaken to take into account the more recent changes to the JET facility and the currently planned future programme;
- Investigation into the treatment and packaging of tritiated waste and detritiation.

Closer to the start of Stage 1 decommissioning:

- Stage 1 decommissioning programme;
- Liaison with the JET project regarding decommissioning, health physics and waste management issues and the disposal of legacy wastes;
- Establishment for the strategy for the management of all wastes once the end condition is known;
- Preparation of necessary plans, safety and environmental documentation and the establishment of the appropriate management and safety-related infrastructure for Stage 1.

6 CONCLUSIONS

One of the interesting challenges in planning the decommissioning of JET is the way the 'target' keeps moving and changing.

Once a fission reactor starts operating and it becomes highly radioactive (the core region) then unless there is an accident the decommissioning plan will change very little with the time the reactor operates. Once it is activated then no matter how long it operates the decommissioning methodology will not change significantly. This would be the same with a power generating fusion plant. However, JET being an experimental facility has not been activated to the design level because of the changes in the experimental programme.

There are a number of uncertainties when planning the decommissioning of the JET facility, including the experimental end date; changes to the machine and associated equipment and the experimental programme itself; and hence the radiological condition of the facility and extent of the liability at the start of decommissioning.

The UKAEA's Fusion Decommissioning Group are undertaking work in preparation for the eventual decommissioning of the JET facility. This work involves a number of areas, including information collection on the facility; the formulation of decommissioning plans; study and development work into decommissioning techniques; and collaboration work with other organisations to learn from their experience in related fields.

Decommissioning the JET facility represents a unique and challenging project.

ACKNOWLEDGEMENT

Funding of this work by the Department of Trade and Industry (DTI) is acknowledged.

Figure 1. JET facility

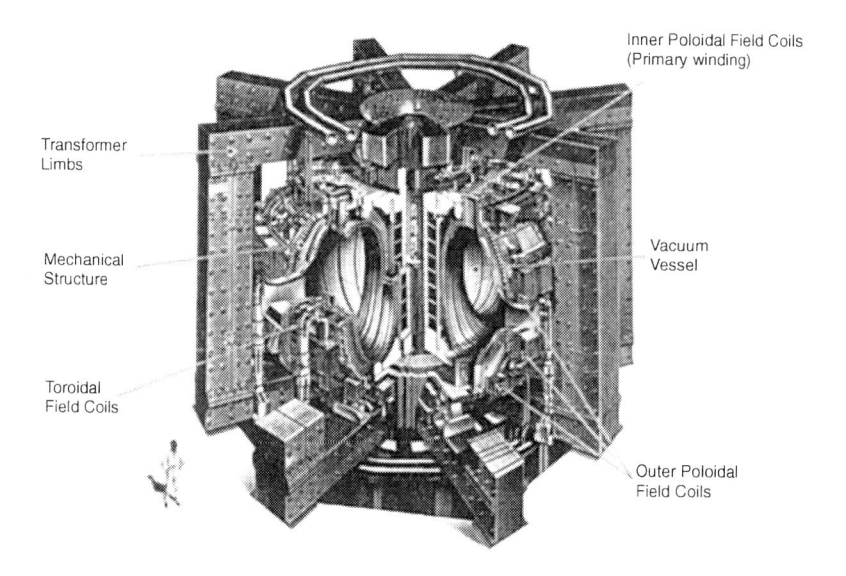

Figure 2. JET device

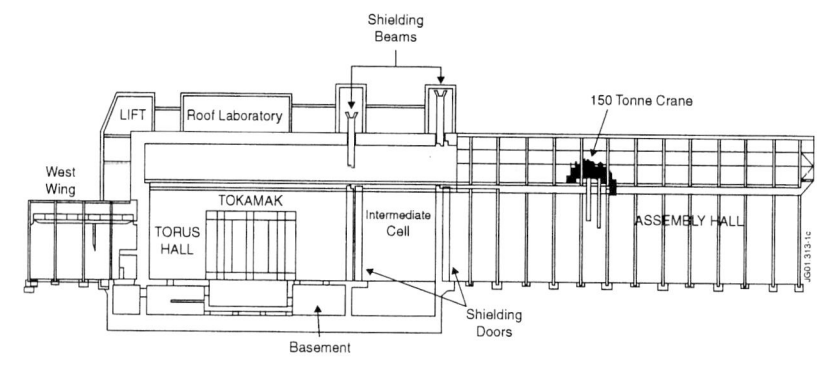

Figure 3. Building J1

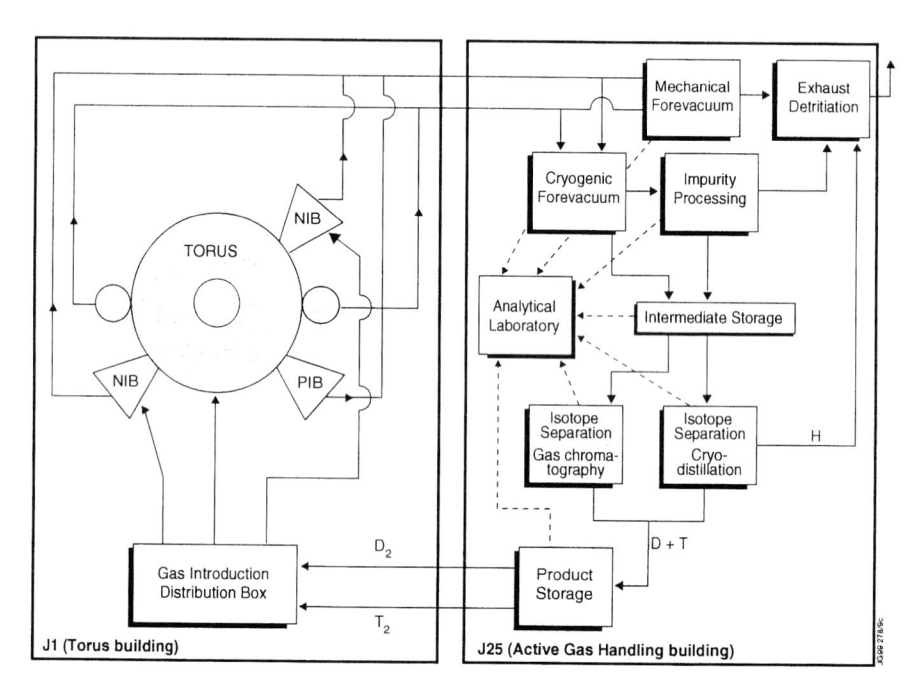

Figure 4. Schematic of the AGHS and Torus device interfaces

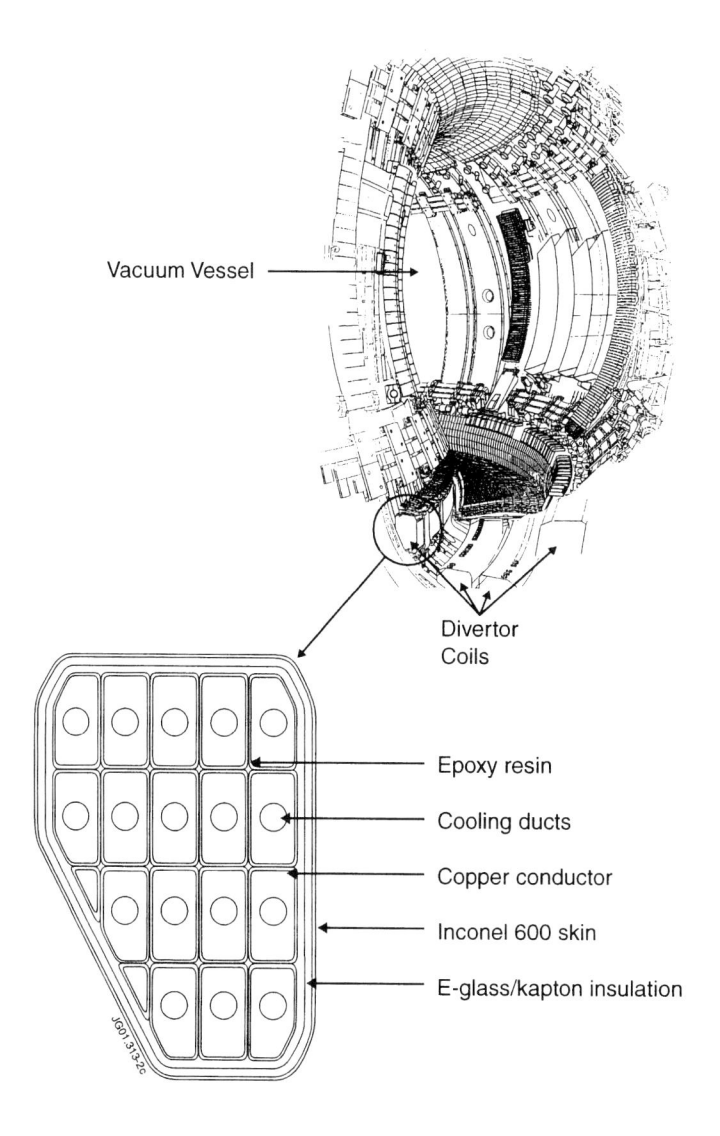

Figure 5. Divertor coils

Additional Paper

C596/028/2001

Regulation of decommissioning projects in a Scottish context

H S FEARN, J GEMMILL, and M KEEP
Scottish Environment Protection Agency, UK

ABSTRACT

The paper sets out current regulatory responsibility in Scotland for the management and disposal of radioactive wastes. The current expectations of the Scottish Environment Protection Agency of operators seeking to utilise the Substances of Low Activity Exemption Order is set out. The need for applications for the disposal of radioactive waste to be supported by Best Practicable Environmental Option studies is confirmed. A possible framework that could be developed for identifying the appropriate level of stakeholder involvement in such studies is discussed. Views on the term Best Practicable Means, and Best Practicable Means are briefly reviewed. The views set out in this paper are those of the authors and do not necessarily represent those of SEPA but do reflect the current regulatory advice being given by the authors.

REGULATORY RESPONSIBILITY IN SCOTLAND

The creation of SEPA (in 1996) and the transfer of responsibility for regulation of radioactive substances to the Agency was particularly significant. Up until 1996 regulation had been carried out by central government and latterly in Scotland by Her Majesty's Industrial Pollution Inspectorate a unit within the Scottish Office. These changes in regulatory structure to "arms length from Government", were essential in order to ensure that the UK Government was (and is) in a position to demonstrate the requirements for independent regulation set out in the Joint Convention on the Safety of Spent Fuel Management and on the Safety of

Radioactive Waste Management, to which the UK is a signatory. A key element of SEPA's approach, indeed SEPA's ethos is to work in an open and transparent way, which is of course is an approach espoused by the European Union and also part of the Government's strategy on modernising government. SEPA's approach has led to a more open environment for debating matters and greater accessibility to information. The result, we believe, has been a step change in the media coverage of issues under discussion and much increased public confidence. SEPA is fully accountable to the First Minister for Scotland who has the responsibility in Scotland for formulating radioactive waste management policy. Duties in this regard were first set down in a Government White Paper Cm 6820 where the Secretary of State for Scotland's duties were recorded, including amongst other things, the responsibility to *"secure the disposal of waste in appropriate ways and in appropriate time and in appropriate places"* and to secure *"the programmed disposal of waste accumulated at nuclear sites"*.

In the wider European context the Treaty of the European Atomic Energy Community (Euratom) gave the European Community the task of establishing uniform safety standards to protect the health of workers and the general public in all Member States for activities concerning ionising radiation. On 13 May 1996 the European Council issued *Council Directive 96/29/EURATOM, laying down basic safety standards for the protection of the health of workers and the general public against the dangers arising from ionising radiation*. This council directive takes account of the recommendations contained in publication 60 of the International Commission on Radiological Protection (ICRP). The ICRP are an independent body of experts, that provide guidance on a range of topics relating to the protection of man against radiation. In their publication *1990 Recommendations of the International Commission on Radiological Protection* (ICRP 60) the health effects of exposure to radiation were set out revising previous guidance taking account of the views of the international scientific community. The system of protection recommended by the ICRP for practices involving radioactive substances is based on the following principles:

> 1) justification of a practice;
> 2) optimisation of protection; and
> 3) the application of individual dose and risk limits.

To implement the relevant parts of this Directive in Scotland, Scottish Ministers exercising powers conferred upon them by section 40(2) of the Environment Act 1995 issued a Direction (in May 2000) *"The Radioactive Substances (Basic Safety Standards) (Scotland) Direction 2000"* to SEPA. This requires SEPA when discharging its functions in relation to the disposal of radioactive waste under the Radio-active Substances Act (RSA) 93 to ensure that the dose limits for members of the public set out in Article 13 of Council Directive 96/29/EURATOM are not exceeded. The dose limit is set at 1 milliSievert in a year. The Direction to SEPA also requires that the contribution to public dose arising from the authorised radioactive discharges of any one new nuclear installation should be constrained to a maximum of 0.3 milliSievert in a year. In addition where a number of nuclear facilities are adjacent, possibly owned by different organisations, an overall site constraint of 0.5 milliSievert will be applied. Additionally SEPA is required to ensure that reasonable steps are taken such that the contribution to the exposure of the population as a whole from practices that involve the disposal of radioactive waste is kept as low as reasonably achievable, economic and social factors being taken into account. The Direction also requires SEPA to observe the requirements of the Council Directive with regard to the estimation of effective and equivalent doses, the estimate of population doses and of the requirement for the holder of an authorisation issued by SEPA to appoint a qualified expert.

SEPA regulates the disposal of radioactive waste under the Radioactive Substances Act 1993 and issues authorisations with appropriate conditions for the disposal of radioactive waste. Authorisations are only issued or refused after due consideration of an application made by an operator which in the case of nuclear sites is subject to public consultation. In addition, for nuclear sites SEPA is required to consult with the Nuclear Installations Inspectorate (NII), who are responsible for nuclear safety and the recently created Food Standards Agency (Scotland), which is responsible for ensuring the protection of the food chain. Both organisations are statutory consultees under RSA 93 for applications relating to the disposal of radioactive waste, thus allowing the consideration (by SEPA) of the implications for the public from all pathways before considering whether to grant an authorisation. Consultation is also carried out with Scottish Ministers. Scottish Ministers can require applications to be determined by them and have powers under RSA 93 to cause a local inquiry to be held in relation to any application. Scottish Ministers can also issue directions to SEPA in relation to the terms and conditions of any authorisation that SEPA is minded to grant. In any authorisation that SEPA issues the practice that gives rise to the disposal of radioactive waste will be specified based upon the information supplied by the applicant. For the systematic decommissioning for a large scale facility or large structure which would inevitably take place over a large timescale, SEPA would expect the applicant to give detailed information on activities to be carried out and waste to be generated within the first five years of any authorisation and as much detail as possible for the next five years. Any application should be supported by a Best Practicable Environmental Option (BPEO) study to underpin the waste management strategy that is to be adopted and a report containing the detailed consideration of how the applicant has addressed the issue of best practicable means (BPM). SEPA will attach conditions in any authorisation that it issues requiring operators to use BPM to reduce the activity in all the waste disposed and where relevant its physical volume (to ensure waste minimisation), to maintain in good repair all equipment and plant, to keep records of all waste disposed, set limits for the overall disposal of radioactive waste and attach appropriate conditions to satisfy the requirements of the BSS Direction.

After setting the regulatory framework we will considering in a little more detail SEPA's developing thoughts on BPEO, BPM and the disposal of decommissioning wastes at activity concentrations below regulatory control.

UNCONDITIONAL EXEMPTION FROM RSA 93

In managing the disposal or future disposal of radioactive waste it is imperative that suitable waste quality assurance procedures are in place to firstly ensure that waste minimisation is achieved and secondly that any wastes that are created are properly categorised at source so that the most appropriate available disposal routes can be utilised. For large scale decommissioning operations, particularly the stripping out of plant and demolition of buildings, large volumes of radioactive waste containing trace quantities of radionuclides can be anticipated. Such wastes may be suitable for unconditional exemption from the requirements of RSA 93 under the 1986 Substances of Low Activity Exemption Order (SOLA) providing the wastes are substantially insoluble and have a specific activity of

<0.4Bq/g. The UK Government has reviewed this exemption order against the clearance levels set out in the EU Basic Safety Standards and have informed the commission that waste disposed under SOLA would comply with the BSS criterion that the resultant dose to any individual would be less than the 10 μSv. SEPA does not anticipate any immediate changes to this exemption order or the issue of an exemption order specific to the decommissioning of nuclear plant.

Before waste is disposed under SOLA or segregated for disposal as non radioactive waste, any site operators must have in place suitable quality assurance (QA) arrangements for the management of its wastes including its segregation and be able to demonstrate their effectiveness to SEPA. This will inevitably require waste to be tested and monitored but reasoned arguments, where the full history of the item is known and is well documented, may limit the amount of monitoring deemed necessary by an operator under his QA arrangements. SEPA would expect a site operator to keep records of how they apply SOLA conditions to exempt any waste and to record the quantities of waste disposed. SEPA is likely to audit the QA and management procedures being used by a site operator to segregate radioactive wastes that meet the requirement of SOLA from other wastes in order to form a view on their adequacy. This may involve the independent analysis of samples of waste using the Environment Agency (for England and Wales) Quality Checking laboratory at Winfrith in Dorset. When considering an operator's QA system SEPA will also need to be satisfied that the best practicable means (BPM) principle has been embodied in the system.

For bulk wastes such as contaminated soils or cement based rubble, SEPA would expect operators to carry out adequate sampling of the waste as dictated by the operators chosen robust statistical approach to ensure that the results are representative of the bulk waste. Methods used for quality assurance of the purity of bulk commodities may be suitable for radioactive waste and are worth investigating, indeed an appropriate British Standard may exist. For bulk structures it will be necessary before any demolition is carried out to investigate and quantify the depth profile of any radioactivity. This is likely to involve core sampling, the extent of which will be dictated by the adopted statistical approach ensuring that the sampling is representative of the bulk. Any scrutiny of operators procedures will as a minimum consider the items identified in the table below.

Are appropriate QA and management procedures in place for waste segregation, classification and monitoring;
was every consignment of waste monitored or was statistical sampling methods used;
what number of samples were taken and what statistical method was adopted;
is the contamination homogeneous or is the waste profile such that activity is being concentrated in the depth profile;
what is the profile through bulk material (core sampling);
is the sample representative of the bulk material;
what sample volumes were taken;

> how was the radionuclide content identified;
>
> were the measurements appropriate to define the nature and extent of the radioactivity; and
>
> are the results readily reproducible?

For items where there is no discernible depth profile to the radioactivity (surface contaminated objects) then it would be reasonable to expect operators to take all reasonable steps to remove the contamination before applying SOLA. In carrying out any decontamination the concentrate and contain principle must be adopted. SEPA is aware that many operators carry out surface contamination measurements for comparison with the levels set in the 1968 unsealed Ionising Radiation Regulations for controlled areas (4Bq/cm^2 for $\beta\gamma$ and 0.4Bq/cm^2 for α) but this, in its self, does not necessarily demonstrate compliance with SOLA. Where suitable and rigorous decontamination has taken place SEPA is likely to accept that any remaining radioactive contamination can be considered as fixed. In these circumstances providing that the waste item had been broken down to its smallest components averaging of the contamination over the weight of the constituent parts appears to SEPA to be acceptable providing that the operators QA system rejects "high surface dose rate" objects from disposal under SOLA. The rejection criteria should be set to ensure that the any resultant dose to any individual would be less than the 10 µSv for any conceivable reuse of the waste being disposed of. The table below details some areas of an operators protocol that SEPA would wish to examine.

> Is contamination fixed or ready removable/loose (if loose contamination, waste cannot be released) and how was this determined;
>
> was contamination removed where practicable and then SOLA applied;
>
> was clearance carried out by the measurement of surface contamination or other sampling techniques;
>
> were all areas of the waste monitorable;
>
> what measurement probes were used and their limits of detection;
>
> has the whole surface of the waste been monitored;
>
> details of the methodology used to relate surface contamination (Bq/cm^2) measurements to specific activity (Bq/g);
>
> where sampling techniques were used are they statistically representative; and
>
> if any contamination was found by surface monitoring, was the contamination analysed separately?

For any decommissioning project the operator should produce a waste characterisation document which sets out the likely radioactive wastes (in terms of activity and quantities) that will require storage or disposal and identifies the most appropriate disposal routes. This is likely to be based on a consideration of the history of the facility to be decommissioned including a review of past health physics monitoring followed by a full health physics survey of the facility. Decommissioning could then commence segregating the waste into appropriate categories as the work progressed. The waste that has been segregated for disposal under SOLA should then go to a final monitoring and sentencing facility where the tests and sampling carried out should be reviewed before a final measurement of activity is made using a suitable drum or bag monitor to confirm its suitability for release under SOLA. A number of companies, particularly scrap metal recycling yards have installed gate detectors at their premises and these have proved effective in a number of circumstances. There is increased interest within the EU of such devices being installed at boarder crossings. Such devices could be usefully employed within a decommissioning project but would never replace the rigorous arrangements for monitoring and segregation of waste that SEPA would expect operators to have in place.

BEST PRACTICABLE MEANS

There is a fundamental requirement to keep all exposure to radiation as low as reasonably achievable, taking into account economic and social factors- the ALARA requirement. This is used as part of the optimisation principle and is given effect within authorisations issued by SEPA by the inclusion of conditions requiring the operator to use the best practicable means (BPM) to reduce the activity in all the waste discharged. The requirements of the precautionary principle in protecting the wider environment are also given effect by the BPM requirement. The concept of BPM is well established in the UK in relation to the protection of the environment and has been used where appropriate in authorisations granted under the Radioactive Substances Act 1960 since the Act, now superseded by the 1993 Act, came into force. Various definitions of BPM have been incorporated into authorisations over the years. Currently SEPA bases its definition on that given in the Government White Paper Cm2191. SEPA's definition reflects the need, for any particular aspect of an operators process within any undertaking carried out by that operator, to represents the best practicable means.

> *"best practicable means" within any particular undertaking, means that level of management and engineering control that minimises, as far as practicable, the release of radioactivity to the environment whilst taking account of a wider range of factors, including cost-effectiveness, technological status, operational safety and social and environmental factors. In determining whether a particular aspect of a process within any undertaking represents the best practicable means SEPA will not require the company to incur expenditure, whether in money, time or trouble, which is disproportionate to the benefit likely to be derived.*

This makes clear that the duty applies not just to the treatment of wastes prior to discharge but to the method of operation of the process by which the waste arises. It applies also to the maintenance and supervision of the plant and equipment as well as to its provision.

SEPA is aware that what constitutes BPM may vary from one site to another and from time to time at a given site according to particular circumstances and developments in science, technology and policy. The determination of what constitutes BPM therefore has an ongoing element. SEPA currently addresses this in its authorisations by requiring not only the use of BPM at all times but also by requiring periodic studies by an operator of the application of BPM to its waste generating activities. However if a company only undertakes BPM studies to justify that its existing plants and processes are BPM then such a study is of no benefit and is likely to be rejected by SEPA. Any BPM study of older plant and processes is likely to identify a range of plant and operational improvements that could be undertaken, indeed if this was not the case then SEPA would question the basis and validity of the study. Any identified improvements will need to be reviewed against the life span of the particular undertaking and the cost of implementing the improvements and their practicability. New plants and processes will have been designed taking account of the results of a detailed study of BPM. At periodic intervals the underlying process for "producing the product" will need to be reviewed within a BPEO framework. This is considered later in this paper. It is clear that for the application of BPM a hierarchy exists with most reliance being placed on "best" for new plant and for plant nearing the end of their useful life great reliance being placed on "means". In respect of the decommissioning of any plant or facility, there will be no operating experience within this plant or facility for this new type of work and hence nothing to underpin the "means" in the BPM requirement. It is likely therefore that plant and facilities will require significant upgrading to meet modern standards in respect of containment, ventilation systems, monitoring systems etc. before they are decommissioned as the decommissioning projects will need to be reliant on "best".

There would be benefit to operators if BPM was defined in the legislation controlling the disposal of radioactive waste and in which case the definition should be consistent across the different legislation for which SEPA is the regulatory authority. This could be usefully explored in any relevant statutory guidance which could clarify the relationship between BPM and the use of "Best Available Technology" and other such terms used in different environmental legislation and under international obligations such as OSPAR.

BEST PRACTICABLE ENVIRONMENTAL OPTION

Any undertaking that generates radioactive waste should be underpinned by a BPEO study. The main purpose of such a study is to provide a transparent decision making process to justify that the process selected provides the most benefit or causes least damage to the environment as a whole, at acceptable cost, in the long term as well as in the short term. The Environment Agencies are in the process of having guidance prepared on their expectations of such studies. This will be available in due course and will refine the advice that SEPA currently gives to the nuclear industry. Any BPEO procedure that is adopted should follow the guidance in the Twelfth Report of the Royal Commission on Environmental Pollution (RCEP). It is however recognised by SEPA that full and early stakeholder involvement may not be necessary in all circumstances.

It is clear that the need for early involvement of stakeholders should be proportionate to the likely level of public interest and to the likely environmental detriment. If this were not the case then an undue burden would be placed on all stakeholders possibly leading to "consultation fatigue". However where stakeholder involvement is necessary or desirable then, as recommended by the RCEP, all stakeholders should be involved at an early stage. Such early involvement of stakeholders in similar processes has been noted recently by the House of Lords Select Committee on Science and Technology in its report on the Management of Nuclear Waste and by the Radioactive Waste Management Advisory Committee (RWMAC) in its response to this report. Both organisations cited the uncertainties that could result when early involvement is not undertaken.

In deciding what level of stakeholder involvement is appropriate for any decommissioning project a methodology based on the categorisation of potential risk could be developed by operators. SEPA is aware that at operational sites decommissioning activities are periodically carried out and that these are usually undertaken as small scale projects. SEPA would not however countenance the splitting of a site wide decommissioning plan into separate projects or a site wide approach being taken over time as a myriad of small projects simply to avoid the undertaking of a BPEO study that involved meaningful stakeholder involvement. It is possible that decommissioning projects could be investigated with regard to stakeholder involvement under a "potential risk" methodology. Under such a methodology the first step would be to attempt to define the scope of the project in the broadest terms possible. This would prevent potentially acceptable options being excluded at an early stage. References to specific processes or technologies should be avoided. So for example in terms of decommissioning the projects aims may be - 'to manage a material in such a way that it no longer poses a threat to workers, the public or the environment' or 'to dispose of an item in an environmentally acceptable manner'.

The second stage is to assess the environmental risks related to the project. When carrying out this assessment no account of any controls, whether existing or planned, should be taken into account. Factors that should be considered include:

Material inventory

Toxicity of all feed-stocks, products, intermediates, catalysts etc.

Potential for accumulation in both humans and the wider environment.

Physical state of the materials. i.e. How easy is it for the material to get into humans and the wider environment.

Public interest

The output from the assessment will allow the project to be assigned an environmental category for example high, medium or low. The environmental categorisation could then be used to define the extent to which stakeholders should be involved in the preparation and review of the BPEO study or whether a simpler optioneering study would be more appropriate.

Low Hazard. Internal stakeholders and regulators should be involved in the preparation and consideration of the options.

Medium Hazard. In addition to the above the project should be subjected to independent external review by a body outside the industry.

High Hazard. In addition to the above the project should be subjected to full public consultation and all stakeholders should be involved in the preparation and considerations of the options.

Finally the study would proceed to :

Define the minimum requirements for the project in terms of outputs, environment, safety, cost etc.

Generate as many options as practicable. Brain storming is a useful tool for this exercise.

Undertake a coarse screening to reject the options that fail to meet minimum requirements.

Undertake a detailed study of the options brought forward to determine that which achieves most benefit or least detriment to the environment.

The study must be carried out in such a way that allows all interested parties to understand how the decisions have been made and the information that has contributed to the decision making process. For facilities where the decommissioning will take several decades or where decommissioning as been differed then it will be inevitable to reaffirm previous decisions and strategies with new generations of "stakeholders".

Authors' Index

A

Allyson, J D 303–314
Arnold, H-U 153–160

B

Banford, H M 303–314
Batchelor, S J 265–276
Bayliss, C R 61–70
Bigg, P 3–12
Bischoff, H 31–38
Blakeway, S J 13–22
Bossart, S 153–160
Bredell, P J 337–344
Bull, F 123–138
Burnett, B 139–152

C

Carr, T J 217–222
Christ, B G 369–374
Conner, C 161–168
Coppins, G J 61–70
Curtis, C 81–94

F

Fairhall, G 123–138
Fearn, H S 405–414
Forrester, K 287–302
Fouquet, R 171–182
Froschauer, K 369–374

G

Gemmill, J 405–414
Gregson, K P 223–232

H

Hudson, I D 123–138

I

Ingham, B 73–80

K

Keep, M 405–414

L

Laraia, M 345–354
Lauridsen, K 315–324
Li, L 103–122
Lunt, M 183–194
Luyten, J-P 49–60

M

McCool, T M 303–314
Meservey, R 161–168
Murray, A 325–336

N

Nicol, R D 207–216

P

Parkinson, S J 223–232
Paul, R 49–60
Pearl, M 61–70
Peckitt, R A 277–286
Petrasch, P 49–60

R

Riley, K 123–138
Rittscher, D 235–254
Robertson, I 303–314
Rosenberger, S 153–160, 161–168

S

Salmon, N P 277–286
Sanders, M J 153–160, 161–168
Scott, R D 303–314
Sellers, R M 41–48
Shaw, I 255–264
Simon, G G 369–374
Smith, P K J 277–286
Staples, A 223–232
Starkey, J 355–368
Steele, M J 265–276
Sterner, H 235–254

T

Taylor, F E 13–22

W

Wall, S 255–264
Walters, G J 265–276
Warner, D R T 41–48

Welsh, P 23–28
Western, R E J 95–100
Williams, D M 375–384
Wilson, K A 387–402

Y

Young, T 195–204